Springer Proceedings in Mathematics & Statistics

Volume 185

More information about this series at http://www.springer.com/series/10533

Springer Proceedings in Mathematics & Statistics

This book series features volumes composed of select contributions from workshops and conferences in all areas of current research in mathematics and statistics, including OR and optimization. In addition to an overall evaluation of the interest, scientific quality, and timeliness of each proposal at the hands of the publisher, individual contributions are all refereed to the high quality standards of leading journals in the field. Thus, this series provides the research community with well-edited, authoritative reports on developments in the most exciting areas of mathematical and statistical research today.

Ilias S. Kotsireas • Anna Nagurney
Panos M. Pardalos
Editors

Dynamics of Disasters–Key Concepts, Models, Algorithms, and Insights

Kalamata, Greece, June–July 2015

Springer

Editors
Ilias S. Kotsireas
Department of Physics
and Computer Science
Wilfrid Laurier University
Waterloo, ON, Canada

Anna Nagurney
Department of Operations
and Information Management
University of Massachusetts
Amherst, MA, USA

Panos M. Pardalos
Department of Industrial
and Systems Engineering
University of Florida
Gainesville, FL, USA

ISSN 2194-1009 ISSN 2194-1017 (electronic)
Springer Proceedings in Mathematics & Statistics
ISBN 978-3-319-43707-1 ISBN 978-3-319-43709-5 (eBook)
DOI 10.1007/978-3-319-43709-5

Library of Congress Control Number: 2016948727

Mathematics Subject Classification (2010): 90, 91, 65, 93

Printed on acid-free paper

This Springer imprint is published by Springer Nature
The registered company is Springer International Publishing AG
The registered company address is: Gewerbestrasse 11, 6330 Cham, Switzerland

Preface

This volume is a collection of papers presented at the 2nd International Conference on Dynamics of Disasters held in Kalamata, Greece, June 29–July 2, 2015, with additional invited papers, all of which were reviewed. The conference was organized by Ilias S. Kotsireas, Anna Nagurney, and Panos M. Pardalos and brought together academics and practitioners to discuss their latest research on some of the most challenging problems associated with disasters from mitigation and preparedness to response and recovery. The collection of 18 papers is organized alphabetically by the first initial of the last name of the first author of each paper with highlights of each paper given below.

The coeditors of this volume thank the authors of the papers that appear in this volume and also thank the referees for their constructive reports on the papers. We hope that this volume demonstrates the breadth and depth of challenges associated with disasters and the underlying dynamics and also illustrates the different methodological as well as conceptual frameworks to address some of these challenges.

Fuad Aleskerov and Sergey Demin in their paper, "An assessment of the impact of natural and technological disasters using a DEA approach," focus on disaster mitigation and prevention. The authors emphasize that regions differ in terms of their resistance to different disasters since they are characterized by, among other features, different sizes of populations as well as the distribution of sources of potential disasters, whether natural or technological. They propose a Data Envelopment Analysis (DEA) approach, consisting of two methods, a standard one, and one new one with sequential exclusion of alternatives, in order to determine reasonable rankings of regions in terms of preventive measures. The approach takes into account the risks of the implementation of different preventive measures, their cost, and the heterogeneity of the regions. The application of their framework is illustrated through numerical examples for regions of the Russian Federation.

Burcu Balcik in her paper, "Selective routing for post-disaster needs assessment," turns to rapid needs assessment that relief agencies conduct immediately following a disaster, in order to determine the effects of the disaster and the needs of the affected communities. The topic of needs assessment has not received much attention in the humanitarian logistics literature. Balcik identifies and introduces a new problem, the Selective Assessment Routing Problem (SARP), that addresses site selection and routing decisions for the needs assessment teams. She uses a purposive sampling strategy with assessment teams aiming to select sites that involve different community groups with distinct characteristics. She proposes both a mathematical model and greedy heuristics for SARP, accompanied by numerical analysis that reveals that the heuristic version that balances the trade-off between coverage and travel times provides reasonable solutions for realistic problem instances.

Rasmus Dahlberg in his paper in this volume, "Preparing for long-term infrastructure disruptions," provides an overview of infrastructure and what is meant by European Critical Infrastructure and then discusses the preparedness plans and alternative routes for passengers and freight in the case of a more than 30-day disruption of a fixed link. Specifically, the fixed link consists of the combined road and rail Oresund Bridge and the Drogden tunnel, which more than 70,000 people traverse on a daily basis and which links Denmark and Sweden. A long-term disruption there could happen, for example, because of a ship collision with the bridge or a plane crash, and although such events may be considered extreme, the Eurotunnel that connects France and England has already closed twice. Rasmus then discusses the preparedness plans of a working group and their findings and compares the possible fixed link closure scenario to the impacts of the Champlain Bridge closure in the eastern United States and the Forth Road Bridge closure in Scotland. He describes how the closures of the latter two bridges disrupted immediately local communities and affected them severely. He argues that swift and affirmative action from infrastructure operators and authorities is required in the case of such disruptive events as bridge closures that are important infrastructure.

Emil-Sever Georgescu, Cristina Olga Gociman, Iolanda-Gabriela Craifaleanu, Mihaela Stela Georgescu, Cristian Iosif Moscu, Claudiu Sorin Dragomir, and Daniela Dobre are the authors of the next paper in this volume. Their paper, "Multi-hazard scenarios and impact mapping for a protected built area in Bucharest, as a base for emergency planning," is inspired by earthquake disasters that have repeatedly struck Romania but with changes in patterns and dynamics, with increases in losses from 1940 to 1977, in contradiction with the positive developments in earthquake engineering, architecture, and urban planning. This paper presents multi-hazard scenarios of an area of Bucharest that suffered trauma in the 1980s and considers the effects of earthquakes, flooding, and terrorist attacks on public institutions. The authors used spatial databases and online tools in order to identify locations of shelters for evacuation purposes and security centers, in the case of an earthquake, as well as in the case of flooding, and other hazards.

Sulejman Halilagic and Dimitris Folinas in their paper, "Lean thinking and UN field operations: a successful coexistence?" strive to identify the supply chain management and lean thinking principles of the business world for humanitarian operations and the expected benefits of the application of both principles and best practices. They note that the "One-UN" culture is one of the biggest challenges in humanitarian operations for the United Nations, with major associated organizational challenges that include lower costs of operation, offering better service to beneficiaries by bringing all of its operations under a single umbrella. The authors argue that humanitarian operations, especially in slow-onset/man-made disasters, have a major potential of benefiting from lean principles and practices.

Georgios Marios Karagiannis and Costas E. Synolakis in the paper, "Collaborative incident planning and the common operational picture," describe the common operational picture in disaster response, with a goal of reducing the gap between the technological and operational aspects. Disasters typically extend across jurisdictions, and the disaster response operational picture is filled with multiple agencies with different objectives. The authors use the incident planning process as a means to determine the information requirements of emergency managers during disaster response operations. In addition, they provide a typology of current capabilities and report on the major types of existing software, such as hazard modeling software, hazard mapping, vulnerability and risk mapping, and alert notifications, among others, and discuss how software supports disaster response coordination. They emphasize that, although situational awareness is a necessity in disaster response, it is impossible to achieve without effective coordination and communication. They also identify the unmet modern operational needs in existing software products and suggest directions for product development of common operational picture software.

Thomai Korkou, Dimitris Souravlias, Konstantinos Parsopoulos, Konstantina Skouri in their paper "Metaheuristic optimization for logistics in natural disasters" recognize that logistics in natural disasters or emergencies involve highly complicated optimization problems with diverse characteristics and introduce a multi-period model aiming to minimize the shortages of different relief products in a number of affected areas during a disaster relief effort. The relief products are transported via multiple modes of transportation from dispatch centers to these areas while adhering to traffic restrictions. A test suite of benchmark problems is generated from the proposed model and solved to optimality with CPLEX. The aforementioned test suite is used for benchmarking a number of established metaheuristics. In addition, they provide the necessary modifications in the algorithms, in order to fit the special requirements of the specific problem type. Algorithmic performance is assessed in terms of solution accuracy with respect to the optimal solutions. Comparisons among these metaheuristics offer valuable insight regarding their ability to tackle humanitarian logistics problems.

Andrey Kozelkov and Efim Pelinovsky in their paper, "Tsunami of the meteoric origin," address the modeling of a tsunami caused by an asteroid-meteorite. They first provide a review of such a hazard to planet earth. They note that the main danger on a global scale is from bodies of more than 1 km in diameter, while major continental or regional destruction can be caused by falling bodies of much smaller diameters. They highlight that the first documented catastrophe on a regional scale was the Tunguska meteorite, after which the forest was destroyed over an area of 2000 km^2. The mathematical model that they construct is based on Navier-Stokes equations and captures the generation of disturbances in the water and the surface. The authors derive formulas that assess the parameters of such a tsunami and conduct numerical simulations of the effect of the angle of entry of the body into the water and the characteristics of the resulting waves.

Emmett J. Lodree, Derek Carter, and Emily Barbee in their paper, "The donation collections routing problem," rigorously study a problem that arises immediately following a large-scale disaster in the response phase and is motivated by the surge of donations of relief items as well as volunteers that often occurs after a disaster strikes. The problem, which they introduce a mathematical formalism for using integer programming, is the donation collections problem (DCP). It is a network routing problem that seeks to address a practical alternative to post-disaster logistics operations associated with materiel and volunteer convergence. They propose common sense heuristic procedures and evaluate their performance through computational experimentation. Simple heuristic procedures in humanitarian contexts may be warranted since these can more easily be implemented in practice. Their findings may seem, at first glance, counterintuitive, but are, indeed, plausible, since postponing collections at nodes (collection sites) with large accumulation rates of donations produces larger stockpiles of donated goods as compared to servicing modes with lower rates at the end of the collection route. The donation collection problem and the methodology described in this chapter to tackle it are also useful for food banks.

Evangelos Mitsakis, Josep Maria Salanova, Iraklis Stamos, and Emmanouil Chaniotakis in their seventh paper, "Network criticality and network complexity indicators for the assessment of critical infrastructures during disasters," focus on the improvement of disaster management through the use of network analytics/criticality indicators. Network criticality indicators provide powerful tools for the assessment of parts of networks, notably, transportation networks, the closure of which would affect the overall performance of the network to the greatest degree. The authors overview centrality indicators for disaster management and provide an extended application of those metrics for the case of the Peloponnese region in Greece, which was subject to catastrophic fires in 2007. Their findings show that adopting interdisciplinary advances, as their synthesis of different network concepts and metrics reveals, can yield useful insights to decision-makers involved in all phases of disaster management: mitigation, preparedness, response, and recovery/reconstruction.

The next paper, "Freight service provision for disaster relief: a competitive network model with computations," by Anna Nagurney, presents a mathematical model in which freight service providers compete in order to deliver relief supplies provided by a humanitarian relief organization to points of demand post-disaster. Hence, this model focuses on the response phase. The behavior of the disaster relief organization assumes cost minimization, since the organization must manage its budget wisely and also report to donors and stakeholders. Freight service providers, in turn, are profit maximizers. The multitiered disaster relief supply chain network problem is formulated as a variational inequality problem, and a path-based projection method proposed for the determination of the product relief item flows. The algorithm resolves the problem into specially structured network problems for which a special purpose exact equilibration algorithm is provided. Qualitative results in terms of existence and uniqueness of the solution pattern are also given. The author also introduces a cooperative system-optimized model for freight service provision for disaster relief and describes a price of anarchy in the disaster relief setting. In addition to illustrative numerical examples, a case study, focusing on the recent immense Ebola healthcare crisis in Western Africa, is presented for which the relief item flows as well as freight service provision prices are reported. This paper adds to the literature on game theory and disasters.

Anna Nagurney and Ladimer S. Nagurney in their paper, "A mean-variance disaster relief supply chain network model for risk reduction with stochastic link costs, time targets, and demand uncertainty," construct an integrated computable model that incorporates features of disaster preparedness and mitigation as well as response. The model is an optimization model in which the humanitarian organization seeks to minimize its total expected operational costs and the total risk in operations with an associated individual weight, as well as to minimize expected costs of shortages and surpluses and tardiness penalties associated with time target goals at demand points. The model handles both the pre-positioning of relief items, whether local or nonlocal, and their procurement (also local or nonlocal), and the transport and distribution post-disaster. The proposed algorithm yields closed form expressions at each iteration for the variables. The numerical examples include smaller examples and a case study focusing on hurricanes striking Mexico, which is a country ranked as one of the world's 30 most exposed countries to three or more types of natural disasters.

Gabriela Perez-Fuentes, Enrica Verrucci, and Helene Joffe in the paper, "A review of current earthquakes and fire preparedness campaigns: what works?," focus on preparedness campaigns in the case of natural disasters. They note that most of the natural disaster campaigns rely primarily on information delivery, although studies have consistently shown that this is not sufficient to affect behavior. In addition, many of the campaigns lack evaluation, and, hence, their effectiveness cannot be ascertained. The authors report the findings of an online search conducted to determine major earthquake and fire preparedness campaigns and analyze the content, design, and theoretical background of the campaigns and the results of their evaluation. The authors argue that there is a need for a multi-hazard approach to emergency preparedness interventions since a public that is better prepared for multiple hazards is better prepared for specific as well as unpredictable ones.

Denise Sumpf, Vladimir Isaila, and Kristine Najjar tackle an ongoing disaster in their paper, "The impact of the Syria crisis on Lebanon." Their paper vividly explores the impacts of the war in Syria and the dynamics and spillovers on its immediate neighbor, Lebanon, including the effects of the Syrian refugee crisis on Lebanon, which currently is hosting more than one million Syrian refugees, adding about 25 % to the existing Lebanese population. The authors identify economic issues as well as social costs of the Syria crisis, specifically, on health and education. They also delineate the environmental costs and the impacts on agriculture and provide potential solutions. They discuss food security and waste management challenges and potential solutions. The authors note that the humanitarian crisis response has prioritized addressing immediate needs, but for that to be effective in damping the impact of the continuing conflict, the integration of a development perspective is essential in finding a medium- to long-term solution.

The next paper, "Absenteeism impact on local economy during a pandemic via hybrid SIR dynamics," is by Edward W. Thommes, Monica Gabriela Cojocaru, and Safia Athar. The authors construct a hybrid model by combining a Susceptible-Infected-Recovered (SIR) model, with roots in epidemiology, and discrete probability transition matrices, to explore the costs of absenteeism and presenteeism (going to work while sick) during a pandemic in a local economy. Workers consist of those who live and work in the same city as well as those who are commuters between cities. The authors make use of projected dynamical systems theory for the dynamics. Thommes, Cojocaru, and Athar also conduct simulations to investigate the effects of a fear factor and the severity of the disease on the number of missed work days in the region, which they then translate into loss of productivity costs. The results reveal that higher fear parameter values lead to high absenteeism and lower infection levels. Nevertheless, in the case of severe pandemics, there exist scenarios in which there is a unique value of the fear parameter which results in minimum economic costs for the regional economy. The implication for policy-makers is that "stay at home" policies during a pandemic could be implemented for workers, without resulting in a state of emergency.

Theodore B. Trafalis, Budi Santosa, and Michael B. Richman in the paper, "Tornado detection with kernel-based classifiers from WSR-88D radar data," focus on the important problem of detection of tornadoes in a timely manner so that warnings for evasive action are possible, which has been an objective of weather forecasters. The prediction of tornadoes is challenging due to the small scale of their circulation and their rapid production in the atmosphere. Through the use of technology, there has been progress in increasing the lead time of tornado warnings, which translates into lives saved. The authors use machine learning for weather prediction of tornadoes and introduce and apply two types of kernel-based methods, Support Vector Machines (SVM) and Minimax Probability Machines (MPM), to detect tornadoes through the utilization of attributes from radar-derived velocity data. They compare their results to those obtained using neural networks (NN) and demonstrate that kernel approaches are more accurate for tornado detection.

Chrysafis Vogiatzis and Panos M. Pardalos in their paper, "Evacuation modeling and betweenness centrality," consider the problem of evacuating people in an urban area from danger zones to safe zones. This is a large-scale problem, which has received much attention in the literature and has been subject to formulation and solution using various methodological approaches. The authors' approach falls into the category of heuristics, and their heuristic decomposes the large-scale evacuation problems into a series of smaller, scalable integer linear programming problems. In addition, Vogiatzis and Pardalos incorporate information on the transportation network using ideas from graph centrality, in order to ensure a more robust decomposition. Specifically, the decomposition approach takes into consideration the betweenness of a set of nodes in the transportation network and tries to obtain clusters from those nodes that can be easily solved so as to divide the flow of evacuees more evenly toward multiple paths to safety. The authors provide results for their approach on both synthetic and real-life networks, including a large-scale network representation of Jacksonville, Florida. The authors' approach yields solutions to such challenging problems, which have, heretofore, been proven impossible to be solved via commercial solvers.

The final paper in this volume, by Deborah Wilson, is entitled "Ode to the humanitarian logistician: humanistic logistics through a nurse's eye." It provides a firsthand, gripping, account of the Ebola crisis in Liberia through the lens of a courageous nurse (the author), who was deployed there for 6 weeks at a 120-bed Ebola Treatment Unit (ETU) with the humanitarian organization Medecins Sans Frontieres (MSF), known in English as Doctors Without Borders. The focus of the paper is on the role of the humanitarian logisticians in battling, together with the healthcare providers, the greatest outbreak of this disease to date. The tons of supplies that were provided to battle this terrible disease had to be effectively managed, recorded, transported, and set up by the logisticians. In addition, communications were established by the logisticians or "logs" as well as the procedures for the testing of blood samples, which needed to be transported, while the patients awaited the results, and this took 5 days. The logs also oversaw the destruction of all infectious waste and were responsible for restoring the electric power generators. Their skillset and expertise were essential to the effectiveness of the work of the medical personnel.

The disaster research area requires the development of multidisciplinary theories, tools, techniques, and methodologies that span a wide spectrum of disciplines, from the social and behavioral sciences, humanities, and government to management, engineering, medicine, mathematics, and computer science. These characteristics make disaster research challenging and exciting, and we hope that this volume contributes to exemplifying these aspects. Disasters come in various forms, each with their own particular issues. Whether one studies hurricanes, floods, earthquakes, tsunamis, tornadoes, volcanic eruptions, hazardous chemical incidents/attacks, plane crashes, fires, civil disturbances, riots, and so forth, researchers have identified important commonalities, such as organizational and community preparation, response, recovery, and others. The interplay between researchers and policy-makers is another important aspect of disaster research. We hope that the DOD 2015 *Book of Proceedings* will stimulate further interest in disaster research.

Acknowledgements Anna Nagurney thanks Oxford University for its support through the visiting fellowship program at All Souls College to complete the coediting of this volume.

Waterloo, ON, Canada — Ilias S. Kotsireas
Amherst, MA, USA — Anna Nagurney
Gainesville, FL, USA — Panos M. Pardalos

Contents

Contributors

Fuad Aleskerov National Research University Higher School of Economics, Institute of Control Sciences of Russian Academy of Sciences, Moscow, Russia

Safia Athar University of Guelph, Guelph, ON, Canada

Burcu Balcik Industrial Engineering Department, Ozyegin University, Istanbul, Turkey

Emily Barbee Department of Information Systems, Statistics, and Management Science, Culverhouse College of Commerce, The University of Alabama, Tuscaloosa, AL, USA

Derek Carter Department of Information Systems, Statistics, and Management Science, Culverhouse College of Commerce, The University of Alabama, Tuscaloosa, AL, USA

Emmanouil Chaniotakis Centre for Research and Technology Hellas – Hellenic Institute of Transport, Thermi, Thessaloniki, Greece

M.G. Cojocaru University of Guelph, Guelph, ON, Canada

Iolanda-Gabriela Craifaleanu Technical University of Civil Engineering, Bucharest, Romania

Institute for Research and Development URBAN-INCERC, Bucharest, Romania

Rasmus Dahlberg Copenhagen Center for Disaster Research, University of Copenhagen, Copenhagen, Denmark

Danish Emergency Management Agency, Birkerød, Denmark

Sergey Demin National Research University Higher School of Economics, Moscow, Russia

Claudiu Sorin Dragomir Faculty of Land Reclamation and Environment Engineering, University of Agronomic Sciences and Veterinary Medicine, Bucharest, Romania

National Institute for Research and Development URBAN-INCERC, Bucharest, Romania

Daniela Dobre Technical University of Civil Engineering Bucharest, Bucharest, Romania

National Institute for Research and Development URBAN-INCERC, Bucharest, Romania

Dimitris Folinas Department of Logistics, TEI-KM, Katerini, Greece

Emil-Sever Georgescu National Institute for Research and Development URBAN-INCERC, Bucharest, Romania

Mihaela Stela Georgescu "Ion Mincu" University of Architecture and Urban, Planning, Bucharest, Romania

Cristina Olga Gociman "Ion Mincu" University of Architecture and Urban, Planning, Bucharest, Romania

Sulejman Halilagic Office of Central Support Service, FF Building, New York City, NY, USA

Vladimir Isaila Economic Governance and Planning Section, Economic Development and Integration Division, United Nations Economic and Social Commission for Western Asia, Beirut, Lebanon

Helene Joffe Department of Clinical, Educational & Health Psychology, UCL, London, UK

Georgios Marios Karagiannis Technical University of Crete, School of Environmental Engineering, Chania, Greece

Thomai Korkoua Department of Computer Science and Engineering, University of Ioannina, Ioannina, Greece

Andrey Kozelkov Russian Federal Nuclear Center, All-Russian Research Institute of Experimental Physics, Sarov, Russia

Alekseev Nizhny Novgorod State Technical University, Nizhny Novgorod, Russia

Emmett J. Lodree Department of Information Systems, Statistics, and Management Science, Culverhouse College of Commerce, The University of Alabama, Tuscaloosa, AL, USA

Evangelos Mitsakis Centre for Research and Technology Hellas – Hellenic Institute of Transport, Thermi, Thessaloniki, Greece

Cristian Iosif Moscu "Ion Mincu" University of Architecture and Urban, Planning, Bucharest, Romania

Anna Nagurney Department of Operations and Information Management, Isenberg School of Management, University of Massachusetts, Amherst, MA, USA

Ladimer S. Nagurney Department of Electrical and Computer Engineering, University of Hartford, West Hartford, CT, USA

Kristine Najjar Economic Governance and Planning Section, Economic Development and Integration Division, United Nations Economic and Social Commission for Western Asia, Beirut, Lebanon

Panos M. Pardalos Department of Industrial and Systems Engineering, University of Florida, Gainesville, FL, USA

Konstantinos Parsopoulos Department of Computer Science and Engineering, University of Ioannina, Ioannina, Greece

Efim Pelinovsky Alekseev Nizhny Novgorod State Technical University, Nizhny Novgorod, Russia

Institute of Applied Physics, Nizhny Novgorod, Russia

National Research University – Higher School of Economics, Moscow, Russia

Gabriela Perez-Fuentes Department of Clinical, Educational & Health Psychology, UCL, London, UK

Michael B. Richman School of Meteorology, University of Oklahoma, Norman, OK, USA

Josep Maria Salanova Centre for Research and Technology Hellas – Hellenic Institute of Transport, Thermi, Thessaloniki, Greece

Budi Santosa Industrial Engineering Department, Institut Teknologi Sepuluh Nopember, Sukolilo, Indonesia

Konstantina Skouri Department of Mathematics, University of Ioannina, Ioannina, Greece

Dimitris Souravlias Department of Computer Science and Engineering, University of Ioannina, Ioannina, Greece

Iraklis Stamos Centre for Research and Technology Hellas – Hellenic Institute of Transport, Thermi, Thessaloniki, Greece

Denise Sumpf Economic Governance and Planning Section, Economic Development and Integration Division, United Nations Economic and Social Commission for Western Asia, Beirut, Lebanon

Costas E. Synolakis University of Southern California, Viterbi School of Engineering, Los Angeles, CA, USA

E.W. Thommes University of Guelph, Guelph, ON, Canada

Theodore B. Trafalis School of Industrial and Systems Engineering, University of Oklahoma, Stillwater, OK, USA

Enrica Verrucci Department of Civil, Environmental & Geomatic Engineering, UCL, London, UK

Chrysafis Vogiatzis Department of Industrial and Manufacturing Engineering, North Dakota State University, Fargo, ND, USA

Deborah Wilson RN, BSN, CRNI

An Assessment of the Impact of Natural and Technological Disasters Using a DEA Approach

Fuad Aleskerov and Sergey Demin

Abstract We consider a model of regions' ranking in terms of their vulnerability to natural and technological disasters. Regions are different in terms of their resistance to different disasters, by their population, by the distribution of the sources of potential disasters, etc. We consider different models of a data envelopment analysis (DEA) approach taking into account the risks of the implementation of different measures, their cost as well as the heterogeneity of regions. The numerical examples demonstrate the application of the constructed model for the regions of Russian Federation.

Keywords Technological and natural disasters • DEA • Ranking of regions

1 Introduction

Nowadays natural and man-made disasters occur and threaten people more and more often and with increasing impact, and cause a great amount of damage in terms of the numbers of killed and affected and the economic losses (Makhutov 2006; Guha-Sapir and Hoyols 2012; Kates et al. 2001).

The average damage, as well as the average number of killed and injured people, rose during the period from 1970 to 2010. Meanwhile, it is proved statistically that prevention of disasters can not only save people's lives, but also is cost-efficient (The World Bank 2010).

In these studies, pre-disaster spendings include expenditures on identifying risks, risk reduction by invention of new technologies and their implementation, risk transfer by using insurance, upgrading early warning systems, and educating the

F. Aleskerov (✉)
National Research University Higher School of Economics, Institute of Control Sciences of Russian Academy of Sciences, Moscow, Russia
e-mail: alesk@hse.ru

S. Demin
National Research University Higher School of Economics, Moscow, Russia
e-mail: ssdemin@edu.hse.ru

I.S. Kotsireas et al. (eds.), *Dynamics of Disasters—Key Concepts, Models, Algorithms, and Insights*, Springer Proceedings in Mathematics & Statistics 185,
DOI 10.1007/978-3-319-43709-5_1

public. In turn, post-disaster expenditures consist of expenditures on restoring resources, liquidating pollution and other ecological consequences, search and rescue operations, and rehabilitation and reconstruction.

Above all, in almost all countries, except Colombia, post-disaster expenditures are generally higher than pre-disaster expenditures (The World Bank 2010). In addition, the fluctuations in post-disaster spendings also have larger range, which can be explained by the fact that pre-disaster spendings are planned not only for 1 year, but also for a long period of time, which makes it more balanced and removes any significant fluctuations.

While analyzing the consequences of devastating disasters, certain systems of preventive measures, which can help to reduce disaster risk, were developed. It is convenient to divide all preventive measures into three groups: the first one—risk assessment and measures that reduce vulnerability of potentially dangerous objects; the second—monitoring, which can give indicators of impending disaster, and the last is early warning systems and educating the public.

It is important to point out that all types of precautionary measures play great role and none can be underestimated. This fact is supported by many studies. For instance, some researches give the list of factors that must be observed in order to prevent any negative consequences (Modoi et al. 2009). In turn, Coldewey (2009) studied parameters, which play great role in tailing security provision, and highlighted the following attributes: water level in reservoirs, potential precipitation, which can raise the water level, seismological activity, which can damage the reservoir, etc. Tracking of these parameters may detect potential threats in advance and, as a result, prevent disaster or at least mitigate its repercussions.

Meanwhile, some other studies pay more attention to early warning systems and educating the public (Akimov et al. 2004).

At the same time it is important to know what actions should be implemented because there may be a financial shortfall to fund all possible safety measures. There are several models, which can represent the results of precautionary measures and, as a result, will help in choosing the best strategy for disaster prevention or at least mitigation of its consequences.

Li et al. (2013) categorized the evaluation methods of disaster vulnerability by highlighting quantitative and qualitative approaches. The first is based on the relationship between preventive measures and consequences of potential disaster and, as a result, can show the best list of precautionary measures. These models can provide a great amount of information about a potential disaster; however, they require a great volume of input data with lots of details.

In turn, qualitative methods are based on expert assessments, and here lies the main problem of this approach—it makes assessment subjective. However, these methods help to solve the problem of data shortage. Moreover, some qualitative methods may become semi-quantitative. For instance, Wang et al. (2011a, b) applied a fuzzy analytic hierarchy process for assessment of flood risk.

One of the quantitative methods is introduced by Yanenko et al. (2008). This model, which evaluates the probability of a disaster occurrence for a nuclear power reactor, is based on the assumption that the progress of every disaster can be described as a model with three steady states. The first state conforms to a common situation at the object. The second one corresponds to extreme situations (for example, an electricity blackout) that can cause a disaster. And the third state is an occurrence of a disaster with devastating consequences.

Then, all parameters characterizing the condition of the reactor were divided into four groups. Subsequently, taking into account the safety values of parameters and their weights, the authors evaluate the risk assessment of the reactor. Moreover, they also obtain the vulnerabilities connected to different features of a reactor, such as the efficiency of the protective system, the activity of radioactive waste, or the human factor (e.g., staff qualification).

In that paper, an open system with intake means and resources, and outside—products and contaminations—is studied. This mathematical model represents the dynamics of many parameters connected to the disaster, including money flows for different aims, the power of the pollution, the speed of distribution of the contamination, the speed of resource depreciation, and many other indicators. Moreover, taking into account that this is a dynamic model, we can know not only the final consequences of the disaster, but also the details of the process.

Another study, connected to the technological disasters, describes the mathematical model that helps to fix the size of taxation for the level of air pollution by enterprises (Aleskerov 1990). For this aim, we simulate the contamination from a group of polluters, taking into account the geographical situation in the region and the location of the factories and find the value of tax on each polluter. Moreover, afterwards, considering limited resources and opportunities of the single enterprise, this study discusses preventive measures for contamination reduction.

The next model explains how to estimate the damage and the casualties (Aleskerov et al. 2005). In this study, the authors at the first step divide the region in question into the clusters according to the features of the buildings (the construction, number of stories, and the construction year). Afterwards they estimate the potential damage to the buildings in case of earthquakes of different intensity, according to the statistics. Subsequently, this data helps to estimate the casualties.

Finally, the model that helps to choose the best list of precautionary measures for the region with application to the Yaroslavl region is given in Aleskerov et al. (1988). In this study, all measures were divided into several classes according to two main characteristics (value of prevented damage per one cost unit and reduction of the acuteness value for the implementation area) by highlighting Pareto-optimal alternatives. Subsequently, the decision-maker should just chooses the measures, and the computer program will show the features of the selected choice, such as the percentage of the prevented damage and cost of measures.

However, Huang et al. (2011) denoted that many methods of quantitative assessment are very sensitive to the weight indices. The authors mentioned that the main shortcoming is the relative contribution of variables used for assessment, and the variables should be weighted differently, while some other scientists, on the contrary, prefer to neglect these mutual contributions (Cutter et al. 2000). This is why we apply a method based on another approach to the problem—data envelopment analysis (DEA).

We use DEA as a mathematical tool to compare different decision-making units (DMUs). For this purpose, all objects get some input and output parameters, which describe these objects. Subsequently, based on the efficiency, which is represented as the comparison of parameters, we find the best bound, where the efficiency is equal to 100 %. Afterwards, we rank the objects according to the position regarding the efficiency frontier and choose the best alternative taking into account constructed ranging scale.

The main advantage of this method is the fact that DEA does not only show the efficiency of all DMUs, but also identifies the benchmark elements for inefficient DMUs. For this reason, DEA has been used in many fields. For instance, Abankina et al. (2012) used this method for ranking universities. Further applications of this method include analysis of national innovation systems efficiency (Shao and Lin 2002; Kotsemir 2013), bank efficiency and productivity growth (Andries 2010), etc.

One more advantage is the fact that efficiency value is independent of the measuring unit (Wang and Tsai 2009). Therefore, it is not necessary to assess inputs and output in one scale.

However, turning back to the main aim of our research, it is crucial to examine applications of DEA in the sphere of disaster prevention.

Furthermore, this method can be combined with other approaches that can improve the model. For instance, Saein and Saen (2012) used an improved DEA model for the assessment of the region vulnerability to earthquakes. This model, introduced by Saen (2011), is based on cross-efficiency approach. The main idea of this method implies the use of DMUs cross-comparison instead of self-evaluation. For this purpose, each time one determines the highest efficiency of a certain DMU one should take into account all other DMUs' efficiency. Finally, the result for every concrete DMU is calculated as the mean of all cross-efficiencies.

In turn, De Almada Garcia Adriano et al. (2013) used DEA for the assessment of the security level at a nuclear power plant. For this purpose, the authors offered to use the improved and more realistic DEA model. They took into consideration the effect of weight indices restriction. For instance, taking into account the expert assessments, it was assumed that the severity of the failure mode is more important than other criteria (occurrence and detectability). As a result, there was added one more restriction (v_S, v_D, v_O—weight coefficients of severity, occurrence, and detectability accordingly):

$$v_S - (v_D + v_O) \geq 0$$

This approach allows constructing of a more realistic and more precise method, which will pay attention to the ratio of importance of different criteria.

Another application of DEA is offered by Zhang and Fu (2012), who introduced a model for the evaluation of emergency logistics performance. The main idea of this method is the combination of two basic approaches—DEA and analytical hierarchy process (AHP). At the first stage, AHP calculates the weight of each part of the emergency logistics system (Zhang and Fu 2012). Subsequently, the relative efficiency of different parts is developed by the DEA method. Finally, one should calculate the overall efficiency of the emergency logistics system by summarizing the results of different parts.

In addition, it is important to point out that in this work we provide a new approach to the application of DEA method to an assessment of natural and technological disasters based on the new model of DEA with sequential exclusion of alternatives.

2 Framework

As was mentioned above, the main method that will be used in our research is DEA, specifically the CCR approach, named by first letters of the authors' surnames: Charnes, Cooper, and Rhodes (Charnes et al. 1978). The key idea of this model is presentation of the efficiency of preventive measures as a fraction over resources spent. We will analyze the following goal functions:

$$\max_{u_i, v_j} \left(e_k = \frac{\sum_{i=1}^{M} u_{ik} x_{ik}}{\sum_{j=1}^{N} v_{jk} y_{jk}} \right), \tag{1}$$

under the constraints:

$$\begin{cases} \dfrac{\sum_{i=1}^{M} u_{ik} x_{ik}}{\sum_{j=1}^{N} v_{jk} y_{jk}} \leq 1, \; k = 1, \ldots, R \\ \forall i \; u_i > 0 \\ \forall j \; v_j > 0 \end{cases} \tag{2}$$

In these inequalities we introduced the following variables: e_k—efficiency of k-th safety measures; u_{ik}, v_{jk}—weight coefficients, that illustrate the importance of appropriate parameters; x_{ik}—output parameters, which show achieved results; y_{jk}—input parameters, which show spent resources; M—the number of output parameters; N—the number of input parameters; and R—the number of preventive measures.

The main advantage of this technique is automatic selection of u_{ik}, v_{jk}, based on criterion (1). In addition, this model can be transformed to the linear programming task by the conversion proposed by Charnes and Cooper (1962):

$$\min_{\theta_k, \lambda} \theta_k \tag{3}$$

In this way, the restrictions will also change:

$$\begin{cases} -q_k + Q * \lambda \geq 0, \\ \theta_k x_k - X * \lambda \geq 0, \\ \lambda \geq 0, \end{cases} \tag{4}$$

where θ_k is a scalar, indicative the efficiency of k-th safety measures, λ—vector of constants $(R * 1)$, Q—matrix of output indices of all preventive measures $(M * R)$, X—matrix of input indices of all precautionary measures $(N * R)$. $\theta_k \in [0; 1]$, $\theta_k = 1$ shows the 100 % efficiency of k-th safety measure relative to others.

We will analyze now all available precautionary measures and construct the efficiency frontier, which shows us the best possible result. Subsequently, we rank all variants of preventive measures taking into account that the closer to the borderline the alternative is, the more efficient it is.

Another method, which will be used for an assessment, was introduced by Aleskerov and Petrushchenko (2015). The main improvement of this model is a less strict assessment of the inefficient DMUs. It is achieved by exchanging the benchmark for inefficient alternatives taking into account the heterogeneity of the evaluated sample.

The authors propose to generate a new efficiency frontier using the best alternative and the barycenter of all DMUs. For this purpose, they introduce a heterogeneity coefficient μ, which is equal to the ratio of the mean value of the distance between the alternative and the barycenter to the maximum value. Then the benchmark is constructed by combining the best alternative and the barycenter with weights $(1 - \mu)$ and μ, respectively. Afterwards, the efficiency of inefficient DMUs is evaluated. Then all alternatives, which has an efficiency less than 1, are excluded from the sample, and the assessment continues for inefficient alternatives, which get an efficiency equal to 1.

It is important to choose which elements will be compared. There might be two approaches to this task. In the first one, different regions (including all enterprises in them) are compared. In the second one, the elements are enterprises themselves.

3 Input and Output Parameters

In many DEA studies, it is mentioned that the selection of input and output parameters is one of the most important stages in this method (e.g., Golan and Roll 1989). Moreover, at this stage some problems with evaluation of parameters could take place, because sometimes it is very difficult to estimate the value of a parameter (Saein and Saen 2012). The authors point out that the best solution in such cases is to use the value of a parameter with respect to some classes on a scale.

We will apply our model for both types of disasters: technological and natural. For more accurate results, in each case, we will examine a certain source of danger, because different catastrophes require different input and output parameters. Thus, we will consider industrial accidents at a chemical or nuclear plant as an example of technological disaster and earthquakes as an example of a natural one.

3.1 Parameters for Technological Disasters

The main target of our project is to rank different safety measures according to their efficiency. Any precautionary measures demand financial funds, and it is one of the most crucial input parameters. In addition, it is important to divide all expenditures into money flows according to the aims.

As a result, we will identify the following money flows:

1. Expenditures on equipment upgrading. This is the main use of financial funds, because it helps us to keep the object in a steady state. For instance, in 2011 the Fukushima disaster caused a great amount of damage (Lipscy et al. 2013). One of the main reasons for enormous negative consequences is the unavailability of qualitative protective equipment.
2. Expenditures on a disaster alert system and educating the public. It is important to point out that in many cases the number of victims could be much less if the population would be warned in time and made aware of disaster-safe behavior. For instance, in 1984 in Bhopal (Dutta 2002), residents were not warned, and, as a result, about 18,000 people died and more than 150,000 affected. Therefore, this money flow plays an important role in reducing the number of victims from a disaster.
3. Expenditures on scientific research. This part of financial funds influences the efficiency of the first money flow (updating equipment) because by using new technologies we can decrease certain hazards.

However, there is a great difference between two cases: a technological disaster somewhere in the middle of a desert with nobody around the epicenter, and the same technological disaster, for instance, in the center of a big city. Thus, we use the number of people, not far from the epicenter of the disaster.

However, we should parameterize not only the source data, but also the results of precautionary measures. In this part, we will point out one parameter—the probability of a devastating consequence of the disaster. This parameter will depend on expenditures on updating equipment. Here we will use the assumption that all financial funds have decreasing returns to scale.

Therefore the probability is evaluated as $p = ae^{-be_1}$, where p is the probability of the occurrence of extreme situations with devastating consequences, e_1—expenditures on updating equipment, and a, b—rate-setting coefficients. These coefficients should be evaluated separately for each situation (possibly by expert assessment). We will use two assumptions. The first one is that with no expenditures on updating equipment p will be high (e.g., 90 %). And vice versa, huge expenditures will decrease p to 10 %.

Table 1 Input and output parameters of DEA model for technological disasters

Input parameters	Output parameters
e_1—expenditures on equipment upgrading	p—probability of devastating consequences of the disaster
e_2—expenditures on disaster alert system and educating the public	y_1—number of killed
	y_2—number of affected
e_3—expenditures on scientific research	y_3—economic losses in spite of production stoppage
x—the number of people, who live not far from the epicenter of the disaster	y_4—spendings on recovery from the disaster
	s—defeat area—area, wherein people have a probability of injury

However, in addition to the probability, we should highlight one more group of parameters, which will show the consequences of a potential disaster: the number of killed, the number of affected people, the economic losses, in spite of production stoppage, spendings on recovery from the disaster, and contamination of the territory. For the contamination of the territory, we have the same problems as with the population of the territory. So, the solution will be the same: we will use defeat area—area, wherein people have a probability of injury.

As a result, we get the following list of input and output parameters for our DEA model (Table 1).

3.2 *Parameters for Natural Disasters*

As in the case of technological disasters, again the majority of input parameters of the model are different types of expenditures. We will identify the following flows of investment:

1. Expenditures on scientific research,
2. Expenditures on disaster alert system and educating the public,
3. Expenditures on construction of new earthquake-proof buildings and seismic improvement of those already existing.

Another input parameter is the construction typology of houses in regions because an earthquake can have varied influence on buildings with different types of structure. Four main groups of buildings in question are: masonry, reinforced concrete, wooden masonry composite, and steel frame.

Each type of construction has the unique rate of seismic resistance and an average number of people living in one house of this type (Aleskerov et al. 2005). We also assume that the population in the houses of the type is the same for all regions. This assumption is not applicable for more exact analysis, but we do not have access to this information from regions.

Table 2 Input and output parameters of DEA model for natural disasters

Input parameters	Output parameters
e_1—expenditures on scientific research	y_1—economic losses
e_2—expenditures on disaster alert system and educating the public	y_2—number of killed
	y_3—number of affected
e_3—expenditures on construction of new earthquake-proof buildings and seismic improvement of already existing construction typology of houses	y_4—number of homeless

As for output parameters to illustrate the consequences of a potential disaster in case of earthquake occurrence we will use expected economic losses, the number of deaths, the number of affected people, and the number of homeless.

As a result, we get the following list of input and output parameters for our DEA model (Table 2).

4 Application of the Model

We apply our method for comparing the efficiency of 27 Russian regions, which have the highest Seismic Risk Index (value more than 0.1). These regions constitute 65 % of Russia's territory. As it was mentioned by Li et al. (2013) there should be used at least twice the number of input and output parameters. Since the exact data on security expenditures are not open to the public, we had to use a simulation technique.

4.1 Application to Technological Disasters

We used main parameters of the chosen regions, such as population and gross domestic product (GDP). Then, we make a list of assumptions.

Firstly, the data about the number of killed and the number of affected have been obtained from the comparative analysis of losses and injures in some previous situations. Secondly, we believe that GDP shows the production volume, which influences the economic losses in spite of production stoppage and spendings on recovery from the disaster.

As a result, after application of our DEA model for assessment of efficiency in regions we get the following results (efficiency 1 is the efficiency according to the basic DEA approach, while efficiency 2 is the efficiency according to the improved DEA approach) (Table 3).

Table 3 Results of DEA model application for technological disasters

Region	Efficiency 1	Efficiency 2
Altai Krai	0.67	0.71
Altai Republic	0.64	0.77
Amur Oblast	1	1
Chechen Republic	1	1
Chukotka Autonomous Okrug	1	1
Irkutsk Oblast	0.84	0.87
Jewish Autonomous Oblast	1	1
Kabardino-Balkar Republic	0.6	0.67
Kamchatka Krai	1	1
Karachay-Cherkess Republic	0.41	0.63
Kemerovo Oblast	0.76	0.8
Khabarovsk Krai	0.77	0.93
Krasnodar Krai	1	1
Krasnoyarsk Krai	1	1
Magadan Oblast	0.6	0.69
Primorsky Krai	0.75	0.8
Sakha (Yakutia) Republic	0.76	0.9
Sakhalin Oblast	1	1
Stavropol Krai	0.71	0.74
The Republic of Adygea	0.41	0.69
The Republic of Buryatia	0.44	0.54
The Republic of Dagestan	0.83	0.86
The Republic of Ingushetia	0.97	1
The Republic of Khakassia	0.9	0.92
The Republic of Northern Ossetia—Alania	0.41	0.6
Tuva Republic	0.46	0.89
Zabaykalsky Krai	1	1

4.2 Application to Natural Disasters

In the case of natural disasters we used almost the same data and assumptions as in Sect. 4.1. The only important difference is the assessment of the number of killed, injured, and affected people.

For this purpose, we used data, which was highlighted above—the unique rate of seismic resistance and the average number of people living in one house of this type (Aleskerov et al. 2005).

As a result, after application of our DEA model for assessment of efficiency in regions we get the following results (efficiency 1 is the efficiency according to the basic DEA approach, while efficiency 2 is the efficiency according to the improved DEA approach) (Table 4).

Table 4 Results of DEA model application for natural disasters

Region	Efficiency 1	Efficiency 2
Altai Krai	0.71	0.86
Altai Republic	0.81	0.83
Amur Oblast	0.84	0.94
Chechen Republic	1	1
Chukotka Autonomous Okrug	0.88	0.96
Irkutsk Oblast	0.81	0.91
Jewish Autonomous Oblast	0.7	0.75
Kabardino-Balkar Republic	0.83	0.91
Kamchatka Krai	1	1
Karachay-Cherkess Republic	0.75	0.82
Kemerovo Oblast	0.71	0.86
Khabarovsk Krai	0.71	0.75
Krasnodar Krai	1	1
Krasnoyarsk Krai	0.71	0.82
Magadan Oblast	0.87	0.96
Primorsky Krai	0.7	0.75
Sakha (Yakutia) Republic	0.69	0.75
Sakhalin Oblast	1	1
Stavropol Krai	0.71	0.76
The Republic of Adygea	1	1
The Republic of Buryatia	0.96	0.97
The Republic of Dagestan	0.96	0.98
The Republic of Ingushetia	0.84	0.94
The Republic of Khakassia	1	1
The Republic of Northern Ossetia—Alania	0.79	0.83
Tuva Republic	1	1
Zabaykalsky Krai	0.89	0.95

5 Analysis

As expected, the results of the two methods give us almost the same efficiency for the regions. The only difference is the fact that the DEA with sequential exclusion of alternatives shows better results (higher efficiency), because it uses the special alternative as a benchmark for inefficient DMUs instead of the best DMU as in the basic DEA approach.

5.1 Results for Technological Disasters

The results of our model highlight 9 regions with 100 % efficiency in disaster prevention. Good results of the majority of these regions can be explained by their specifics—all of them except Krasnodar Krai and Chechen Republic have a low density of population. In this case, it is not necessary to spend a great part of local budget on industrial security.

In turn, the efficiency of Krasnodar Krai can be explained by the occurrence of the Krasnodar Krai flood in 2012, which forced local authorities to raise industrial security and implement a vast range of precautionary measures.

Returning to general results for all regions, it is important to highlight the Republic of Northern Ossetia—Alania, which has the worst result—41 % (60 %). Some data force certain regions to take measures to improve their efficiency.

5.2 Results for Natural Disasters

In case of the assessment of earthquake mitigation efficiency, the model points out 7 regions with 100 % result. Some leaders, such as Krasnodar Krai, Sakhalin Oblast, Kamchatka Krai, and Chechen Republic, are the same. However, there are also some regions, for which results in technological disaster mitigation are not so good, the Republic of Adygea, the Republic of Khakassia, and Tuva Republic.

In turn, Sakha (Yakutia) Republic gets the worst result—just 69 % (75 %). Here we can detect one more fact, which should be highlighted. The range of result distribution is much less—just 31 % (59 % in technological disaster mitigation). However, the results of this application of the model also force certain regions to take measures to improve their efficiency.

6 Conclusion

In our work, we applied two methods based on DEA to the regions of the Russian Federation. The first method is the standard DEA approach, the second one is the new method based on sequential exclusion of alternatives. Both methods give reasonable rankings of regions in terms of preventive measures efficiency. However, the second method gives higher values of efficiency for regions. It can be explained by the fact that this method does not refer to the best practice in the whole sample, but rather takes into account compactness in terms of their evaluations subsamples.

These methods may be applied to similar problems.

Acknowledgements The article was prepared within the framework of the Basic Research Program at the National Research University Higher School of Economics (HSE) and supported within the framework of a subsidy by the Russian Academic Excellence Project "5-100".

We express our appreciation for help in collecting data for the research to Mr. Nikita Kolesnikov.

We are grateful for the comments and suggests of the anonymous reviewers.

References

Abankina, I.V., Aleskerov, F.T., Belousova, V.Yu., Bonch-Osmolovskaya, A., Petruschenko, V.V., Ogorodniychuk, D., Yakuba, V., Zin'kovsky, K.V.: University efficiency evaluation with using its reputational component. In: Proceedings of the 4th International Conference on Applied Operational Research, pp. 244–253. Tadbir Operational Research Group, Bangkok (2012)

Akimov, V.A., Lesnyh, V.V., Radaev, N.N.: Basics of Risk Analysis and Risk Management in Natural and Technological Spheres (in Russian). Delovoy Express, Russia (2004)

Aleskerov, F.T., Petrushchenko, V.V.: An approach to DEA for heterogeneous samples. In: Le Thi, H.A., et al. (eds.) Modelling, Computation and Optimization in Information Systems and Management Sciences, Advances in Intelligent Systems and Computing, vol. 360, pp. 15–21. Springer, Switzerland (2015)

Aleskerov, F.T., Vil'ner, M.Ya., Vol'skiy, V.I., Litvakov, B.M., Novikov, S.G.: Man-machine system for a rational choice of nature conversation strategy of a region. In: Proceedings of III IFAC/IFIP/IFORS/IEA Conference on Man–Machine Systems, Oulu, Finland, pp. 145–150 (1988)

Aleskerov, F.: Taxation for improving regional ecological situation. In: Ecological Economics of Sustainability Conference, World Bank, Washington, DC, May 1990, 12 p

Aleskerov, F.T., Say, A.I., Toker, A., Akin, H.L., Altay, G.: A cluster-based decision support system for estimating earthquake damage and casualties. Disasters **29**(3), 255–276 (2005)

Andries, A.M.: The Determinants of Bank Efficiency and Productivity Growth in the Central and Eastern European Banking Systems. Faculty of Economics and Business Administration, Alexandru Loan Cuza University of Lasi, Romania, Draft (2010)

Charnes, A., Cooper, W.: Programming with linear fractional functional. Nav. Res. Logist. Q. **9**, 181–186 (1962)

Charnes, A., Cooper, W.W., Rhodes, E.: Measuring the efficiency of decision making units. Eur. J. Oper. Res. **2**, 429–444 (1978)

Coldewey, W.G.: Emergency Planning For Tailings Dams, Optimisation of Disaster Forecasting and Prevention Measures in the Context of Human and Social Dynamics, pp. 115–121. IOS Press, Amsterdam (2009)

Cutter, S.L., Mitchell, J.T., Scott, M.S.: Revealing the vulnerability of people and places: a case study of Georgetown county, South Carolina, USA. Ann. Assoc. Am. Geograph. **90**(4), 713–737 (2000)

De Almada Garcia Adriano, P., Curty Leal Junior, I., Alvarenga, O.M.: A weight restricted DEA model for FMEA risk prioritization, Producao **23**(3), 500–507 (2013)

Dutta, S.: The Bhopal Gas Tragedy. ICFAI Center for Management Research, Hyderabad (2002)

Golan, B., Roll, Y.: An application procedure for DEA. OMEGA **17**(3), 237–250 (1989)

Guha-Sapir, D., Hoyois, P.: Measuring the Human and Economic Impact of Disasters. CRED, Government Office for Science, Brussels (2012)

Huang, J., Liu, Y., Ma, L.: Assessment of regional vulnerability to natural hazards in China using a DEA model. Beijing, China. Int. J. Disaster Risk Sci. **2**(2), 41–48 (2011)

Kates, R.W., Clark, W.C., Corell, R., Hall, J.M., Jaeger, C.C., Lowe, I., McCarthy, J.J., Schellnhuber, H.J., Bolin, B., Dickson, N.M.: Sustainability science. Science **292**(5517), 641–642 (2001)

Kotsemir, M.N.: Measuring National Innovation Systems Efficiency – A Review of DEA Approach. Science, Technology and Innovation, WP BRP, The Higher School of Economics, No. 16, Moscow (2013)

Li, C.-H., Li, N., Wu, L.-C., Hu, A.-J.: A relative vulnerability estimation of flood disaster using data envelopment analysis in the Dongting Lake region of Hunan. Nat. Hazards Earth Syst. Sci. **13**, 1723–1734 (2013)

Lipscy, P.Y., Kushida, K.E., Incerti, T.: The Fukushima disaster and Japan's nuclear plant vulnerability in comparative perspective. Environ. Sci. Technol. **47**, 6082–6088 (2013)

Makhutov, N.A.: Methodology for Assessing the Risks of Terrorism, Countering Urban Terrorism in Russia and the United States, pp. 207–222. The National Academic Press (NAP), Washington, DC (2006)

Modoi, O-C., Stefanescu, L., Marginean, S., Arghius, C., Ozunu, A.: Management of Risks Associated with Mining Wasted (Tailing Dams and Waste Heaps), Optimisation of Disaster Forecasting and Prevention Measures in the Context of Human and Social Dynamics, pp. 130–143. IOS Press, Amsterdam (2009)

Saein, A.F., Saen, R.F.: Assessment of the site effect vulnerability within urban regions by data envelopment analysis: a case study in Iran. Comput. Geosci. **48**, 280–288 (2012)

Saen, R.F.: A decision model for selecting the best entry modes via data envelopment analysis. Int. J. Appl. Decis. Sci. **4**(3), 213–229 (2011)

Shao, B.B.M., Lin, W.T.: Technical efficiency analysis of information technology investments: a two-stage empirical investigation. Inf. Manage. **39**, 391–401 (2002)

The World Bank: Natural Hazards, Unnatural Disasters: The Economics of Effective Prevention (2010)

Wang, L.-C., Tsai, H.-Y.: Evaluation of highway maintenance performance using data envelopment analysis (DEA) in Taiwan, Taiwan. J. Mar. Sci. Technol. **17**(2), 145–155 (2009)

Wang, F., Cao, Y., Liu, M.: Risk early-warning method for natural disasters based on integration of entropy and DEA model, USA. Appl. Math. **2**, 23–32 (2011a)

Wang, Y., Li, Z., Tang, Z., Zeng, G.: A GIS-based spatial multi-criteria approach for flood risk assessment in the Dongting Lake region, Hunan, Central China. Water Resour. Manage. **25**, 3465–3484 (2011b)

Yanenko, V.M., Rykhtovsky, V.O., Yanenko, N.V.: Environmental, medical, technogenic and computer technology: modeling, risk assessment and cost/benefit analysis of the accidents. Environ. Technol. New Dev. **15**, 245–264 (2008)

Zhang, J., Fu, S.: An effective DEA-AHP algorithm for evaluation of emergency logistics performance. AISS **4**(12), 1–8 (2012)

Selective Routing for Post-disaster Needs Assessments

Burcu Balcik

Abstract In the immediate aftermath of a disaster, relief agencies perform needs assessment operations to investigate the effects of the disaster and understand the needs of the affected communities. Since assessments must be performed quickly, it may not be possible to visit each site in the affected region. In practice, sites to be visited during the assessment period are selected considering the characteristics of the target communities. In this study, we address site selection and routing decisions of the rapid needs assessment teams that aim to evaluate the post-disaster conditions of a diverse set of community groups with different characteristics (e.g., ethnicity, income level, etc.) within a limited period of time. In particular, we study the Selective Assessment Routing Problem (SARP) that determines sites to be visited and the order of site visits for each team while ensuring sufficient coverage of the given set of characteristics. We present a mathematical model and greedy heuristics for the SARP. We perform numerical analysis to evaluate the performance of the greedy heuristics and show that the heuristic version that balances the tradeoff between coverage and travel times provides reasonable solutions for realistic problem instances.

Keywords Needs assessment • Selective routing • Coverage • Purposive sampling • Greedy heuristics

1 Introduction

Once a disaster occurs, humanitarian relief organizations and/or government agencies first perform rapid assessment of the affected region. A rapid needs assessment process is performed by experienced experts, who have knowledge of the local context and may have specialties in different areas such as logistics, food, hygiene, health, and shelter (ACAPS, 2011). During the rapid assessment stage, which usually starts within 24 h after the disaster (IFRC, 2008), assessment teams visit

---

B. Balcik (✉)
Industrial Engineering Department, Ozyegin University, Istanbul, Turkey
e-mail: burcu.balcik@ozyegin.edu.tr

I.S. Kotsireas et al. (eds.), *Dynamics of Disasters—Key Concepts, Models, Algorithms, and Insights*, Springer Proceedings in Mathematics & Statistics 185,
DOI 10.1007/978-3-319-43709-5_2

a number of sites to investigate the impact of the disaster on people and understand their urgent needs. The findings of the needs assessment are critical for making effective funding appeals and prioritizing needs. However, since assessments must be completed quickly, assessment teams cannot usually visit all of the affected sites. In practice, sites to be visited during the rapid needs assessment stage are selected by sampling. After the rapid assessment stage, humanitarian organizations may continue assessing the needs through the relief horizon; in later stages of assessment, organizations typically visit more sites and spend more time at each site to evaluate community needs in more detail (IFRC, 2008). In this study, we focus on the rapid needs assessment stage where time plays a critical role and site selection decisions are particularly important.

The general aim of sampling in the rapid needs assessment stage is to choose the sites that will allow the assessment teams to observe the post-disaster conditions of different community groups in a short period of time. In general, the selected sites should not be limited to the worst-affected regions and/or easy-to-reach areas (ACAPS, 2011). Also, visiting sites that are similar to each other in terms of regional and/or community characteristics may not lead to new information that would contribute to assessment findings. In practice, two sampling methods are typically used to select sites during the rapid need assessment stage (IFRC, 2008). In *random sampling*, sites to be visited are selected randomly, which may be reasonable if different community groups are expected to be affected by the disaster similarly. However, random sampling may not lead to accurate assessments if the community needs change throughout the affected region. For instance, communities in mountainous and coastal sites may be affected by the disaster in different ways; in such a case, random sampling may not ensure that different community groups are observed. Therefore, relief organizations often use *purposive sampling*, which allows them to assess and compare post-disaster conditions of a diverse group of community groups with different vulnerabilities and needs. This study focuses on a purposive sampling strategy for site selection. In practice, relief organizations apply purposive sampling by following the steps below (ACAPS, 2011):

- *Specify the critical characteristics of interest.* First, the critical characteristics that will be used to define different community groups are identified (ACAPS, 2011). These characteristics may depend on the specific disaster situation; for instance, they may be related to physical aspects of the region (e.g., geography, topography, and closeness to the disaster epicenter) and/or demographical and socio-economical characteristics of the affected people (e.g., age, gender, ethnicity, religion, disability, minority or immigrant status, income level, etc.) (IASC, 2012; ACAPS, 2011). In this step, it is important to avoid using many characteristics and only focus on the ones that will make the most difference in terms of the assessment results (ACAPS, 2011).
- *Site selection.* Second, sites to be visited by the assessment teams are selected. In general, as the heterogeneity of the affected area increases in terms of the selected critical characteristics, more sites may need to be visited. On the other hand, selecting large samples may not be possible since the assessments must

be completed quickly by using scarce resources (such as staff and vehicles). In general, it is important to select sites such that sufficient information about different community groups can be obtained.
- *Scheduling/Routing*. Finally, an assessment schedule is developed, which involves assigning sites to the teams and specifying the order of site visits for each team, considering the length of the assessment period.

Since the above steps are applied sequentially in practice, it may not be possible to obtain a feasible assessment schedule in the last step that visits all of the selected sites within the given limited time period. In such cases, relief organizations can either decrease the number of critical characteristics of interest and/or consider visiting a smaller number of sites (ACAPS, 2011). This trial-and-error process may be time-consuming and is not guaranteed to produce an effective assessment schedule.

In this study, we focus on developing a systematic method that will allow selection of sites and scheduling site visits simultaneously. Our study assumes that the critical characteristics that define community groups are specified in advance. Given the set of critical characteristics carried by each site, we study the Selective Assessment Routing Problem (SARP) that focuses on determining the sites to visit and the assessment routes. We develop a mathematical model that ensures a balanced coverage of the critical characteristics in the objective function, and time duration concerns are captured via the constraints. We present greedy heuristics to solve the SARP and present numerical results to evaluate the performance of the heuristics.

In Sect. 2, we review the relevant literature. In Sect. 3, we describe the SARP and present the mathematical model. We present our greedy heuristics in Sect. 4 and numerical results in Sect. 5. Finally, we conclude and discuss future work in Sect. 6.

2 Literature Review

In Sect. 2.1 we review the related studies from the humanitarian logistics literature, and in Sect. 2.2 we discuss the related routing literature.

2.1 Needs Assessment in Humanitarian Logistics

The increasing number and impact of disasters in recent years have led to a growing body of Operations Research (OR) studies that focus on humanitarian operations. Many studies address problems that involve routing decisions in various humanitarian settings (e.g., see review by de la Torre et al. 2012); most of the existing studies focus on post-disaster relief distribution (e.g., Tzeng et al., 2007; Balcik et al., 2008; Lin et al., 2011; Ozdamar and Demir, 2012; Huang et al., 2012).

We observe that problems that address routing decisions in the needs assessment stage have not received much attention. To the best of our knowledge, the only study that considers routing of assessment teams is presented by Huang et al. (2013). The authors develop a model that captures time sensitivity of the needs assessment process by minimizing the sum of the node arrival times. They apply continuous approximation to obtain routing policies that can be easily implemented in the field. Huang et al. (2013) assume that all nodes in the network are visited; that is, sampling issues and site selection decisions are not considered. The tools presented in Huang et al. (2013) may be more relevant to the detailed needs assessment stage, where time is less of a concern. Our study contributes to the humanitarian logistics literature by focusing on the rapid needs assessment operations and introducing a new problem that involves site selection and routing decisions.

The characteristics of the needs assessment process are described in various practical resources and conceptual studies (e.g., IFRC, 2000; WHO, 2004; IFRC, 2008; ACAPS, 2011; Darcy and Hofman, 2003; IASC, 2012). Several studies present assessment reports related to past real-world disasters (e.g., BNPB et al., 2009; The Government of the Republic of Haiti, 2010; GFDRR, 2013). Tatham (2009) discusses the use of unmanned aerial vehicles for conducting rapid needs assessments. Several articles focus on the sampling methods used after real-world disasters [e.g., Texas hurricane (Waring et al., 2005)], Indonesian tsunami (Garces et al., 2010), and Turkey earthquake (Daley et al., 2001). ACAPS (2011) and ACAPS (2014) specifically focus on the purposive sampling method and describe site selection process in detail.

2.2 Related Routing Problems

The traditional Vehicle Routing Problem finds the cost-minimizing routes that visit all nodes in a given network. However, completing routes within a limited time period is critical in some applications, which makes node selection another decision. There are many studies in the OR literature that consider site selection and vehicle routing decisions simultaneously. The SARP in this study is mostly similar to the Selective Routing Problem (also known as the Team Orienteering Problem (TOP)), which aims to maximize the total profit collected from the visited locations while completing each route within a given time duration; the problem is first introduced by Butt and Cavalier (1994) as the Multiple Tour Maximum Collection Problem, and then named by Chao et al. (1996a) as the TOP.

A variety of modeling and solution approaches are developed for the TOP and the Orienteering Problem (OP), which is the single-vehicle version of the TOP [see, e.g., review studies by Feillet et al. (2005), Vansteenwegen et al. (2011), and Archetti et al. (2013)]. There are a few studies that focus on developing exact methods for the TOP; for instance, Butt and Ryan (1999) and Boussier et al. (2007) present branch and price algorithms. Since the OP and the TOP are NP-hard, most studies focus on developing heuristics. There are several studies that focus on developing greedy/constructive heuristics for the OP and the TOP. Some of these heuristics assign an attractiveness/desirability score to each node and attempt to select attractive nodes for inserting to the routes (e.g., Tsiligirides, 1984; Golden et al., 1987). The attractiveness scores are usually defined as a function of profits and travel distances. Chao et al. (1996b) develop a greedy constructive algorithm for the OP, which is adapted to the TOP by Chao et al. (1996a); in these heuristics, several paths are generated by applying the cheapest insertion heuristic, and then the path(s) with the largest total profit is(are) selected to generate an initial solution. The authors apply exchange and remove/insert procedures to improve the initial solution. More recently, the literature focuses on developing metaheuristic algorithms for the TOP. Tang and Miller-Hooks (2005) develop a tabu search heuristic that involves an adaptive memory procedure. Archetti et al. (2007) present a tabu search algorithm and a variable neighborhood search algorithm. Other metaheuristics developed for the TOP involve ant colony optimization (Ke et al., 2008), guided local search (Vansteenwegen et al., 2009), greedy randomized adaptive search procedure with path relinking (Souffriau et al., 2010), particle swarm optimization (Dang et al., 2013), and memetic algorithm (Bouly et al., 2010).

Although the recent literature mostly focuses on developing advanced algorithms that can find near-optimal solutions for the TOP, the aforementioned greedy heuristics are still attractive as they can find good solutions quickly; indeed, the greedy heuristics developed for the TOP are widely used as benchmarks in the studies that present metaheuristics for the TOP. We are inspired by the competitive performance of the greedy algorithms that can effectively solve the TOP and focus on developing a greedy heuristic for the SARP.

The SARP is a variant of the TOP with a different objective. While the TOP maximizes total collected profits, the SARP maximizes the minimum coverage ratio across the characteristics. Further, TOP and SARP differ in terms of the benefits that can be collected by visiting a node; in TOP, the benefit from visiting a site is explicitly defined by a single measure (e.g., profit, revenue, etc.), whereas in SARP, the benefit of visiting a site depends on the set of characteristics carried by the site as well as the characteristics of the other sites that will be included in the assessment schedule. Due to these differences, the existing solution algorithms developed for the TOP cannot be directly applied to solve the SARP. More specifically, the greedy algorithms developed for the TOP focus on identifying the most profitable node(s) to include in the solution, while in SARP one needs to develop a procedure that can identify the most promising node(s) for improving the minimum coverage ratio. In this study, we take a first step and develop an easy-to-implement greedy heuristic that can quickly find reasonable solutions for the SARP.

3 The Selective Assessment Routing Problem

We focus on the rapid needs assessment operations that start immediately after a disaster and must be completed within a few days. Assessment teams travel around the affected region during the assessment period and visit a number of sites to collect information about the impact of disaster on different communities. Since assessments must be completed quickly, teams can visit a limited number of sites. We focus on the purposive sampling method, which is widely applied in the rapid needs assessment stage for selecting sites to visit (ACAPS, 2011). To implement purposive sampling, first, critical characteristics (such as topography, ethnicity, and income sources) that may affect post-disaster conditions of the communities are determined. Then assessment teams select a number of sites that will allow them to evaluate and compare the post-disaster conditions of different community groups; for instance, to understand how a flood might have affected people in fishermen villages versus people in agricultural regions, visiting at least one coastal site and one rural site may be necessary. Finally, the selected sites are assigned to the teams, and the order of visits to sites is determined for each team.

The SARP involves site selection and vehicle routing decisions of the rapid needs assessment teams that apply a purposive sampling strategy. We assume that the critical characteristics of interest are chosen in advance. Also, the critical characteristics that can be observed at each site are known. Each site may carry a different set of critical characteristics. Each team can travel through the region for a specific time period. For simplicity, we focus on travel times and ignore the service durations at the visited sites; if necessary, estimated service times can be easily incorporated into the travel time matrix.

Example routes for the SARP are illustrated in Fig. 1; in this example, there are four characteristics of interest, two of which correspond to geographical aspects (mountainous and coastal), and the other two are related to disaster impact (high and medium). According to the figure, the two teams visit a total of seven sites, observing the effects of disaster in mountainous and coastal regions once and three times, respectively. Also, the impact of the disaster on communities that live in high impact and medium impact areas is observed three times and two times, respectively.

In SARP, we define a coverage type objective to ensure a balanced coverage of the selected critical characteristics. More specifically, we assume each characteristic

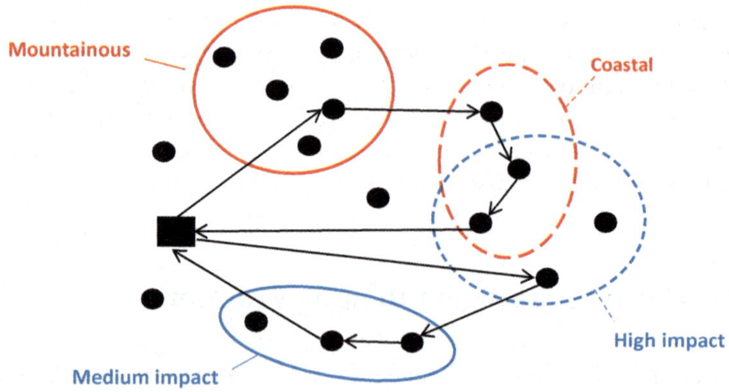

Fig. 1 Example routes for the SARP

must be observed at least once; further, if the duration of assessment period allows one to observe a characteristic multiple times, the characteristic that exists in the network most often is given the priority for additional observations. For instance, while the assessment routes in Fig. 1 visit all of the three coastal sites in the network, only one of the five mountainous sites is visited. In this case, it may be desirable to increase the number of visits to the mountainous sites, while decreasing the number of visits to the coastal sites. To ensure a balanced coverage of the critical characteristics, we define an objective that *maximizes the minimum coverage ratio* across the critical characteristics, where coverage ratio is calculated by dividing the number of times that a characteristic is covered by the number of times that characteristic exists in the network.

To model the SARP, let $\mathcal{N}$ represent the set of sites in the affected region. There are K teams (vehicles); each team $k \in \mathcal{K}$ departs from the origin node $\{0\}$ and returns back to the origin node after completing visits to the sites. Let $\mathcal{N}_0 = \mathcal{N} \cup \{0\}$. The travel time between nodes is represented by t_{ij} $\forall i, j \in \mathcal{N}_0$. Each vehicle is allowed to travel at most T_{max} time units. Let $\mathcal{C}$ denote the set of critical characteristics of interest. A coverage parameter α_{ic} is defined, which takes the value 1 if node $i \in \mathcal{N}$ carries characteristic $c \in \mathcal{C}$, and 0 otherwise. The number of sites that carry characteristic $c \in \mathcal{C}$ is represented by τ_c.

There are site selection variables y_{ik}, which take the value of 1 if team $k \in \mathcal{K}$ visits site $i \in \mathcal{N}$, and 0 otherwise. Finally, x_{ijk} are routing variables, which take the value of 1 if team $k \in \mathcal{K}$ visits site $j \in \mathcal{N}_0$ after site $i \in \mathcal{N}_0$, and 0 otherwise. The formulation for the SARP is

$$\text{maximize} \quad Z, \tag{1}$$

$$\text{subjectto}: \quad Z \leq \sum_{i \in \mathcal{N}} \sum_{k \in \mathcal{K}} \alpha_{ic} y_{ik} / \tau_c \qquad \forall c \in \mathcal{C}, \tag{2}$$

$$\sum_{j \in \mathcal{N}_0} x_{ijk} = y_{ik} \qquad \forall i \in \mathcal{N}_0, k \in \mathcal{K}, \tag{3}$$

$$\sum_{j \in \mathcal{N}_0} x_{jik} = y_{ik} \qquad \forall i \in \mathcal{N}_0, k \in \mathcal{K}, \tag{4}$$

$$\sum_{k \in \mathcal{K}} y_{ik} \leq 1 \qquad \forall i \in \mathcal{N}, \tag{5}$$

$$\sum_{k \in \mathcal{K}} y_{0k} \leq K, \tag{6}$$

$$\sum_{i \in \mathcal{N}_0} \sum_{j \in \mathcal{N}_0} t_{ij} x_{ijk} \leq T_{max} \qquad \forall k \in \mathcal{K}, \tag{7}$$

$$u_i - u_j + N x_{ijk} \leq N - 1 \qquad \forall i \in \mathcal{N}, j \in \mathcal{N} (i \neq j), k \in \mathcal{K}, \tag{8}$$

$$Z \geq 0, \tag{9}$$

$$u_i \geq 0 \qquad \forall i \in \mathcal{N}, \tag{10}$$

$$x_{ijk} \in \{0, 1\} \qquad \forall i \in \mathcal{N}_0, j \in \mathcal{N}_0, k \in \mathcal{K}, \tag{11}$$

$$y_{ik} \in \{0, 1\} \qquad \forall i \in \mathcal{N}_0, k \in \mathcal{K}. \tag{12}$$

The objective function (1) maximizes the minimum coverage ratio, which is defined through constraints (2). Constraints (3) and (4) ensure that an arc enters and leaves each selected site. Constraints (5) guarantee that each site is visited at most once. Constraint (6) limits the number of routes by the available number of teams. Constraints (7) ensure that each route is completed within the allowed duration. Constraints (8) are for eliminating subtours (adapted from Miller et al., 1960). Constraints (9) and (10) define positive variables, and constraints (11) and (12) define the binary routing and site selection variables, respectively.

The SARP is a variant of the TOP, which is NP-hard. Humanitarian organizations usually do not have access to computational resources (such as software and staff) that would be necessary to apply advanced algorithms for solving difficult optimization problems. Moreover, needs assessments need to be planned immediately after the disaster in an inherently chaotic environment. Therefore, simple tools that would allow relief agencies to make effective and quick needs assessment decisions are valuable. We present an easy-to-implement greedy heuristic that can find reasonable solutions for the SARP.

4 A Greedy Heuristic for the SARP

As discussed in Sect. 2, there are various greedy algorithms in the literature that can effectively solve the TOP. The existing algorithms usually involve sorting the nodes with respect to a measure (such as profits or attractiveness scores). In SARP, it is difficult to measure and fix the benefit associated with visiting a specific node in advance. Since the SARP maximizes the minimum coverage ratio across the characteristics, the benefit of visiting a particular node not only depends on the set of characteristics carried by the node but also on the characteristics of the other nodes that will appear in the solution.

We design a constructive greedy heuristic (GH) that re-evaluates the benefit to be obtained from each unvisited node in each iteration. More specifically, in each iteration, the GH determines the characteristics with a low coverage ratio and specifies one or more of these characteristics as "*critical characteristic(s).*" Then the algorithm checks whether a node that carries a critical characteristic can be inserted in a new or an existing route feasibly. Among the candidate nodes that carry the critical characteristic(s), the node that gives the cheapest insertion is considered as the next node to be inserted in an existing route. The algorithm continues until no feasible insertions could be obtained.

We develop three versions of the GH, which slightly differ in terms of the method used to determine critical characteristics. Depending on the method used, a different number of candidate nodes are evaluated for cheapest insertion in each iteration. Specifically, in the first version, called the GHa, the characteristic(s) with the lowest coverage ratio and the largest τ_c value is(are) selected as the critical characteristic(s). Different than the GHa, the second version of the heuristic, the GHb, considers all characteristics that have the lowest coverage ratio as critical without checking the value of τ_c; hence, a larger set of candidate nodes may enter the cheapest insertion evaluation in each iteration of the GHb compared to the GHa. Finally, in the third version, GHc, the lowest and the second lowest coverage ratios are used to define the critical characteristics, which may further increase the size of the candidate node set. The pseudocode for the GHa is provided in Appendix 1; note that GHb and GHc follow the same structure; the only difference is in the definition of the set $\mathcal{A}$, which involves the critical characteristics.

By focusing only on the characteristic(s) with the lowest coverage ratio and the largest τ_c value, the GHa evaluates a relatively small number of nodes for cheapest insertion in each iteration and hence acts more greedily than the other versions in improving the coverage ratio. On the other hand, the GHc gives more importance to keeping travel times low by evaluating the largest set of candidate nodes for cheapest insertion in each iteration. The GHb evaluates a moderate number of candidate nodes in each iteration, and hence acts in between GHa and GHc in balancing the tradeoff between travel times and coverage ratio. We compare the performance of the GH versions in Sect. 5.

5 Numerical Analysis

We describe test instances in Sect. 5.1, present example SARP solutions in Sect. 5.2, and evaluate the performance of the GH versions in Sect. 5.3.

5.1 Test Instances

We develop a set of test instances involving 25, 50, 75, and 100 nodes. All of our instances are generated by modifying Solomon's 100 node random (R) and random-clustered (RC) instances (Solomon, 1987). We directly use Solomon's 100 node instances to generate our 100 node R and RC instances. To obtain 25, 50, and 75 node R instances, we select the first 25, 50, and 75 nodes from the Solomon's 100 node R instance, respectively. To obtain 25, 50, and 75 node RC instances, we select a subset of nodes from the Solomon's 100 node RC instance in a way that the random-clustered network structure is maintained; for instance, we get the first 15 and the last 10 nodes of the Solomon's 100 node RC network to obtain a 25 node RC network. Example 25 node R and RC networks are presented in Appendix 2.

We specify 12 critical characteristics for each test instance. Three of these characteristics are assumed to be related to disaster impact (high, medium, and low) and these characteristics are assigned to the nodes based on their proximity to the disaster epicenter. Specifically, if the Euclidean distance between a node and the disaster epicenter is less than 35 units, that node is assumed to be affected by the disaster at the highest level, and the nodes that are farther than 70 units from the disaster epicenter are assumed to be affected by the disaster at the lowest level. The remaining nodes are assigned to the medium-impact zone.

In real-world settings, geographically close sites may be similar to each other in some of the characteristics of interest. For example, there may be several close villages that involve communities with the same ethnicity, income source, or language. In such networks, assessment teams may avoid visiting close sites since the majority of information that can be obtained from the close sites would be similar. To address such settings, we apply the following procedure that considers geographical proximity while assigning site characteristics to the nodes. Specifically, our procedure assumes that close nodes are identical to each other in terms of three randomly chosen characteristics, and the remaining six characteristics are assigned to the nodes at random. Appendix 2 presents characteristic matrices (i.e., α_{ic} values), for example, 25 node networks; the last three columns of the tables in Appendix 2 correspond to the disaster impact characteristics, and the remaining ones are assigned by applying the procedure below.

Step 0. Calculate the average distance D across all nodes in the network.

Step 1. Select a site randomly and denote it as node i. Assign each characteristic to node i randomly; that is,

draw a random number for each characteristic and decide whether the node carries the characteristic or not.

Step 2. Consider node j that is closest to node i.

Step 3. If the distance between nodes i and j is less than D, go to 3.1; otherwise go to Step 3.2.

Step 3.1. Pick three characteristics carried by node i randomly and assign these characteristics to node j.

Assign the remaining six characteristics to node j randomly.

Step 3.2. Assign all nine characteristics to node j randomly.

Step 4. Set $i := j$ and go to Step 2.

We consider different values for K and T_{max} for each network, and as a result we obtain 48 instances. We discuss example SARP solutions in the next subsection.

5.2 SARP Solutions

We solve each of the 48 instances by IBM ILOG OPL/CPLEX by limiting the solution time to 2 h. All computational experiments are carried out on a 64-bit

Windows Server with two 2.0 GHz Intel Xeon CPU's and 32 GB RAM. The best integer solution obtained by CPLEX (Z(CPLEX)), the optimality gaps, and the solution times (in seconds) are provided in Table 1. As observed from the table, only three of the 25 node R type instances could be solved optimally by CPLEX. For other instances, optimal solutions could not be guaranteed within 2 h and large optimality gaps are observed in general.

The SARP solutions obtained for different K and T_{max} values are examined. Since CPLEX could not guarantee optimality for most of the problem instances, our observations are based on the best solutions obtained by CPLEX. Example solutions for the 25 node instances involving two vehicles are illustrated in Fig. 2; specifically, Fig. 2a, b illustrates routes obtained by using different T_{max} values on R and RC type networks, respectively. In Fig. 2b, one of the routes is not explicitly illustrated for keeping the figure clear as the route is long and exhibits zig zagging behavior; the selected nodes are circled without specifying the exact visiting order for that instance.

As expected, route lengths increase with increasing T_{max} for both network types. For small T_{max} values, routes cover a restricted geographical area. As illustrated in Fig. 2b, the teams may not have the opportunity to visit some node clusters when time is limited.

When we examine route composition, we observe that some nodes are consistently visited in all solutions. For instance, nodes 1, 2, and 22 are visited in all solutions of the R instances in Fig. 2a. Note that these nodes are relatively close to the depot. Additionally, we observe that these nodes carry at least one characteristic with a relatively small or large τ_c value, and the other nodes carrying those characteristics are widely dispersed in the network. Specifically, nodes one and two in Fig. 2a carry characteristic four with $\tau_4 = 12$, and most of the other nodes that carry this characteristic are distant to the origin (see Appendix 2 for node characteristics). Additionally, node 22 carries the characteristic 12 with $\tau_{12} = 6$, and most of the other nodes carrying characteristic 12 are far from the depot. We make similar observations for the RC network solutions in Fig. 2b. Again, we observe that there are some nodes common to all solutions; specifically, nodes 16, 19, 21, 23, and 25 are visited in all of the solutions with different T_{max} values. As before, we observe that these nodes are generally advantageous in terms of travel times. Also, nodes that include characteristics with relatively small or large τ_c values (e.g., characteristics 2, 3, and 11) are preferred.

According to the best solutions obtained, we observe that routes do not necessarily expand by adding nodes to the current solution when T_{max} is increased. That is, the visited node set may significantly change for different T_{max} values. For instance, in the solutions of the R instances in Fig. 2a, nodes 3, 4, 13, and 21, which are visited when T_{max} is two or three, are not chosen when T_{max} is increased to four. Similarly, in the RC instances, nodes 18 and 20 are visited when T_{max} is two; however, these nodes do not appear in the solution when T_{max} is three.

In summary, example SARP solutions indicate that the assessment routes may be highly affected by the network structure, assessment duration, and how the characteristics of interest are dispersed through the nodes of the network. Our GH

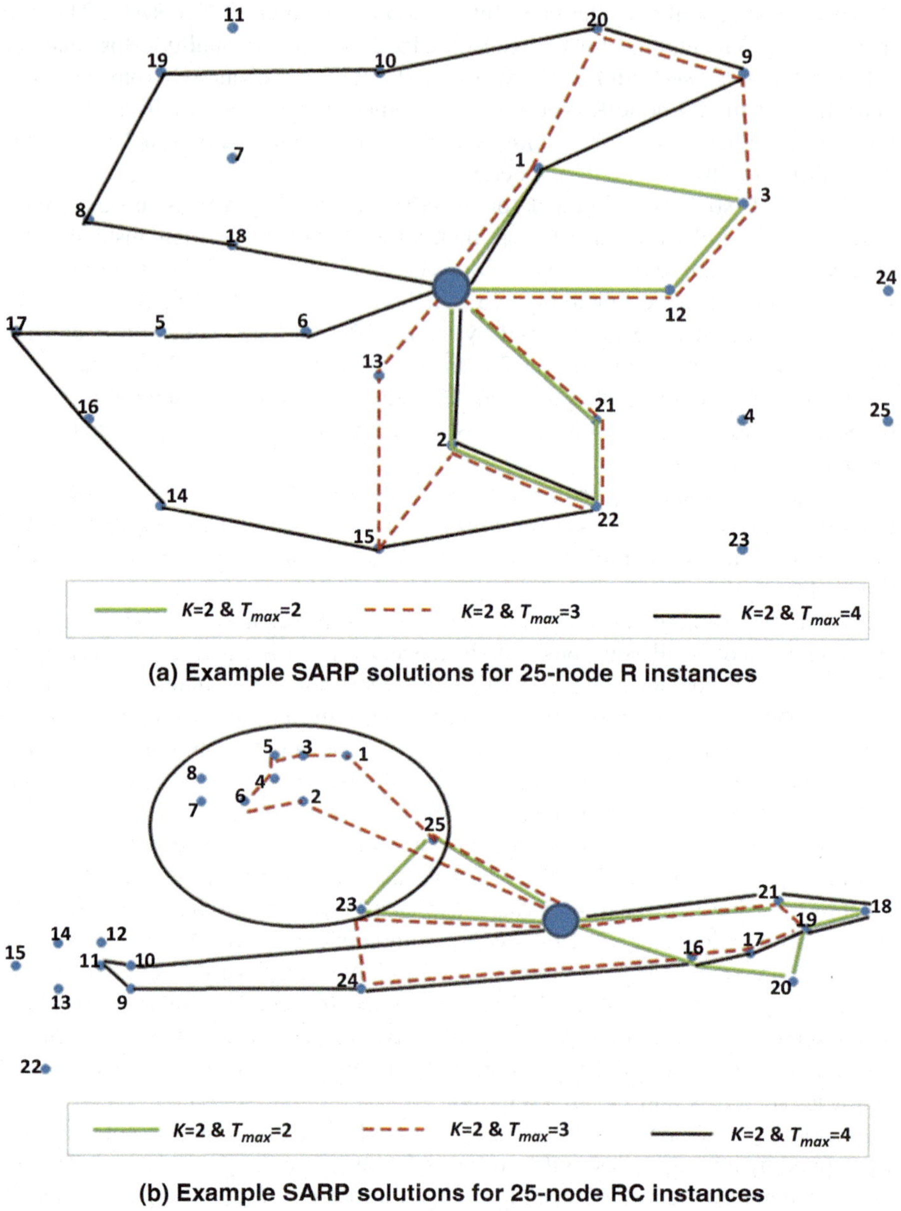

Fig. 2 Example SARP solutions. (**a**) For 25-node R instances. (**b**) For 25-node RC instances

algorithm constructs assessment routes by identifying a set of promising nodes in each iteration, whose insertion may improve the minimum coverage ratio value. In the next subsection, we evaluate the performance of the GH in solving the SARP.

5.3 Heuristic Performance

We solve each of the 48 problem instances by applying the three versions of the GH. The objective function values obtained by the GH versions (Z(GHa), Z(GHb), and Z(GHc)) are provided in Table 1. Each heuristic solution is within a second.

When we compare the solutions obtained by different GH versions, we observe that GHb finds the same solution or a better solution compared to other GH versions in 77 % of the instances, whereas GHa and GHc provide an equivalent or superior solution than the other GH versions in 42 and 50 % of the instances, respectively. These results suggest that GHb tends to perform better than other GH versions. This may be because GHb balances the objectives of improving the coverage ratio and reducing traveling time; recall that in each iteration of the algorithm, GHa focuses on improving the coverage ratio in a more greedy way, where GHc gives more importance to route efficiency. Although GHb solutions are more promising, since solving an instance by applying any GH version takes less than a second, applying three versions of the heuristic simultaneously and choosing the best solution could be practical. We refer to the best solution obtained by the GH versions as the "best GH solution" and its objective value as the GH_{best}. The last column of Table 1 presents the percentage gap between the GH_{best} and the Z(CPLEX); the bold numbers in this column indicate the instances for which the best GH solution value is equivalent to or better than the best CPLEX solution.

When the best CPLEX solutions are compared with the best GH solutions, we observe that CPLEX solutions are dominated by the best GH solutions in 56.25 % of the instances; also, CPLEX and the GH achieve the same solution in 18.75 % of the instances. Furthermore, when medium/large problem instances with 50 or more nodes are considered, we observe that GH achieves a better solution than the CPLEX in about 89 % of the instances. For the 100 node instances, while GHb generally gives the best results, any version of the GH generally performs better than CPLEX, and the percentage gap between GH_{best} and Z(CPLEX) can reach 65.74 % for the R instances and 73.61 % for the RC instances.

In summary, numerical results show that the GH provides competitive solutions for the SARP. The GHb can be applied to obtain quick and reasonable solutions for the SARP in practice. Further, it is possible to improve solution quality by applying the three versions of the GH and choosing the best solution among them.

6 Conclusion

This study focuses on the rapid needs assessment process in humanitarian relief and introduces a new problem that addresses site selection and routing decisions for the assessment teams. A purposive sampling strategy is considered; accordingly, assessment teams aim to choose and visit sites that involve different community groups with distinct characteristics. We develop a mathematical model that ensures

Table 1 Comparison of solutions obtained by CPLEX and GH versions

Network	N	K	T_{max}	O. Gap (%)	S. time (s)	Z(CPLEX)	Z(GHa)	Z(GHb)	Z(GHc)	$\frac{GH_{best} - Z(CPLEX)}{Z(CPLEX)} * 100(\%)$
R	25	2	2	0.0	44.7	0.167	0.000	0.000	0.000	−100.00
R	25	2	3	0.0	1009.9	0.333	0.167	0.143	0.000	−49.85
R	25	2	4	35.2	7208.9	0.500	0.333	0.333	0.167	−33.40
R	25	3	2	0.0	304.2	0.167	0.000	0.000	0.000	−100.00
R	25	3	3	75.8	7259.4	0.444	0.286	0.500	0.500	**12.61**
R	25	3	4	20.0	7223.8	0.833	0.571	0.833	0.667	**0.00**
R	50	3	3	86.3	7272.9	0.333	0.250	0.250	0.333	**0.00**
R	50	3	4	62.0	7237.3	0.500	0.476	0.476	0.476	−4.80
R	50	3	5	38.8	7274.6	0.714	0.667	0.714	0.583	**0.00**
R	50	4	3	140.6	7268.8	0.333	0.389	0.444	0.429	**33.33**
R	50	4	4	71.4	7281.5	0.583	0.611	0.667	0.556	**14.41**
R	50	4	5	16.7	7274.5	0.857	0.833	0.917	0.833	**7.00**
R	75	3	3	89.2	7271.3	0.294	0.222	0.294	0.304	**3.40**
R	75	3	4	101.6	7294.5	0.353	0.429	0.471	0.444	**33.43**
R	75	3	6	48.4	7267.0	0.652	0.652	0.739	0.714	**13.34**
R	75	5	3	120.1	7339.9	0.389	0.391	0.471	0.471	**21.08**
R	75	5	4	70.0	7249.2	0.588	0.619	0.706	0.690	**20.07**
R	75	5	6	12.5	7210.1	0.889	0.950	1.000	1.000	**12.49**
R	100	3	4	59.1	7210.1	0.435	0.435	0.444	0.435	**2.07**
R	100	3	6	33.7	7218.8	0.690	0.706	0.703	0.696	**2.32**
R	100	3	8	35.3	7208.8	0.739	0.882	0.913	0.913	**23.55**
R	100	6	4	76.2	7207.8	0.568	0.680	0.786	0.652	**38.38**
R	100	6	6	64.7	7245.0	0.607	1.000	1.000	1.000	**64.74**
R	100	6	8	38.9	7353.3	0.720	1.000	1.000	1.000	**38.89**
RC	25	2	2	44.4	7203.8	0.154	0.000	0.154	0.154	**0.00**
RC	25	2	3	83.2	7209.8	0.437	0.000	0.167	0.154	−61.78

Network	N	K	T_{max}	O. Gap (%)	S. time (s)	Z(CPLEX)	Z(GHa)	Z(GHb)	Z(GHc)	$\frac{GH_{best}-Z(CPLEX)}{Z(CPLEX)} * 100(\%)$
RC	25	2	4	50.0	7235.3	0.667	0.385	0.462	0.462	−30.73
RC	25	3	2	402.1	7231.6	0.154	0.125	0.154	0.154	**0.00**
RC	25	3	3	50.0	7203.5	0.667	0.385	0.231	0.231	−42.28
RC	25	3	4	12.5	7208.6	0.889	0.667	0.778	0.833	−6.30
RC	50	3	3	207.1	7250.4	0.286	0.286	0.286	0.286	**0.00**
RC	50	3	4	150.0	7243.1	0.400	0.480	0.500	0.500	**25.00**
RC	50	3	5	100.0	7264.4	0.500	0.643	0.714	0.714	**42.80**
RC	50	4	3	180.0	7285.8	0.357	0.286	0.321	0.321	−10.08
RC	50	4	4	55.6	7227.5	0.643	0.643	0.500	0.500	**0.00**
RC	50	4	5	33.3	7257.9	0.750	0.786	0.714	0.714	**4.80**
RC	75	3	3	1111.4	7431.0	0.063	0.063	0.063	0.063	**0.00**
RC	75	3	4	182.3	7249.3	0.313	0.333	0.154	0.125	**6.39**
RC	75	3	6	65.0	7235.5	0.606	0.576	0.591	0.576	−2.48
RC	75	5	3	1500.0	7244.5	0.063	0.063	0.063	0.063	**0.00**
RC	75	5	4	73.7	7288.5	0.576	0.500	0.438	0.250	−13.19
RC	75	5	6	10.0	7225.4	0.909	0.909	1.000	0.929	**10.01**
RC	100	3	4	208.3	7222.6	0.242	0.360	0.360	0.240	**48.76**
RC	100	3	6	87.4	7442.0	0.515	0.560	0.560	0.545	**8.74**
RC	100	3	8	35.9	7220.2	0.736	0.788	0.758	0.788	**7.07**
RC	100	6	4	177.8	7353.4	0.360	0.560	0.571	0.400	**58.61**
RC	100	6	6	73.7	7319.0	0.576	0.935	1.000	1.000	**73.61**
RC	100	6	8	56.3	7207.4	0.640	1.000	1.000	1.000	**56.25**
% of instances with $GH_{best} > Z(CPLEX)$										56.25 %
% of instances with $GH_{best} = Z(CPLEX)$										18.75 %
% of instances with $GH_{best} < Z(CPLEX)$										25.00 %

coverage of the selected characteristics by maximizing the minimum coverage ratio in the objective function; and the necessity of completing assessments within the given deadline is captured via travel duration constraints. We develop an iterative greedy heuristic for the problem, which inserts a new node to the solution in each iteration. We evaluate three versions of the greedy heuristic, which differ from each other in terms of the number of candidate nodes evaluated for insertion to routes in each iteration. Our numerical results show that all versions of the greedy heuristics can provide reasonable solutions for the SARP especially for medium/large size instances; furthermore, the heuristic version that balances the tradeoff between coverage ratio and travel time gives the most competitive results.

Needs assessment operations have not received much attention in the humanitarian logistics literature before; hence, there may be several avenues for future research. For instance, accessibility may be a major concern in the post-disaster environment; therefore, it may be valuable to incorporate travel time uncertainty in making site selection and routing decisions. Also, uncertainties may arise during the sampling process; for instance, the characteristics that the sites carry may not be known in advance and site selection decision may need to be updated as more information is collected. Finally, future research can focus on developing solution algorithms that can solve the SARP more effectively; even though computational resources may not be available to run advanced algorithms in the field, it would be valuable to obtain better benchmark solutions to evaluate the performance of practical heuristics.

Acknowledgements This work was presented at the 2nd International Conference on Dynamics of Disasters, Kalamata, Greece, June 29–July 2, 2015. This research has been funded by the Scientific and Technological Research Council of Turkey (TUBITAK) Career Award [213M414]. The author would like to thank Ilknur Singin, Alperen Talaslioglu, Busra Uydasoglu, Burak Guragac, and Yasin Dogan for their help with various phases of this ongoing project. The author would also thank to the Science Academy of Turkey for the BAGEP research award.

Appendix 1

The pseudocode for the GHa is provided in Algorithm 1. The pseudocodes for GHb and GHc are not provided as they are the same as Algorithm 1 except the definition of the set $\mathcal{A}$. Specifically, to adapt the following pseudocode for the GHb, one can define set $\mathcal{A}$ as the characteristic(s) with the smallest value of r_c. Finally, in GHc, set $\mathcal{A}$ includes all characteristics that have the smallest or the second smallest values of r_c.

Algorithm 1: Greedy Heuristic (GHa)

// Initialization
r_c = coverage ratio for characteristic c
Set $\mathcal{A}$ = characteristic(s) with the smallest value of r_c and the largest value of τ_c
Set $\mathcal{S}_A$ = set of nodes that carry a characteristic from $\mathcal{A}$
Set $\mathcal{N}_k$ = set of nodes visited by vehicle k
boolean b := true;

// Main loop
while *(b true)* **do**
  // Check whether there is an unused vehicle or not:
  if $\exists \mathcal{N}_k == \varnothing$ **then**
    // Start a new route:
    k:= an unused vehicle;
    z:= the closest node $i \in \mathcal{S}_A$ to the origin node;
    if *(insertion of z to the route of vehicle k is time-feasible)* **then**
      Add node z to the route of vehicle k; $\mathcal{N}_k := \mathcal{N}_k \cup \{z\}$;
      Update: r_c, $\mathcal{A}$, $\mathcal{S}_A$
    end
    else
      b := false;
    end
  end
  else
    //Insert to an existing route:
    Calculate the cost of cheapest insertion $\forall i \in \mathcal{S}_A$ to all existing routes;
    z:= the node that gives the cheapest insertion;
    k:= the vehicle that gives the cheapest insertion;
    if *(insertion of z to the route of vehicle k is time-feasible)* **then**
      Add node z to the route of vehicle k; $\mathcal{N}_k := \mathcal{N}_k \cup \{z\}$;
      Update: r_c, $\mathcal{A}$, $\mathcal{S}_A$
    end
    else
      b := false;
    end
  end
end

Appendix 2

Example networks and characteristic matrices are presented in Fig. 3 and Tables 2 and 3.

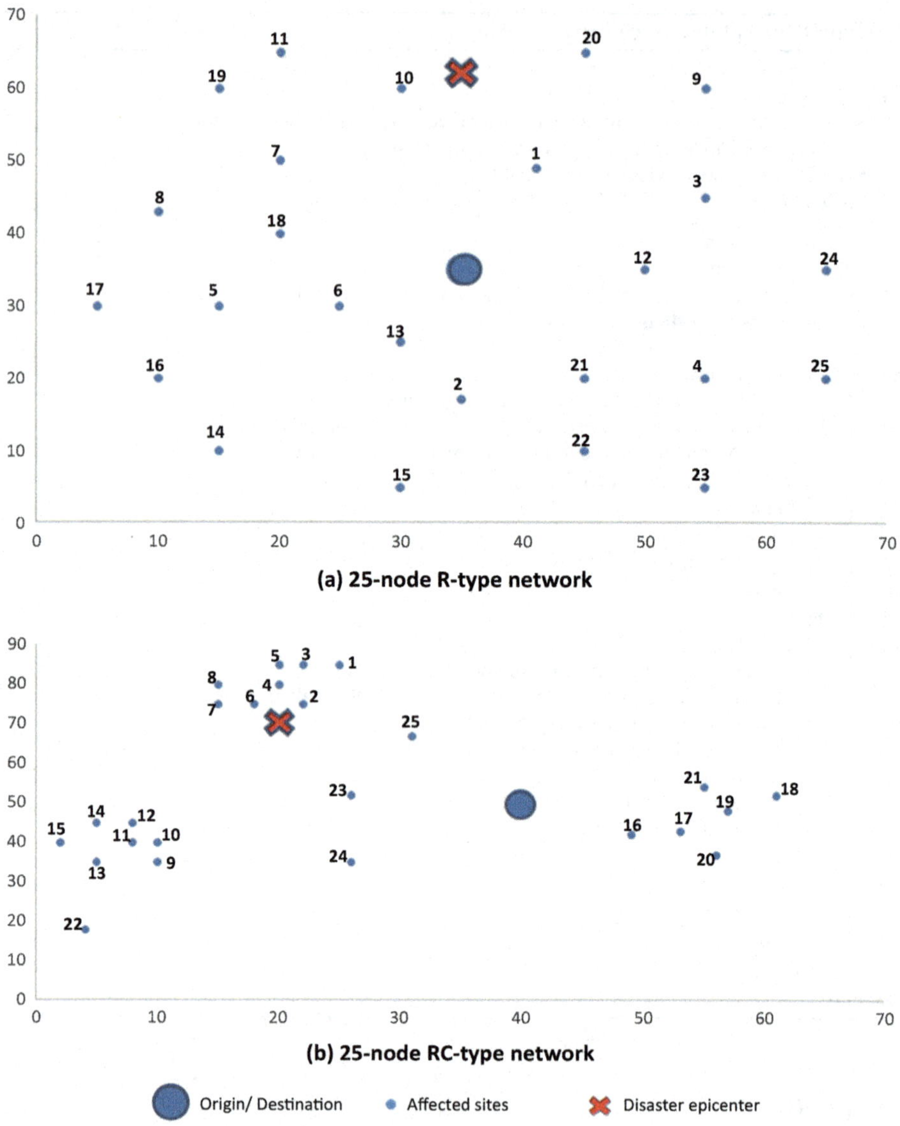

Fig. 3 Example 25 node networks used in problem instances. (**a**) 25-node R-type network. (**b**) 25-node RC-type network

Table 2 Node characteristics (α_{ic} values) for the 25 node R type network

	Characteristics											
Node	1	2	3	4	5	6	7	8	9	10	11	12
0	0	0	0	0	0	0	0	0	0	0	0	0
1	0	0	1	1	0	0	0	1	0	1	0	0
2	1	0	0	1	0	0	0	0	1	0	1	0
3	1	0	0	0	1	0	1	0	0	0	1	0
4	0	0	1	0	0	1	1	0	0	0	1	0
5	0	0	1	1	0	0	0	1	0	0	1	0
6	1	0	0	1	0	0	0	1	0	0	1	0
7	1	0	0	1	0	0	0	1	0	0	1	0
8	0	1	0	1	0	0	1	0	0	0	1	0
9	0	0	1	0	1	0	1	0	0	1	0	0
10	1	0	0	1	0	0	0	1	0	1	0	0
11	1	0	0	1	0	0	0	0	1	1	0	0
12	0	1	0	0	1	0	0	1	0	0	1	0
13	0	0	1	1	0	0	0	0	1	0	1	0
14	0	0	1	0	1	0	1	0	0	0	0	1
15	0	1	0	1	0	0	0	1	0	0	0	1
16	1	0	0	0	1	0	1	0	0	0	0	1
17	0	0	1	0	0	1	1	0	0	0	1	0
18	1	0	0	0	0	1	0	0	1	0	1	0
19	1	0	0	1	0	0	0	1	0	1	0	0
20	0	1	0	0	1	0	0	0	1	1	0	0
21	0	0	1	0	0	1	1	0	0	0	1	0
22	1	0	0	0	0	1	0	0	1	0	0	1
23	1	0	0	0	1	0	0	1	0	0	0	1
24	1	0	0	1	0	0	0	0	1	0	1	0
25	0	0	1	0	0	1	1	0	0	0	0	1
τ_c	12	4	9	12	7	6	9	9	7	6	13	6

Table 3 Node characteristics (α_{ic} values) for the 25 node RC type network

	Characteristics											
Node	1	2	3	4	5	6	7	8	9	10	11	12
0	0	0	0	0	0	0	0	0	0	0	0	0
1	0	1	0	1	0	0	0	0	1	0	1	0
2	1	0	0	0	0	1	0	1	0	1	0	0
3	0	0	1	1	0	0	0	0	1	0	1	0
4	0	0	1	0	0	1	0	0	1	1	0	0
5	0	0	1	1	0	0	1	0	0	0	1	0
6	0	0	1	0	1	0	0	1	0	1	0	0
7	0	0	1	0	1	0	0	1	0	1	0	0
8	0	0	1	1	0	0	1	0	0	1	0	0
9	0	0	1	1	0	0	0	0	1	0	1	0
10	1	0	0	0	1	0	0	0	1	0	1	0
11	1	0	0	0	0	1	0	1	0	0	1	0
12	0	0	1	0	0	1	0	1	0	0	1	0
13	0	0	1	1	0	0	1	0	0	0	1	0
14	0	0	1	0	0	1	0	0	1	0	1	0
15	0	0	1	1	0	0	0	0	1	0	1	0
16	0	1	0	0	0	1	1	0	0	0	0	1
17	1	0	0	0	1	0	1	0	0	0	0	1
18	0	0	1	0	0	1	0	0	1	0	0	1
19	0	0	1	1	0	0	1	0	0	0	0	1
20	0	1	0	1	0	0	0	1	0	0	0	1
21	0	1	0	0	0	1	0	0	1	0	1	0
22	0	0	1	0	1	0	0	1	0	0	0	1
23	1	0	0	0	1	0	0	1	0	0	1	0
24	0	0	1	0	1	0	0	1	0	0	1	0
25	0	0	1	0	1	0	0	0	1	1	0	0
τ_c	5	4	16	9	8	8	6	9	10	6	13	6

References

ACAPS (The Assessment Capacities Project): Technical Brief, Purposive sampling and site selection in phase 2. http://www.acaps.org/sites/acaps/files/resources/files/purposive_sampling_and_site_selection_in_phase_2.pdf (2011). Accessed 5 Sept 2016

ACAPS: Humanitarian Needs Assessment: The Good Enough Guide. ACAPS, Emergency Capacity Building Project (ECB) and Practical Action Publishing, Rugby (2014). http://reliefweb.int/sites/reliefweb.int/files/resources/h-humanitarian-needs-assessment-the-good-enough-guide.pdf. Accessed 5 Sept 2016

Archetti, C., Hertz, A., Speranza, M.G.: Metaheuristics for the team orienteering problem. J. Heuristics **13**, 49–76 (2007)

Archetti, C., Speranza, M.G., Vigo, D.: Vehicle routing problems with profits. Working Papers, Department of Economics and Management, University of Brescia (2013)

Balcik, B., Beamon, B.M., Smilowitz, K.: Last mile distribution in humanitarian relief. J. Intell. Transp. Syst. **12**, 51–63 (2008)

BNPB (Badan Nasional Penanggulangan Bencana): West Sumatra and Jambi natural disasters: damage, loss and preliminary needs assessment. http://reliefweb.int/report/indonesia/indonesia-west-sumatra-and-jambi-natural-disasters-damage-loss-and-preliminary (2009). Accessed 5 Sept 2016

Bouly, H., Dang, D.-C., Moukrim, A.: A memetic algorithm for the team orienteering problem. 4OR **8**, 49–70 (2010)

Boussier, S., Feillet, D., Gendreau, M.: An exact algorithm for the team orienteering problem. 4OR **5**, 211–230 (2007)

Butt, S.E., Cavalier, T.M.: A heuristic for the multiple tour maximum collection problem. Comput. Oper. Res. **21**, 101–111 (1994)

Butt, S.E., Ryan, D.M.: An optimal solution procedure for the multiple tour maximum collection problem using column generation. Comput. Oper. Res. **26**, 427–441 (1999)

Chao, I.M., Golden, B.L., Wasil, E.A.: The team orienteering problem. Eur. J. Oper. Res. **88**, 464–474 (1996a)

Chao, I.M., Golden, B.L., Wasil, E.A.: A fast and effective heuristic for the orienteering problem. Eur. J. Oper. Res. **88**, 475–489 (1996b)

Daley, W.R., Karpati, A., Sheik, M.: Needs assessment of the displaced population following the August 1999 earthquake in Turkey. Disasters **25**, 67–75 (2001)

Dang, D.-C., Guibadj, R.N., Moukrim, A.: An effective PSO-inspired algorithm for the team orienteering problem. Eur. J. Oper. Res. **229**, 332–344 (2013)

Darcy, J., Hofman, C.A.: Humanitarian needs assessment and decision-making. Humanitarian Policy Group Briefing Paper 13 (2003)

de la Torre, L.E., Dolinskaya, I.S., Smilowitz, K.R.: Disaster relief routing: integrating research and practice. Socio Econ. Plan. Sci. **46**, 88–97 (2012)

Feillet, D., Dejax, P., Gendreau, M.: Travelling salesman problems with profits. Transport. Sci. **39**, 188–205 (2005)

Garces, L.R., Pido, M.D., Pomeroy, R.S., Koeshendrajana, S., Prisantoso, B.I., Fatan, N.A., Adhuri, D., Raiful, T., Rizal, S., Tewfik, A., Dey, M.: Rapid assessment of community needs and fisheries status in tsunami-affected communities in Aceh Province, Indonesia. Ocean Coast. Manag. **53**, 69–79 (2010)

GFDRR (Global Facility for Disaster Reduction and Recovery): Post-disaster needs assessments. https://www.gfdrr.org/post-disaster-needs-assessments (2013). Accessed 5 Sept 2016

Golden, B.L., Levy, L., Vohra, R.: The orienteering problem. Nav. Res. Logist. **34**, 307–318 (1987)

Huang, M., Smilowitz, K.R., Balcik, B.: Models for relief routing: equity, efficiency and efficacy. Transport. Res. Part E: Logist. Transport. Rev. **48**, 2–18 (2012)

Huang, M., Smilowitz, K.R., Balcik, B.: A continuous approximation approach for assessment routing in disaster relief. Transport. Res. Part B: Methodol. **50**, 20–41 (2013)

IASC (Inter-Agency Standing Committee): Multi-cluster/sector initial rapid assessment (MIRA). https://docs.unocha.org/sites/dms/Documents/mira_final_version2012.pdf (2012). Accessed 5 Sept 2016

IFRC (International Federation of Red Cross and Red Crescent Societies): Disaster preparedness programme. http://www.ifrc.org/Global/Publications/disasters/all.pdf (2000). Accessed 5 Sept 2016

IFRC (International Federation of Red Cross and Red Crescent Societies): Guidelines for Assessment in Emergencies, Geneva (2008)

Ke, L., Archetti, C., Feng, Z.: Ants can solve the team orienteering problem. Comput. Ind. Eng. **54**, 648–665 (2008)

Lin, Y.H., Batta, R., Rogerson, P., Blatt, A., Flanigan, M.: A logistics model for emergency supply of critical items in the aftermath of a disaster. Socio Econ. Plan. Sci. **45**, 132–145 (2011)

Miller, C.E., Tucker, A.W., Zemlin, R.A.: Integer programming formulation of travelling salesman problems. J. ACM **7**, 326–329 (1960)

Ozdamar, L., Demir, O.: A hierarchical clustering and routing procedure for large scale disaster relief logistics planning. Transport. Res. Part E: Logist. Transport. Rev. **48**(3), 591–602 (2012)

Solomon, M.M.: Algorithms for the vehicle routing and scheduling problem with time window constraints. Oper. Res. **35**, 254–265 (1987)

Souffriau, W., Vansteenwegen, P., Vanden Berghe, G., Van Oudheusden, D.: A path relinking approach for the team orienteering problem. Comput. Oper. Res. **37**(11), 1853–1859 (2010)

Tang, H., Miller-Hooks, E.: A tabu search heuristic for the team orienteering problem. Comput. Oper. Res. **32**, 1379–1407 (2005)

Tatham, P.: An investigation into the suitability of the use of unmanned aerial vehicle systems (UAVS) to support the initial needs assessment process in rapid onset humanitarian disasters. Int. J. Risk Assess. Manag. **13**, 60–78 (2009)

The Government of the Republic of Haiti, Haiti Earthquake PDNA: Assessment of damage, losses, general and sectoral needs. http://siteresources.worldbank.org/INTLAC/Resources/PDNA_Haiti-2010_Working_Document_EN.pdf (2010). Accessed 5 Sept 2016

Tsiligirides, T.: Heuristic methods applied to orienteering. J. Oper. Res. Soc. **35**, 797–809 (1984)

Tzeng, G.H., Cheng, H.J., Huang, T.D.: Multi-objective optimal planning for designing relief delivery systems. Transport. Res. Part E: Logist. Transport. Rev. **43**, 673–686 (2007)

Vansteenwegen, P., Souffriau, W., Vanden Berghe, G., Van Oudheusden, D.: A guided local search metaheuristic for the team orienteering problem. Eur. J. Oper. Res. **196**(1), 118–127 (2009)

Vansteenwegen, P., Souffriau, W., Van Oudheusden, D.: The orienteering problem: a survey. Eur. J. Oper. Res. **209**, 1–10 (2011)

Waring, S., Zakos-Feliberti, A., Wood, R., Stone, M., Padgett, P., Arafat, R.: The utility of geographic information systems (GIS) in rapid epidemiological assessments following weather-related disasters: methodological issues based on Tropical Storm Allison experience. Int. J. Hyg. Environ. Health **208**, 109–116 (2005)

WHO (World Health Organization): Rapid needs assessment for water, sanitation and hygiene. http://globalhandwashing.org/wp-content/uploads/2015/04/WHO-SEARO-Rapid-Needs-Assessment.pdf (2004). Accessed 5 Sept 2016

Bridging the Gap

Preparing for Long-Term Infrastructure Disruptions

Rasmus Dahlberg

Abstract The fixed link between Denmark and Sweden connects two busy cities and a large international airport with many of its travelers and employees. 18,000 vehicles and 160 passenger trains transport each day more than 70,000 people across the combined road and rail Øresund Bridge and through the Øresund Tunnel, approximately 25,000 of them critical to the regional work market. Even though the risk analysis states that the likelihood of a long-term closure (100+ days) is very low Danish and Swedish transport authorities have demanded that the infrastructure operator conducts a survey of the preparedness plans already in place and map possible alternate travel routes for people and freight in case of long-term disruptions. This paper (a) delineates the concept of infrastructure, (b) describes the proceedings of the Work Group for Øresund Preparedness 2014–2016, and (c) discusses the findings presented in its final report to the Danish and Swedish transport authorities while drawing upon experiences from two recent comparable cases of infrastructure disruptions: The Champlain Bridge (2009) and the Forth Road Bridge (2015).

Keywords Infrastructure • Disruption • Resilience • Contingencies • Preparedness • Transport • Possibilism

1 Introducing Infrastructure

A bridge or a tunnel connecting two areas of land across a stretch of water is in daily speech an "infrastructure," as it allows people and goods to cross. A disruption of the infrastructure may occur in the shape of a low frequency, high-impact event such as a ship collision or plane crash that damages the bridge and renders it unusable for a prolonged time. However, demand for the service provided by the infrastructure remains, as people and goods still need to cross the water. After a

R. Dahlberg (✉)
Copenhagen Center for Disaster Research, University of Copenhagen, Copenhagen, Denmark

Danish Emergency Management Agency, Birkerød, Denmark
e-mail: dahlberg@sund.ku.dk

I.S. Kotsireas et al. (eds.), *Dynamics of Disasters—Key Concepts, Models, Algorithms, and Insights*, Springer Proceedings in Mathematics & Statistics 185,
DOI 10.1007/978-3-319-43709-5_3

while the infrastructure is (hopefully) repaired, and the service is restored to its previous state. Now, people and goods may again cross the bridge or pass through the tunnel unobstructed.

From a research point of view, however, an infrastructure has a certain duality to it in that it is at the same time a tangible technology built of concrete and steel or other materials *and* an intangible process involving flows of people, goods, energy, or information. In his 2013 paper, anthropologist Brian Larkin distinguished infrastructures from technologies by stating that "infrastructures are matter that enable the movement of other matter," and when they do so they become systems that "cannot be theorized in terms of the object alone." Systemic operation, in Larkin's terms, means that they are objects that "create the grounds on which other objects operate" (Larkin 2013, p. 329). Applied to a bridge or a tunnel this notion is self-evident: without traffic it is merely a technology, with it, is an infrastructure.

An often-repeated assumption is that infrastructures are by default "invisible," and that they only become visible when they break down (Star 1999; Chang 2009). Seen from an everyday point of view this makes sense as nobody notices the bridge or the tunnel until it fails—but then it will be all over the news. Larkin argues, however, that this notion is only a partial truth: "Invisibility is certainly one aspect of infrastructure, but it is only one and at the extreme edge of a range of visibilities that move from unseen to grand spectacles and everything in between" (Larkin 2013, p. 336). When working with long-term disruptions that have very low probabilities, but potentially huge consequences, Larkin's idea about a scale of visibility is relevant. By addressing the vulnerability of the infrastructure it might be possible to decrease its opaqueness just a little, thus enabling owners, users, and policy makers to better prepare for a contingency.

A subset of the broader concept of infrastructure is the so-called critical infrastructures (CI). These are assets or systems that are critical for the maintenance of vital societal functions, providing services that citizens rely on in their daily life—i.e., power and water supply, healthcare, transport, electronic communication, and banking (Kozine et al. 2015). In other words, a vital societal function delivers a service needed (or at least valued) by society while an infrastructure is a system that enables or supports the delivery of that function. It follows from this definition that a specific vital societal function may be delivered by multiple infrastructures, i.e., a number of power plants all producing electricity to a city interchangeably or two bridges crossing the same body of water. If a vital societal function relies on an infrastructure that has no alternatives, that infrastructure is per definition a CI.

While infrastructure itself has its conceptual roots in the Enlightenment idea of a "world in movement and open to change where the free circulation of goods, ideas, and people created the possibility of progress" (Larkin 2013, p. 332), protection of critical infrastructures only became an important task for the modern industrial state (Brown 2006, p. ix). Traditionally, Critical Infrastructure Protection (CIP) has been very focused on physical protection, but increased interdependency and use of digital systems, especially networks, has since 2000 prompted a turn towards resilience (Chang 2009; Biringer et al. 2013, p. 75; Dahlberg et al. 2015a, b). A resilience approach to CIP acknowledges that all threats from either natural hazards or intentional man-made attacks cannot be avoided or deflected, and therefore, CI must be able to some extent to absorb unexpected perturbations without losing functionality (Boin and McConnell, p. 52). This approach to infrastructure is informed by complexity theory and focuses on the interdependencies of many nodes and actors (Vespignani 2010, p. 984).

Biringer et al. identify three "lines of defense" in CIP: (1) absorptive capacity, (2) adaptive capacity, and (3) restorative capacity (Biringer et al. 2013, pp. 117–123). The first line of defense describes the ability of a system to cushion the effect of an unforeseen impact through endogenous features such as robustness, redundancies, and segregation (de-compartmentalization of vital functions). The second adaptive defense line utilizes alternative ways of maintaining overall performance by substituting, reorganizing, or rerouting processes—or by exploiting basic human ingenuity that can contribute to the adaptive capacity of CI, although in unpredictable ways. The third line of defense seeks to decrease the time and money needed to restore a disrupted CI by installing early warning and monitoring systems in advance as well as prepositioning supplies in key locations.

The acute response phase of critical infrastructure disruptions has been covered elsewhere (for a review of the literature with special emphasis on information sharing, see Petrenj et al. 2013). This paper focuses on what Biringer et al. term "Adaptive Capacity" in CIP: the ability of an infrastructure system to change the way it functions in case of a disruption so the societal function that it delivers is interrupted the least.

2 Crossing the Øresund

The narrow strait of Øresund separates Denmark from Sweden and provides, together with two other Danish straits, access to the Baltic from the Atlantic Ocean. Until 1658 both sides of the water were under the rule of the Danish king, who controlled the passage with fortresses and demanded dues from foreign ships. In modern times Øresund has developed into one of the busiest waterways in the world. Ferries have crossed the strait for centuries linking Copenhagen, the capital of Denmark, with Malmö, the third largest city in Sweden. However, a fixed link comprised of the Øresund Bridge and the Øresund Tunnel was inaugurated in 2000, rendering most of these routes obsolete. Only the ferry connection between Elsinore and Helsingborg 40 km to the north, where the strait is very narrow, maintained service after the bridge was opened. The Øresund Bridge, comprising both the bridge itself and the tunnel as well as the artificial island Peberholm in the middle, is owned by the Danish and the Swedish state through the jointly owned independent Øresund Consortium that also operates the fixed link.

On an average day 70,000 travelers cross the fixed link between Denmark and Sweden, traversing the Øresund Tunnel (4 km) and Europe's longest combined road and rail bridge (8 km), dispersed in approximately 18,000 vehicles and 160 passenger trains. Approximately 25,000 daily travels are estimated to be of critical importance to the local work market. Freight, both regional and local, amounts on average to 18,000 tons daily distributed on 1100 trailers and 20–25 freight trains. An estimated 11,600 people commute on a daily basis, the vast majority of them from Sweden to work in Denmark. A traffic forecast puts the yearly increase towards 2025 at approximately four percent (middle estimate) for passengers as well as freight, testifying to the popular success of the bridge and tunnel. The number of train travelers alone more than doubled from 5 million in 2001 to 11 million in 2014.

Following Larkin's definition of infrastructure as "matter that moves matter," there is no doubt that the fixed link between Denmark and Sweden qualifies as an infrastructure—but is it also a *critical* infrastructure? In December 2008 the Council of the European Union issued its Directive 2008/114 addressing CIP in the member states. Here, CI was defined as:

> an asset, system or part thereof located in Member States which is essential for the maintenance of vital societal functions, health, safety, security, economic or social well-being of people, and the disruption or destruction of which would have a significant impact in a Member State as a result of the failure to maintain those functions (Council Directive 2008/114/EC)

The potential impact of a disruption of such assets, systems, or parts thereof should be estimated with regard to three criteria: (a) casualties, (b) economic effects, and (c) public effects, with any one of these being sufficient to meet the definition. Threshold values were, however, not defined in the directive, but were left up to the member states to decide upon. Each member state was obliged by the directive to identify infrastructures that could be defined as European Critical Infrastructure (ECI), and in 2010 the Øresund Bridge Consortium issued the report *Vurdering af Øresundsbron som Europæisk Kritisk Infrastruktur* (transl. Assessment of the Øresund Bridge as European Critical Infrastructure).

According to this report the Øresund Bridge is *not* an ECI. Using a 100-day total closure of road and rail traffic as the baseline, the report concludes that even in the most pessimistic estimates none of the criteria are met: casualties from increased road traffic on alternate routes would amount to a mere four additional deaths and 59 injured persons, while the economic repercussions would be just 0.03 % of the Danish and Swedish GNP. The potential effects on public trust and societal coherence were also estimated as very low. An important factor for not defining the Øresund Bridge as ECI was the existence of an alternate transportation route (i.e., the ferry link between Elsinore and Helsingborg) that would allow people and goods to keep flowing in case of a closure, although at a higher cost.

Also in 2010 *Länsstyrelsen i Skåne Län* (the regional Swedish authority) published a report on *Beredskapsplanering i samband med ett långvarigt avbrott i den faste Öresundsforbindelsen* (transl. Preparedness planning in connection with long-time disruptions of the Øresund fixed link). This report estimated the necessary means for handling a 100-day total disruption of the fixed link—the scenario that the above-mentioned assessment of the Øresund Bridge as ECI was based upon. The work group behind the report concluded that the available ferry capacity in the region would be insufficient to replace the fixed link in case of a disruption. During the initial phase large build-ups of road and especially rail traffic should be expected on both sides, and in the longer perspective severe disturbances to travel patterns in the entire region would be unavoidable.

With regard to risk assessment, the fixed link is thought to be an extremely safe system. Using the definition from Biringer et al. the absorptive capacity is very high due to the robustness of the bridge and the tunnel, the redundancies and segregation built into management systems and power supply, and the procedures of surveillance and preparedness organizations. The infrastructure operators' Operational Risk Analysis (ORA), revised in 2008, estimates the probability of a closure of the bridge for more than 30 days at 3.7 % for the link's entire expected lifetime (100 years). The probability of a closure of the tunnel that connects the bridge to Denmark is considerably higher (26.3 %) due to the risk of a vessel colliding with the immersed tube tunnel comprised of 20 prefabricated reinforced concrete segments. Overall, however, the risk of a long-term disruption (100+ days) of the infrastructure is deemed very low even though the fixed link altogether, being a tightly coupled system, depends on the bridge as well as the tunnel to function in order to provide its designated service. All probabilities for long-term disruptions caused by either accidents in the tunnel or on the bridge were estimated at below two percent for the link's entire expected lifetime (Fig. 1).

Nonetheless, Danish and Swedish transport authorities called in 2014 for a mapping of preparedness plans and crisis management procedures relevant to short- and long-term disruptions of the fixed link between Denmark and Sweden. To accomplish this the *Arbetsgruppen för Öresundsberedskap* (transl. Workgroup for Øresund Preparedness) was formed and tasked with writing a report that in addition

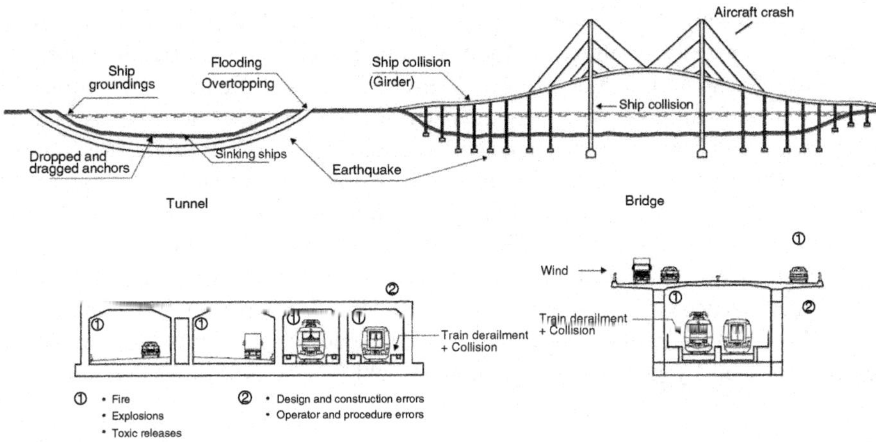

Fig. 1 Cross-section of the Øresund Bridge and Tunnel (not to scale) from the Operational Risk Analysis, revised 2008, with indication of major sources of risk. Copyright: The Øresund Consortium

to mapping the existing plans and procedures would also investigate the possibility of establishing alternate transport routes in case of a disruption.[1]

3 The Impact of the Highly Unlikely

As mentioned above, the likelihood of a total closure of the fixed link due to a ship collision or a plane crash is very low according to the ORA. But so is the calculated likelihood of a closure of the 50-km Eurotunnel that connects England and France—and yet it has already happened twice since its inauguration in 1994. In November 1996 a fire on a train carrying Heavy Goods Vehicles caused a partial closure of the tunnel that lasted until May the following year, and in September 2008 another similar fire resulted in personal injuries and a 5-month partial closure. A third and less severe fire occurred in August 2006.

Such events may be called "extreme" in the sense that the probability of them occurring is very low. They are found in the tail of the normal distribution of probability that governs most modern thinking about risk in general as well as in engineering and social science (Clarke 2008, p. 672, see also Zio and Pedroni 2014 for a more classical risk analytical interpretation of possibilism). The problem with extreme events is that they happen too rarely to allow for meaningful probabilistic risk assessment (PRA)—that is, quantification of occurrences over a time series on which the analyst can apply statistical tools. Lee Clarke proposed in 2006 the so-called *possibilistic* thinking as a complement and antidote to probabilistic thinking. It is an approach that focuses on the consequences instead of the likelihood of a certain event happening and thereby "shifts our gaze away from the center of a normal distribution out to its tails" (Clarke 2008, p. 676).

So, by exposing the potentially huge consequences of a low-probability event the possibilistic way of thinking about risk helps make infrastructure visible to paraphrase Susan Leigh Star (1999). If probability is difficult to determine for infrastructure disruptions, the consequences of such, however, are just as hard to estimate as "too few" have happened in Western societies (Boin and McConnell 2007, p. 51). Clarke advocates for the use of worst case scenarios and points out that thinking possibilisticly does not usually require much "ground truthing" as he calls it—understanding and accounting for all the details of reality. He states that possibilistic or worst case exercises should not try to approximate reality because "their greatest virtue may be their *unreality*" (Clarke 2008, p. 683).

The proceedings of the Work Group on Øresund Preparedness were to a high degree governed by possibilistic thinking. Looking strictly at the ORA, it would

[1]The researcher was allowed to participate in the work of the group as an observer and contributed also to the report with a section on resilience. All data not otherwise referenced in this paper can be found in the final report that was submitted to the Danish and Swedish authorities in Spring 2016 (Arbetsgruppen för Öresundsberedskap 2016).

appear little effort should be invested in preparing for long-term disruptions of the fixed link from a cost-benefit point of view as the likelihood of other kinds of incidents (e.g., traffic accidents, suicide attempts, extreme weather, strikes, and blockades) resulting in short-term closures is probably much higher. But Danish and Swedish authorities nonetheless opted to apply the precautionary principle by establishing the work group so that a thorough mapping and analysis could be carried out. No operational plans, however, resulted from the work; the uncertainties involved are so great that the infrastructure owner, in agreement with the authorities, decided that detailed plans for handling a long-term disruption would be meaningless. Instead, keeping in line with both Danish and Swedish principles for crisis management, an all-hazards approach (focusing on generic capabilities instead of hazard-specific planning) was taken. The group reviewed the procedures for activation of operational staffs and coordination between the responsible sectors as well as mapped the different ways alternate routes could be established in case of a disruption.

A minimum of 30 days of total closure of service was selected as the threshold for long-term disruptions because this time frame would make it necessary to establish temporary alternate means of transportation; at the same time a maximum duration of one year was chosen, as this would be too short a time for a new bridge or tunnel to be built. Rather little attention, however, was paid to the "triggering event" in the long-term disruption scenario during the early meetings in the work group. In the ORA a ship collision with the immersed tunnel was highlighted as the least unlikely scenario, while the risk of a plane crashing into the suspension bridge or a large vessel colliding with the road/rail section was assessed to have extremely low probability. As the waterways in the area are very busy, a large cargo or passenger ship colliding with the bridge's pylons is probably the most likely scenario, but a robust design with underwater barriers is believed to mitigate this risk effectively.

That said, for the possibilistic thinker an extremely low probability is still a probability that needs to be considered. In a study of supply chain flows in and across the Øresund before and after the fixed link was built the following scenario was described:

> There was a heavy fog. A northbound container ship hit one of the protective islands of the high-level bridge pillars. Through the collision some containers fell into the sea, one of them containing carbide. The container, which for security reasons had been placed as far as possible away from the crew and the machine room, was damaged when it fell into the sea. Water came in and acetylene gas was formed, which caught fire through the formation of sparks between the hull, which turned to the north, and the container, which scraped against the side of the hull. A rather powerful explosion followed and fire started in the bow. The bridge pillar was enveloped in flames and it was feared that the concrete would become weakened, so the traffic on the bridge was closed down. (Paulsson 2003, p. 2)

This is a good example of a scenario that utilizes a possibilistic approach to risk. Ask any risk analyst to perform a Quantitative Risk Assessment and calculate the likelihood of exactly this happening using, for example, Fault Tree Analysis, and you will end up with an extremely low probability. But ships with hazardous material *do* traverse the Øresund, so it *could* happen—with potentially huge consequences.

In case of a disruption of the fixed link the response phase will be managed by the standard emergency management organizations on both sides of the bridge. In Denmark the National Operative Staff (NOST) would be activated allowing tight integration between the police, emergency services, the health sector, transport authorities, and other key entities, while the *Länsstyrelsen* (regional authorities) would coordinate the incident on the Swedish side of the Öresund. After the immediate response has been managed, NOST would handover further monitoring and handling of the situation to the *Trafikal genoprettelsesgruppe* (Traffic restoration group), chaired by the Danish transport authorities, which would then be responsible for long-term planning and management of the traffic consequences, in close cooperation with Swedish authorities during the recovery phase.

The traffic restoration group is, however, *not* responsible for restoring the fixed link itself after a disruption; this responsibility rests solely with the infrastructure owner and operator. Reaching back to the before discussed definitions, we may say that the traffic restoration group is concerned with restoring the infrastructure as *process*, while the owner/operator manages the infrastructure as *technology*. This concept, which only is part of the Danish crisis management plan, is aligned with modern resilience thinking in (critical) infrastructure protection as it focuses on adaptive capacities instead of rigid plans and procedures. Overall, contingency planning for the recovery phase resonates with the Biringer et al. concept of second and third lines of defense. Such planning will be the focus of the following two sections of this paper.

4 Contingency Planning

As mentioned above, the fixed link across the Øresund has not been designated as ECI. But such definitions are not wholly independent of politico-economic, but instead depend on context and perspective. Many businesses in the area are to varying degrees dependent on the fixed link. As mentioned earlier, more than 10,000 commuters are traveling daily from the Malmö area in the morning to jobs in Copenhagen, returning late in the afternoon. Some of them will of course be able to work from home or relocate temporarily, but a long-term total closure will have a large impact on many people's daily lives. A disruption will also amplify social inequalities as educated workers will have much more flexibility, for example, to be able to work from home, compared with less highly educated and well-paid workers who must perform their jobs at set locations (e.g., nurses or shop assistants).

Basically, there are two different concerns with long-term disruptions of the fixed link across the Øresund: passengers and freight. Both categories travel on road and rail, and as these means of transportation are tightly coupled, running in the same immersed tunnel and on the same bridge structure, any disruption that could result in a long-term closure is highly likely to affect both travelers and freight. Passengers can be divided into two main groups: commuters and non-commuters, while freight is either local/regional or long-distance (e.g., Volvo cars and spare parts). The work group assumed, based on the findings in *Länsstyrelsens* 2010-report, that the long-distance freight would not be severely affected by a disruption of the fixed link, as there are several rail-ferry connections directly from Southern Sweden to the continent with surplus capacity. These assumptions were confirmed by findings from interviews carried out by the members of the work group with different actors within the sector. Market-driven self-organization is therefore expected to take care of this aspect of future disruptions without interference from the authorities.

The ferries that go between Elsinore and Helsingborg are equipped to carry railcars, but the tracks have been removed from the terminal on the Swedish side, so local and regional freight would have to be reloaded onto lorries. That could result in competition between freight and passengers for the surplus capacity on the ferries, especially during rush hour, so some kind of regulation could be necessary. The 2010-report also describes how the inflexibility of railways very quickly results in build-ups of cars and locomotives in the wrong places, and this is also expected to happen in case of a disruption of the fixed link. However, managing such issues falls outside of the responsibility of the infrastructure operator and the authorities and is a task for the responsible sector and the commercial companies involved. As these actors are professionals with experience in logistics and supply chain management they will, however, quickly adjust to the "new normal" and use the surplus capacity on the ferries to transport goods across the Øresund on lorries.

Individual travelers are the largest challenge, as they are much more difficult to communicate with and do not possess the same tools for coordination and planning as logistics and transport companies. Commuters require special attention, as they rely on the infrastructure service on a daily basis. Some Danish employers are especially dependent on the fixed link as they have many employees residing on the Swedish side: in 2014, Capital Region (Danish regional authority primarily responsible for the health sector), Field's shopping mall on Amager and Copenhagen Airport, Denmark's largest workplace, were some of the major attractors for Swedish labor.

Copenhagen Airport also serves many Swedish customers as an important regional hub for international air travel. As many as 10,000 daily travelers on the bridge are going either to or from Copenhagen Airport, 4000 of them on business trips. In total, Copenhagen Airport served 26.6 million travelers in 2015; four million of those came from Southern Sweden (Magnusson 2016). For commuters as well as for travelers, increased travel times would be, at the best, a nuisance. To many as much as five additional hours of daily travel time via Elsinore–Helsingborg would be unacceptable in the long run (Fig. 2).

In case of a long-term disruption, it would be possible for commercial actors such as shipping companies to set up temporary ferry connections between Copenhagen and Malmö ports. Both are large commercial harbors able to accommodate RO/PAX vessels (ships that can carry both vehicles and passengers), although the available parking space for vehicles is limited. Establishment of such a temporary connection

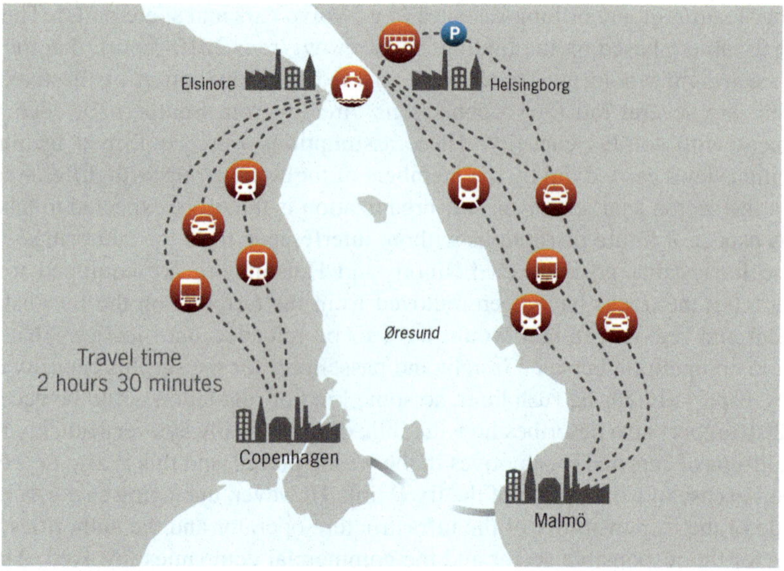

Fig. 2 The alternate route via the ferry connection between Elsinore and Helsingborg increases travel time significantly. And limited rolling stock, congested freeways and lack of parking space close to terminals may create additional bottlenecks during peak hours. Copyright: The Øresund Consortium and BGRAPHIC

is feasible—the authorities state in the report that it would probably take a longer time to identify and negotiate the use of the vessels needed than to obtain the necessary permits. The real challenge, however, is to move vehicles and passengers from the closest train station or freeway through the busy streets of a city like Copenhagen. Many commuters do not live in central Malmö or work in central Copenhagen, adding even more extra travel time to their daily commute, which speaks against setting up a temporary ferry connection between the two ports.

If the closure lasts more than 30 days, the time window from 3 to 6 months will probably pose the most challenges, as this is long enough for workers to wear out the patience of their employers with regard to flexibility but too short to attract commercial actors to a market for alternate transport routes. It is also a challenge that, due to the many daily commutes, there will be an unevenly distributed demand for transportation—if an alternate route, say a high-speed ferry between Copenhagen and Malmö, should be able to accommodate the demands at peak hour, there would be surplus capacity outside of rush hour, which would render such a service commercially problematic unless the carriers were subsidized as part of emergency measures (Fig. 3).

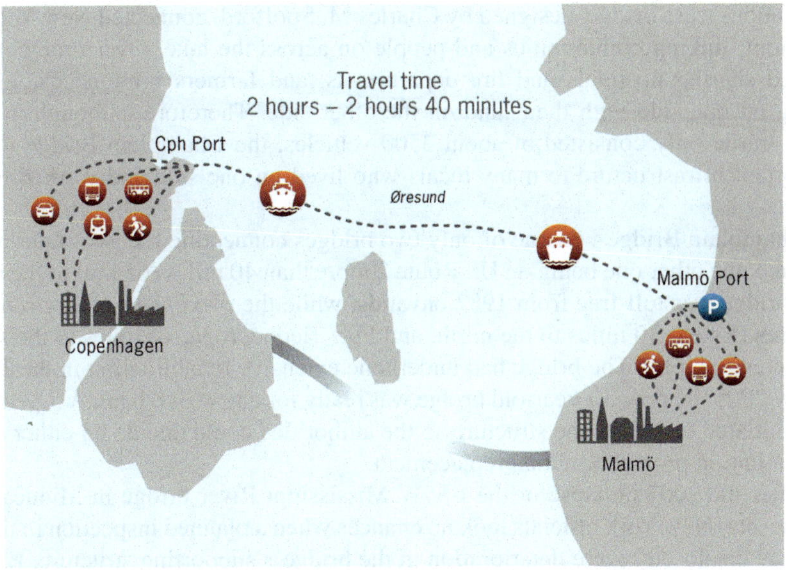

Fig. 3 A temporary ferry connection directly between the ports of Copenhagen and Malmö may seem like a good idea, but travel time still increases significantly due to heavy traffic especially in central Copenhagen. Copyright: The Øresund Consortium and BGRAPHIC

5 The Closure of the Champlain and Forth Road Bridges

After having presented the proceedings and results of the Work Group on Øresund Preparedness it is now appropriate to review two recent incidents that may provide useful insights about disruptions of similar infrastructures. The aim is to investigate the repercussions of two unexpected bridge closures and compare the preparedness plans from the Danish–Swedish context to how those disruptions unfolded.[2]

5.1 The Champlain Bridge

On August 26, 1929, the governor of New York, Franklin D. Roosevelt, cut the ribbon and formally opened the new Crown Point Bridge (known as Champlain Bridge) spanning the big freshwater Lake Champlain. The 2187 ft (666 m)

[2]Unless otherwise referenced, all information about the closure of the Champlain Bridge is taken from the New York State Department of Transportation report about the incident and the new bridge project (NYSDOT 2012), while the description of the Forth Road Bridge closure builds on bridge's official website (accessed February 2016) and Jane Arleen Breed's account of how the events unfolded (Breed 2011).

continuous truss bridge, designed by Charles M. Spofford, connected New York and Vermont, linking communities and people on across the lake. Over time counties started sharing hospitals and fire departments, and farmers grew accustomed to living on one side with their land on the other side. Therefore, although in 2009 daily traffic only consisted of about 3500 vehicles, the Champlain Bridge was an important infrastructure to many locals who lived on one side and worked on the other.

Champlain Bridge was one of only two bridges connecting the two states across the lake, the other one being on US Route 2 more than 40 miles (65 km) to the north. The bridge was toll-free from 1987 onwards, while the two existing ferry routes in the area (Essex, 30 miles to the north, and Fort Ticonderoga, 14 miles to the south) both charged tolls. The bridge had undergone extensive rehabilitation in the 1990s, but by 2009 the now 80-year-old bridge was ready for a new overhaul. A 5-year plan was initiated to survey the structure so the authorities could decide on either a new rehabilitation project or a total replacement.

After the 2007 collapse of the I-35W Mississippi River bridge in Minneapolis, Minnesota, New York officials took no chances when a planned inspection in the fall of 2009 disclosed severe deterioration in the bridge's supporting structure. Experts carried out a number of surveys above as well as below the surface of the lake while traffic on the bridge was restricted to one lane. The condition of the concrete piers was much worse than expected, and on October 16, 2009, the experts concluded that the supports could collapse. On the same day, at 1:30 p.m., NYSDOT closed the bridge to all traffic without any warning—never to reopen it.

The sudden closure of the bridge affected local communities severely. When the lifeline between the communities divided by Lake Champlain were cut, workers, farmers, fire fighters, and paramedics suddenly faced 2 or 3 h increased travel time, and cafes and shops on either side of their crossing lost their customers overnight. The only alternate land route was at least 85 miles (140 km) longer than the direct crossing, and even though the ferries at Essex and Fort Ticonderoga were made free of charge with subsidies from the authorities on October 27, people still had to drive long distances and wait in line to cross the lake. On October 28 a temporary connection for pedestrians was set up using the Basin Harbor tour boat, which ran until November 25, and there were also shuttle bus Park'n Ride services on both sides. From the middle of December, the Ticonderoga Ferry south of the closed bridge only operated sporadically because of the ice conditions on the lake.

The authorities monitored the situation closely. Four days after the disruption the Vermont Secretary of Transportation issued a Declaration of Emergency, and the following day the Governor of New York declared a state of Emergency under an Executive Order. The effects of the disruption were huge. For example:

> The bridge's closure separated residents from employment, medical services, childcare and family members. Farmers with fields and cattle on opposite sides of the lake could not bring in their fall harvests or tend to their livestock. Other residents were leaving home at 3 a.m. to arrive at work on time. (NYSDOT 2012, p. 4)

Exactly as the Work Group for Øresund Preparedness pointed out, the disruption amplified social inequalities. "Hundreds of workers from impoverished upstate New York towns who have low-paying but steady jobs on the Vermont side now face long-distance commutes that add hours to their day and take dollars from their pockets," wrote one newspaper 2 weeks after the closure. Also dairy farmers, already hit hard by declining milk prices, faced potentially fatal unforeseen expenses driving around the lake to feed and milk cows (Filipov 2009).

Public meetings were held on both sides of the lake in late October, and here people demanded a temporary crossing at the location of the now unusable bridge. The local population was furious, but the NYSDOT and VTrans (the Vermont Transportation authority) found that it would be way too expensive to build a temporary bridge, which in any case would take at least 6 months to complete. Instead the authorities decided to set up a temporary ferry connection right next to the closed bridge so the inland infrastructure could still be used (Yanotti 2011).

Setting up a new ferry connection running between two states in an area with many special environmental as well as archeological conditions proved surprisingly demanding. Coordination among the many involved agencies from the Army Corps of Engineers to the Vermont Department of Fish and Wildlife were, however, successful and resulted in a permission from both Vermont and New York on November 11 to set up a ferry service. Then NYSDOT and VTrans could start building the temporary docks and prepare the service, which would be conducted non-stop by two small vessels capable of carrying approximately 20 cars at a time. Due to harsh winter conditions the construction work was difficult, and the temporary ferry connection did not open until February 1, 2010—three and a half months after the disruption of the fixed link. The average daily cost of operation was $24,240, which was covered by NYSDOT and VTrans. Additional costs were carried by the affected residents and business.

While the mitigation efforts were implemented, the authorities also had to manage the long-term perspective. Only two options were possible: either the bridge could be repaired or a new one had to be built. Reinforcement of the fractured supports was considered, but deemed too costly and inefficient, as more permanent repairs would have to be carried out anyway. By the end of 2009 the span of the Chaplain Bridge was gone—it was demolished with explosive charges on December 28. The contracting process was fast-tracked by state and federal agencies, so the contract for building a replacement bridge was signed with the company Flatiron a mere seven and a half months after the closure. On November 7, 2011, the new Lake Chaplain Bridge opened to traffic after more than 2 years of service disruption.[3]

[3]Interestingly, the special situation surrounding the construction of the new Lake Champlain Bridge meant that the building schedule ended up 4 years shorter than if a traditional design-bid-build method had been used and that millions of dollars were saved (APWA 2013, p. 96).

5.2 The Forth Road Bridge Closure

When the Forth Road Bridge, crossing the Firth of Forth near Edinburgh, Scotland, opened on September 4, 1964, it was the longest steel suspension bridge in Europe, with a total length of 8241 ft (2512 m) and a span of 3301 ft (1006 m). It replaced a ferry service that for centuries had transported people and goods back and forth between Quensferry and North Quensferry, complementing the nearby cantilever railway bridge which was inaugurated in 1890. In 2014 approximately 75,000 vehicles crossed the bridge daily on average.

At midnight on Thursday December 3, 2015, the Forth Road Bridge was closed to all traffic after engineers had found a 20 mm wide crack in the supporting structure only 2 days before. An inspection of the bridge in May of that year had not revealed the damage, which was located in one of the most inaccessible parts of the structure. There had been numerous problems with corrosion in the bridge's supporting steel cables over the previous decade, which ultimately led to the decision to build an entirely new bridge adjacent to the Forth Road Bridge, planned to open in late 2016. However, what the engineers had found was actually something completely unrelated: a load-bearing link to the north east tower truss end had fractured. At a media conference one engineer said that an "unprecedented set of circumstances" had forced the Scottish Government's resilience committee to close the bridge to avoid further damage, hoping that repairs could be completed before the end of the year (BBC 2015a).

Already the next morning the bridge closure caused heavy congestion on the alternate routes in the area. Approaching the nearest other bridge spanning the Firth of Forth, the Kincardine Bridge 15 miles (24 km) upstream, were traffic jams over a stretch of 11 miles. The Ministry of Transportation was preparing a full travel plan including busses, trains, and even a temporary ferry (BBC 2015b). ScotRail was treating the closure as a "national emergency," increasing its normal service from 75 to 100 trains a day on the Forth Rail Bridge. Locating enough spare running stock was, however, a challenge, as was manning the many extra trains. This prompted train union leaders to publicly criticize the shortage of capacity now exposed by the current crisis (Carrell 2015).

"We are aware of the potential economic impact for strategic traffic in the east of Scotland and on people living in local communities," said the Scottish transport minister on the first day after the closure, while political opponents called for swift action and full disclosure of Transport Scotland's full contingency plans. A representative of the Scottish Federation of Small Business addressed the need to strike a sound balance between safety and the economy, stating that: "Not only will this closure impact those that use the bridge to bring their goods or services to market, employers of all description will face serious disruption" (Carrell 2015).

Repairs took less time than expected, and, with the exception of Heavy Goods Vehicles, the Forth Road Bridge reopened to traffic on December 23, 2015, after 20 days of total closure. During this period approximately 18,000 seats were added to the local bus capacity, and the police and Transport Scotland worked closely together to ensure that road traffic in the affected areas was managed intelligently so that congestion could be minimized. Authorities engaged in dialogue with communities, business groups, and large employers and encouraged people to use public transportation, consider car sharing and work from home as much as possible.

5.3 Lessons from the Two Cases

Both closure cases serve as examples of how disruptions of infrastructures quite similar to those that are possible with the Øresund Bridge have played out. Even if there are important differences (both were only road bridges, Champlain Bridge had very little traffic compared to Øresund, and Forth Road Bridge was closed for less than 30 days) it is evident that such disruption immediately affects local communities severely, and that swift and affirmative action from infrastructure operators and authorities is required.

What immediately draws attention is that freight is almost non-existing in the documentation of both cases, which could be said to testify to the accuracy of the Work Group for Øresund Preparedness' assumption that professional operators to a large extent will solve the problems themselves. Of course, local/regional cargo transport must have been affected, but long-distance freight is not mentioned in the news coverage. One example of a major actor in this field is Amazon.com, whose biggest UK distribution warehouse is located just north of the Forth Road Bridge. A spokesperson for the major international distributor of books and other items said, when asked by *The Guardian* about the risk of delays of Christmas gift orders, that the company had contingency plans and could cope with the bridge closure by switching operations to its ten other UK fulfillment centers (Carrell 2015). More research is, however, needed to investigate the repercussions of the closures on freight.

The Work Group for Øresund Preparedness has put a lot of effort into meeting with potential stakeholders and partners in order to map where and when and how alternate transportation routes could be established in the event of a disruption. Compared to the apparently rather haphazard process led by the NYSDOT and VTrans in the weeks after the closure of the Champlain Bridge (i.e., having to report complicated information about environmental and wildlife issues in the middle of a transportation crisis) it seems reasonable to at least investigate such matters beforehand.

Lessons learned from the two cases confirm most of the issues identified by the Work Group for Øresund Preparedness: temporary ferry connections are tricky and costly to establish and will likely have to be subsidized heavily by the authorities for many months before they can become commercially feasible to run. Using alternate transportation routes such as existing bridges or ferry connections is preferable, but this requires thorough planning before the event and close coordination and cooperation among the many sectors and actors that will become involved. There are costs associated with these and other adaptive strategies—costs that are borne both by governmental authorities and by residents who are affected. Moreover, negative impacts on residents are likely to fall disproportionately on lower-income and less-well-educated members of the population. Using resilience-related terminology, some groups and sectors of the economy have more adaptive capacity than others and thus will fare better in the event of infrastructure disruptions (Walker et al. 2001; Dahlberg 2015).

One aspect of infrastructure disruption that was present in both cases discussed here, but which the Work Group for Øresund Preparedness only touched briefly upon, is the need for sound public relations and professional crisis communication when disruptions occur. After the Champlain Bridge was closed, local residents gathered in community meeting places and demanded action from the authorities, and after the closure of the Forth Road Bridge, former employees of the Forth Estuary Transport Authority (FETA) went to the press with harsh criticism towards the Scottish transport authorities. This type of outrage, which is understandable, can be mitigated through prompt and forthright communication on the part of authorities, focusing on topics such as how long infrastructure disruptions are expected to last, what options are considered and ultimately chosen to alleviate the impacts of disruptions and why, and how those affected can access information and other resources they need in order to adjust to disruptions (Blom Andersen 2015).

FETA's budget was cut by 58 % in 2011. Subsequently, it was relieved of its responsibilities for the Forth Road Bridge in June 2015, when the private UK company Amey took over as infrastructure operator after winning a 5-year tender from the Scottish government for operation and maintenance of the bridge as well as the new Queensferry Connection that was being built to replace it. Its management had been deeply concerned about handing over the management of the bridge to a private contractor—here expressed by the former convener of FETA:

> There can be no doubt that Transport Scotland were well aware of FETA board's concerns about loss of key staff and the threat that this would have on the future management and maintenance of the bridge (McPherson 2016)

This raises the question of private–public partnerships (PPPs) and their special status with regard to infrastructure protection and disruptions (Dunn-Cavelty and Suter 2009). Since the 1990s many infrastructures have been sold off to or operated by private companies, while the ultimate responsibility for maintaining the vital societal functions still rests with governments. One study suggests that infrastructure resilience should be viewed as an integrated part of Corporate Social Responsibility (Ridley 2011). The Øresund Consortium that owns and operates the fixed link is jointly owned by Denmark and Sweden, but will that information convey well to the public and the media in case of sudden closure?

On a final note should be mentioned that in both of the cases the cause of the closure was NOT the sudden impact from an earthquake, a ship collision or a plane crash, but the result of aging and long-term subtle wear and tear that went by unnoticed by authorities and operators. Another recent case is the combined rail and road Storstrøm Bridge in Southern Denmark (inaugurated 1937), that in October 2011 was closed to rail traffic for a week after the authorities discovered a

25-cm crack in its supporting structure (Rasmussen 2011). Also highly unforeseen socio-political developments with origins far away may severely influence the service of an infrastructure: on January 4, 2016, Swedish authorities introduced identification procedures for travelers going from Denmark to Sweden as a means to control the flow of migrants and refugees, increasing travel time for especially train passengers who were forced to disembark and change trains at Copenhagen Airport (Magnusson 2016).

While the consequences are less sudden and brutal as a ship collision or a plane crash, the root causes of such error trajectories tend to be much more complex and should be sought in the socio-economic-technological systems that surround the infrastructure.

6 Known and Unknown, Knowns and Unknowns

Integrating possibilistic thinking in planning for long-term disruptions of infrastructure should be thought of more as process than an objective. When forced to prepare for low-probability events with potentially huge consequences, the socio-technological system surrounding the infrastructure is exercised on more generic terms, generating awareness, expertise, and knowledge (Boin and McConell 2007, p. 55). A plausible worst case scenario provides excellent opportunities to engage in relevant conversations across sectors and organizations, creating the "chronic state of unease" that is crucial to any High Reliability Organization (Weick and Sutcliffe 2015). Any planning process aimed at catastrophic events at the same time prepares the emergency management and crisis management organizations for more common and trivial events.

No matter how unpopular it might be with quantitative risk experts, possibilistic thinking is a necessary and useful complement to the probabilistic approach. However, there are cognitive limits at work even in possibilistic thinking. Just as Herbert Simon argued that a rational persons rationality is inevitably bounded by the incomplete knowledge he or she possess on which to base decisions (Simon 1955), so is possibilistic thinking limited by our ability to imagine the worst that could happen. In February 2002 then US Secretary of Defense Donald Rumsfeld explained at a Pentagon press briefing that there are "known knowns" (things we know that we know), "known unknowns" (things that we know that we don't know), and "unknown unknowns" (the things that we don't know that we don't know) (youtube.com 2007). Not surprisingly, Rumsfeld found the latter category to be the difficult one.

Applied to risk thinking we may say that PRA is generally well suited to deal with the two first categories—the things that we know we know and those that we know that we don't. PRA requires a thorough understanding of systems, including knowledge of previous events and states over long periods of time. In order to estimate the probability of a certain event happening in the future it is necessary to know the distribution of similar events in the past.

The quantification of risk addresses *aleatory* uncertainties that may be irreducible, but nevertheless can be calculated using probability. Uncertainty caused by randomness such as the tossing of a coin or throwing dice is manageable as long as we understand the behaviors of the system and have access to sufficient past data to describe the probability distribution. That aleatory uncertainty is irreducible means that no matter how much we know about the probability, we'll never be able to say anything more solid about the *next* toss of the coin. Assessing risks over longer time periods and defining acceptable risks are the aims of this approach.

But other kinds of uncertainty are also at play, unfortunately: *epistemic* uncertainties that stem from lack of knowledge about the system and *ontological* uncertainties that resemble Rumsfeld's "unknown unknowns." Risk assessments are based on "world models" that make assumptions about the real-world system that they represent, and if these assumptions are wrong or too simple the result is epistemic uncertainties with potentially catastrophic consequences. Ontological uncertainties constitute a third category that originates not from lack of knowledge but lack of *imagination*. If a risk assessment is based on a world model and that model lacks important factors, then the outcome is of course flawed and dangerous to use for decision-making. Epistemic and ontological uncertainties are usually understood as more dependent on prior assumptions about the world than aleatory uncertainties, although more conventional risk analyses also are based on "subjective" decisions about system boundaries, interpretations of outliers, etc. Therefore, epistemic and ontological uncertainty is often underrepresented in risk assessments done by analysts with a preference for quantifiable "rational" data.

To prepare for disruption, it is necessary to make infrastructure visible before a disruptive event. One approach to this could be to focus more on the infrastructure as *process* than *technology*: if users are made aware of the service that the infrastructure provides instead of thinking about it as a mere stretch of road or rails across the water, that may prompt contingency planning on the individual level—an important element in improving resilience (Rodin 2015). For authorities and infrastructure owners and operators it's about remembering why people buy quarter-inch drill bits. It's because they want quarter-inch holes (Levitt 1986, p. 128). People also use an infrastructure not (only) because they like the view, but because they want to go to the other side of the water.

Making the infrastructure visible before a disruption enables contemplation of not only aleatory but also the epistemic and ontological uncertainties at play. Is cost cutting or other previously unanticipated processes such as climate change slowly undermining an expected infrastructure lifetime of, for example, 100 years, thereby seriously altering the failure probabilities that conventional risk assessments rest upon? Do we take users' and stakeholders' behaviors and opinions into consideration when planning our recovery phase in case of long-term disruption—and if we do not, how can we estimate the costs involved? And do we make sure that we learn the lessons from similar events that have happened elsewhere and incorporate them into our planning processes?

Acknowledgements The author wishes to thank Professor Kathleen Tierney, Natural Hazards Center, University of Colorado Boulder, Professor Henning B. Andersen, Technical University of Denmark, Head of Division Mads Ecklon and Head of Section Maximilian Ritzl, Center for Preparedness Planning and Crisis Management, Danish Emergency Management Agency, Ulla V. Eilersen, Safety Manager, Øresundsbron, and Strategic Consultant Henrik Andersson, Sweco Society AB, for useful comments to a draft of this paper. A special thanks to Ladimer Nagurney and Leif Vincentsen for directing the author's attention towards the two recent cases of infrastructure disruption.

This research was carried out with funding from the READ-project (Resilience Capacities Assessment for Critical Infrastructures Disruptions), funded by the European Commission DG Home.

References

APWA Reporter, July 2013

Arbetsgruppen för Öresundsberedskap: Beredskap för trafiken vid ett långvarigt trafikavbrott på Öresundsbron (2016)

BBC: Forth Road Bridge to be closed until new year, bbc.com, December 4, 2015. http://www.bbc.com/news/uk-scotland-35001277 (2015b). Accessed 2 Feb 2016

BBC: Why is the Forth Road Bridge Closed?, bbc.com, December 4, 2015. http://www.bbc.com/news/uk-scotland-edinburgh-east-fife-35007187 (2015a). Accessed 2 Feb 2016

Biringer, B., Vugrin, E., Warren, D.: Critical Infrastructure System Security and Resiliency. CRC Press, Boca Raton (2013)

Blom Andersen, N.: Analysing communication processes in the disaster cycle: theoretical complementarities and tensions. In: Dahlberg, R., Rubin, O., Vendelo, M.T. (eds.) Disaster Research: Multidisciplinary and International Perspectives. Routledge, New York (2016)

Boin, A., McConnell, A.: Preparing for critical infrastructure breakdowns: the limits of crisis management and the need for resilience. J. Conting. Crisis Manag. **15**(1), 50–59 (2007)

Breed, J.A.: The Loss of the Lake Champlain Bridge: A Traveler's Story. Bloated Toe Publishing, Adirondack (2011)

Brown, K.A.: Critical Path: A Brief History of Critical Infrastructure Protection in the United States, Fairfax. Spectrum, Fairfax, UA (2006)

Carrell, S.: Scotland scrambles for extra trains after Forth Road bridge closure. theguardian.com, December 4, 2015. http://www.theguardian.com/uk-news/2015/dec/04/scotland-scrambles-to-find-extra-trains-after-closure-of-forth-road-bridge (2015). Accessed 17 Feb 2016

Chang, S.E.: Infrastructure resilience to disasters. Bridge **39**, 36–31 (2009)

Clarke, L.: Possibilistic thinking: a new conceptual tool for thinking about extreme events. Soc. Res. **75**(3), 669–690 (2008)

Council Directive 2008/114/EC. http://eur-lex.europa.eu/LexUriServ/LexUriServ.do?uri=OJ:L:2008:345:0075:0082:EN:PDF. Accessed 15 Feb 2016

Dahlberg, R.: Resilience and complexity: conjoining the discourses of two contested concepts. Cult. Unbound **7**, 541–557 (2015)

Dahlberg, R., Johannesen-Henry, C.T., Raju, E., Tulsiani, S.: Resilience in disaster research: three versions. Civil Eng. Environ. Syst. **32**(1–2), 44–54 (2015a)

Dahlberg, R., Rubin, O., Vendelø, M.T.: Disaster Research: Multidisciplinary and International Perspectives. Routledge (2015)

Dunn-Cavelty, M., Suter, M.: Public-private partnerships are no silver bullet: an expanded governance model for critical infrastructure protection. Int. J. Crit. Infrastruct. Protect. **2**, 179–187 (2009)

Filipov, D.: A bridge to nowhere. boston.com, November 1. http://www.boston.com/news/local/vermont/articles/2009/11/01/lake_champlain_bridge_closure_presents_hardships_for_businesses_workers/ (2009). Accessed 26 July 2015

Kozine, I., Andersen, H.B., Trucco, P., Petrenj, B.: Framework to address transboundary Critical Infrastructure Disruptions in the Emergency Management cycle (Deliverable 1). EU Project READ. Available at http://www.read-project.eu/ (2015)

Larkin, B.: The politics and poetics of infrastructure. Ann. Rev. Anthropol. **42**, 327–344 (2013)

Levitt, T.: The Marketing Imagination: New, Expanded Edition. The Free Press (1986)

Magnusson, E.: Gränskoll kan bli bromskloss: 'Kraften i regionen kan gå förlorad', 8till5.se, January 23. http://8till5.se/2016/01/23/granskoll-kan-bli-bromskloss-kraften-i-regionen-kan-ga-forlorad/ (2016). Accessed 25 Feb 2016

McPherson, G.: Vital work shelved due to funding costs, Forth Road Bridge inquiry hears, thecourier.co.uk, January 28. http://www.thecourier.co.uk/news/politics/vital-work-shelved-due-to-funding-cuts-forth-road-bridge-inquiry-hears-1.921835 (2016). Accessed 17 Feb 2016

NYSDOT: The Lake Champlain Bridge Emergency Replacement Project. New York State Department of Transportation, New York (2012)

Paulsson, U.: Supply Chain Flows in and Across Öresund Before and After the Öresund Link – Facts, Risks and a Risk Analysis Model. Lund University, Lund (2003)

Petrenj, E.L., Trucco, P.: Information sharing and collaboration for critical infrastructure resilience – a comprehensive review on barriers and emerging capabilities. Int. J. Crit. Infrastruct. **9**(4), 304–329 (2013)

Rasmussen, D.: 25 centimeter lang revne i brofag lukker Storstrømsboen for tog. ingeniøren.dk, October 19. http://ing.dk/artikel/25-centimeter-lang-revne-i-brofag-lukker-storstromsbroen-tog-123189 (2011). Accessed 25 Feb 2016

Ridley, G.: National security as a corporate social responsibility: critical infrastructure resilience. J. Bus. Ethics **103**, 111–125 (2011)

Rodin, J.: The Resilience Dividend: Managing Disruption, Avoiding Disaster, and Growing Stronger in an Unpredictable World. Profile Books, London (2015)

Simon, H.: A behavioral model of rational choice. Q. J. Econ. **69**, 99–118 (1955)

Star, S.L.: The ethnography of infrastructure. Am. Behav. Sci. **43**(3), 377–391 (1999)

Vespignani, A.: The fragility of interdependency. Nature **464**, 984–985 (2010)

Walker, J., Cooper, M.: Genealogies of resilience: from systems ecology to the political economy of crisis adaptation. Secur. Dialogue **42**(2), 143–160 (2001)

Weick, K., Sutcliffe, K.M.: Managing the Unexpected: Sustained Performance in a Complex World, 3rd edn. Jossey-Bass, San Francisco (2015)

Yanotti, A.P.: Lake Champlain Bridge: decision for closure and response to a crisis. Presentation given at The Department of Civil, Structural and Environmental Engineering, University of Buffalo, 18 April 2011

youtube.com: Donald Rumsfeld Unknown Unknowns! https://www.youtube.com/watch?v=GiPe1OiKQuk (2007). Accessed 23 Feb 2016

Zio, E., Pedroni, N.: Possibilistic Methods for Uncertainty Treatment: An Application to Maintenance Modeling. FonCSI (2014)

Multi-Hazard Scenarios and Impact Mapping for a Protected Built Area in Bucharest, as a Base for Emergency Planning

Emil-Sever Georgescu, Cristina Olga Gociman, Iolanda-Gabriela Craifaleanu, Mihaela Stela Georgescu, Cristian Iosif Moscu, Claudiu Sorin Dragomir, and Daniela Dobre

Abstract In urban areas, there is a need to evaluate the dynamics and impacts of impending disasters, in order to ensure planning for resilience, emergency actions, and humanitarian support. In this respect, a Bucharest area that suffered a deep trauma in the 1980s was studied. The total or partial demolition of the old urban fabric was ordered by the authoritarian regime to build a new Civic Center. The area was declared as a protected built area in the 1990s. The inventory of the building stock was done in the framework of this project, based on data on building's age, height, and materials. Additionally, the cultural value was studied with scoring criteria and methods, based on the six scaling stages of the Romanian Law of Monuments. The functional value, as well as the seismic damage and vulnerability, was evaluated for existing building classes. The paper presents examples of multi-hazard scenarios elaborated for this area, considering earthquakes, flooding, and terrorist attacks on public institutions. The spatial databases and online tools were used for mapping the buildings, as well as hazard impacts. According to the results of the above scenarios, shelters/security centers can be created, for earthquake evacuation, in the nearby "Mihai Eminescu" College while, for high-level flooding,

E.-S. Georgescu (✉)
National Institute for Research and Development URBAN-INCERC, Bucharest, Romania
e-mail: emilsevergeorgescu@gmail.com

C.O. Gociman • M.S. Georgescu • C.I. Moscu
"Ion Mincu" University of Architecture and Urban, Planning, Bucharest, Romania

I.-G. Craifaleanu • D. Dobre
Institute for Research and Development URBAN-INCERC, Bucharest, Romania

Technical University of Civil Engineering, Bucharest, Romania

C.S. Dragomir
National Institute for Research and Development URBAN-INCERC, Bucharest, Romania

Faculty of Land Reclamation and Environment Engineering, University of Agronomic Sciences and Veterinary Medicine, Bucharest, Romania

I.S. Kotsireas et al. (eds.), *Dynamics of Disasters—Key Concepts, Models, Algorithms, and Insights*, Springer Proceedings in Mathematics & Statistics 185,
DOI 10.1007/978-3-319-43709-5_4

the refuge and shelter area should be located in a higher place, near the Academy House or the Romanian Parliament. For other hazards, a center located either at "Mihai Eminescu" College or in Izvor Park is recommended.

Keywords Multi-hazard • Impact mapping • Protected area • Bucharest

1 Rationale for a New Insight into Dynamics of Disasters

At global scale, the need for accelerating disaster risk reduction and building resilience to disasters of natural and built environment of countries and urban fabrics was most recently emphasized during the Third World Conference on Disaster Risk Reduction through the Sendai Framework for Disaster Risk Reduction 2015–2030 (WCDRR 2015). In this context, the dynamics of disaster patterns caused by the impacts of various hazards can be studied as:

- High-speed and low-speed occurrence and/or development;
- Short-term and long-term consequences.

In the case of built environment, the impact of earthquakes, floods, and some man-made hazards is a high-speed process, while materials decay and ageing are low-speed processes. On the other hand, code improvements and enforcement are low-speed processes. Thus, when certain circumstances are met, disasters occur. Urban policies, often enforced in the wake of disasters, may have short-term and long-term consequences, sometimes contradictory.

Short-term dynamics are mainly related to the sheltering and feeding of evacuees and to post-disaster demolitions, propping, etc., and are governed by the need of fast recovery. Long-term dynamics are controlled by urban maintenance, improvement, and rebuilding/transformation, each with specific policies, and are governed by both public policies and market pressure.

The most advanced and trans-disciplinary approaches worldwide consider the dynamics of disasters through a rather complicated mathematical modeling, using the science of networks and operational research for evacuation planning or resource allocation issues (Pardalos 2013; Nagurney and Nagurney 2015). This framework requires a rich database that most Romanian local authorities do not possess at present.

In Europe, several highly populated areas are located in seismic zones. Since Romania is exposed to multiple hazards, the dominant disaster potentials and possible community involvement in preparedness should be foreseen. A large part of Romania is exposed to Vrancea intermediate depth earthquakes. Half of Romania is exposed to high seismic accelerations; in areas exposed to Vrancea earthquakes lives about 35 % of Romania's population, which includes over 66 % of the urban population. The damaging events are those with magnitude over 7 on Richter scale and great Vrancea earthquakes occur at about 30–40 years average interval, as in 1940 and 1977 (Balan et al. 1982; Georgescu and Pomonis 2008). Over 10 % of the urban population of Romania is located in other areas, exposed to shallow earthquakes.

Romania was struck repeatedly by earthquake disasters and there was a change of pattern and dynamics of disasters along its peacetime history. Earthquake damage was frequent in traditional buildings, but casualties were moderate until the 1940 Vrancea earthquake. For the 1977 earthquake the damage and life loss were concentrated in Bucharest, in reinforced concrete buildings. As a consequence of the March 4, 1977 earthquake, the total reported losses were over 2 billion US$; from these, 1.683 billion US$ were direct losses and 0.365 billion US$ were production losses. Damage to constructions was 69.4 % of the total or 84.3 % of the direct losses. In 1977, the losses in the housing sector (US$ 1.0328 billion) represented 71.4 % of construction losses, or 61.4 % of the direct losses, and 50.4 % of the total losses. Casualties were of 1578 deaths (1424 or 90 % in Bucharest) and 11,300 injured (7600 or 68 % in Bucharest) (Georgescu and Pomonis 2008).

This increase of losses, from 1940 to 1977, is in contradiction with the positive development of earthquake engineering, architecture, and urban planning, if regarded separately, since earthquake design codes were enforced in 1963 and 1970. Paramount revisions were enforced in 1978, 1981, 1991–1992, 2006, and 2013.

According to the Emergency Ordinance No. 21/2004, the National System for Emergency Situations Management was established. The National Committee for Emergency Situations, organized under the Ministry of Internal Affairs, and the ministerial committees for emergency situations are responsible for application of the disaster risk reduction policy at national level. At other levels, the Bucharest Municipal Committee for Emergency Situations, county committees for emergency situations and local committees for emergency situations are in charge.

The Government Ordinance on Existing Buildings Risk Reduction No. 20/1994 provided the legal framework for the free evaluation of residential buildings resistance, while for the design and strengthening works the owner may receive a bank credit subsidized for 20 years.

On the other hand, in accordance with the Law no 575/2001, the Plan for National Territory Planning, the maps for flood, landslide, and earthquake risks shall be drafted for every county and locality located in natural risk areas. These maps must be included in the Plans for General Urban Planning in order to implement the specific measures for building and land use. A methodology for post-earthquake emergency investigation of safety of buildings and framework solutions of intervention, ME 003-2007, was enforced in 2007. The immediate inspection is followed by a rapid technical evaluation, with the application of four types of colored placards, in terms of safety and usability. The placing of red placard (unsafe building), blue placard (unsafe zone), and yellow tagged buildings (limited entry) may have significant consequences, as it may require a higher capacity of shelter for evacuated inhabitants. For historical and architectural monuments, the emergency safety assessment will be performed by specialists appointed by the concerned ministry and the authorized public institutions, based on specific rules. The earthquake resistance evaluation for final strengthening is based on the seismic assessment code P100-3/2008.

As it was shown, several national programs for seismic risk mitigation were enforced since 1977, with the obligation to evaluate and, if required, to rehabilitate the existing buildings. Several thousands of evaluation reports and preliminary strengthening projects were already drafted, but the works are costly and delayed because the owners are still reluctant to apply for loans under the clauses of mortgaging their property until the return of debts. The key issue is that of the relationship between the funding provided by government and the actual management of seismic strengthening projects, done by local authorities. Another issue is that owners do not participate in the decision process.

2 Profile of the Study Area

The selected study area is located in Bucharest, in front and around of the present Romanian Parliament building. The area suffered deep trauma in the 1980s, due to the total or partial authoritarian demolition of more than 450 ha of the old urban fabric, aimed for building a new East-West axis and several public institutions (Fig. 1). Subsequently, many high-rise buildings were erected until the end of the 1980s along the new streets frontline, while the lots with low-rise and mid-rise houses, as well as those with major and minor heritage monuments, escaped demolition and remained somehow hidden behind. Following the change of paradigm of 1989, the area was officially listed as a protected built area by the new General Urban Plan (PUG) and the Zonal Urban Plans (PUZ, UAUIM 2000) of the city of Bucharest. However, the already de-structured fabrics (as it is the built protected area under study) need special care and it is very difficult to set the target of reconstruction. Thus, the authors were forced to use less developed but affordable instruments, yet coherent with the risk assessment methods.

The heritage buildings and the vulnerability patterns within the protected area may reveal the roots of hazards impact. Some of them are generic, while others are specific. For seismic hazard, the authors used data on vulnerability adjusted from those of a post-earthquake survey in Bucharest after the 1977 earthquake on a sample exceeding 18,000 buildings, located in various areas of the city (Balan et al. 1982; Sandi et al. 2008). In this framework, high-rise structures erected before 1940 and even some code-designed structures erected before 1977 can be considered as having a built-in vulnerability, due to the lack or the low seismic forces and to the other design requirements used in these historical periods. The INCERC record of the 1977 earthquake showed that the actual accelerations were much higher than prescribed and that Vrancea earthquakes could induce long-period oscillations of the ground in the city area. Moreover, some other code provisions were insufficient (lack of ductility, drift demand, shear reinforcement, etc.).

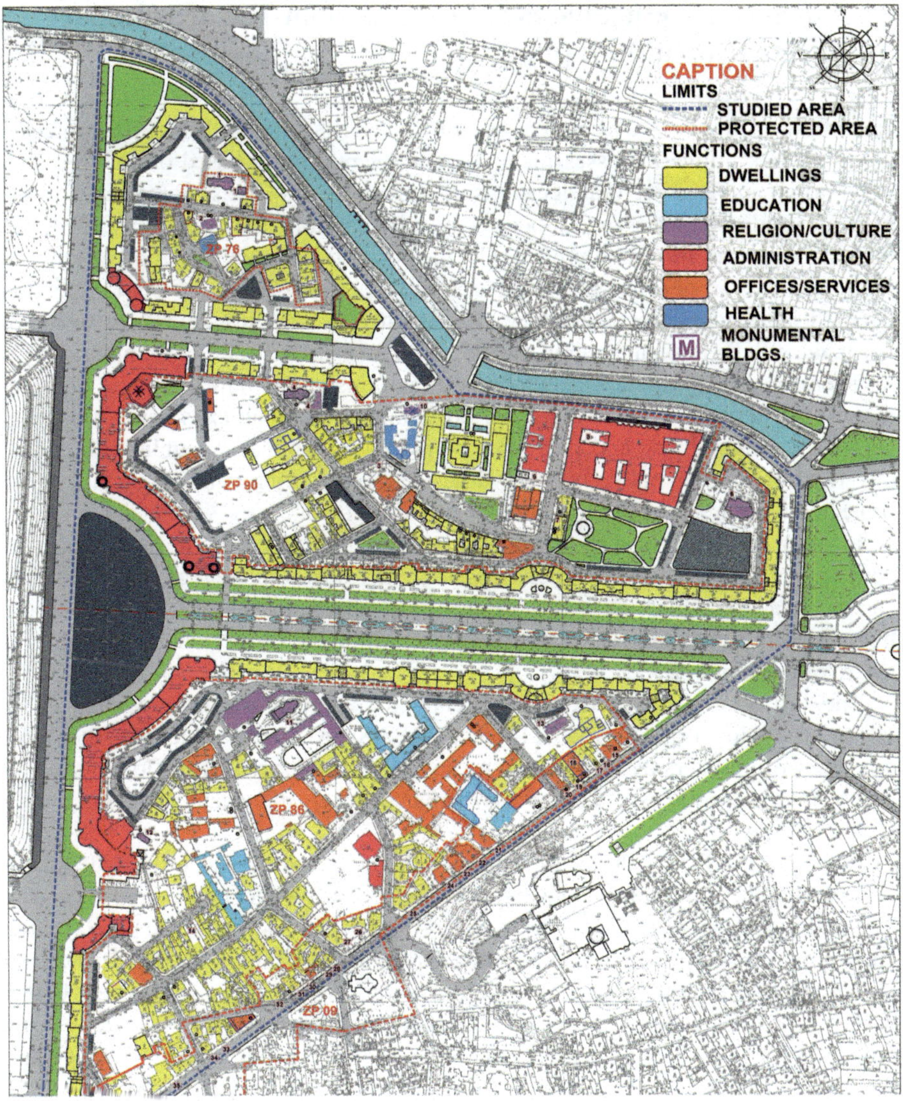

Fig. 1 The study area and the mapping of buildings classification/functional value (Project URBASRISK, 2014)

The extent and degree of damage from past earthquakes in the studied area were moderate; however, after several decades of decay and weathering, the vulnerability may have increased. Moreover, the governmental institutions in the vicinity may create unknown hazards, as terrorist attacks.

3 Advanced Methodological Tools Used for Hazards Impact Assessment and Mapping

According to the laws and regulations in force, it is supposed that, for protected areas, all risk mitigation assessments and interventions are based on rather elaborated urban and architectural decisions on the analyzed objects, in reference to the vicinities and urban areas, and supported by wise administrative decisions. However, the authors are aware that at urban scale, the need of resiliency is hampered by gaps in the application of urban planning, architecture, engineering, and civil protection detailed approaches, as they are devoted mainly to new investments or to major heritage, and the implementation of mitigation measures is time-consuming.

Therefore, a gap exists between knowledge transfer and public policies enforcement. One of its reasons is that each professional community is using a specific approach; vulnerability and risk assessments are complicated, public officers are governed by administrative targets, while the targeted communities, as that in the protected built area, remain in a sort of "no man's land."

If a hazard would occur, the analyzed community would be at great risk. It is necessary to know how and where disasters may occur and spread, depending on causal hazards—which are the dynamics of impending disasters—to prepare the humanitarian support in this sensitive area and to make the community living in it aware of possible risks.

The Project URBASRISK is an attempt to reduce this gap (Gociman et al. 2014a, b; Gociman et al. 2015a, b; Project URBASRISK 2012). Thus, the short-term dynamics of disasters in the area can be positively influenced by the long-term and mid-term urban policies, yet tailored for urban disasters management at community scale.

In the URBASRISK Project, the state of the urban fabric was inspected and the preliminary, though extensive, inventory included data on age, height, and materials. The cultural value was studied with 13 scoring criteria and methods, based on the six scaling stages of the Romanian Law of Monuments. The Law No. 51/1991 on the construction authorization and the provisions of the General (Master) Urban Plan (PUG) of Bucharest provides a classification of the functional areas (Law No. 51, 1991). Thus, the functional value, as well as the past seismic damage and vulnerability, was evaluated for building classes (Florescu et al. 2014; Georgescu et al. 2014, 2015; Gociman et al. 2014a, b; Gociman et al. 2015a, b) (Fig. 1).

The URBASRISKdb geodatabase was created for storing attributes of the buildings in the study area, based on the past experience. These attributes are used within the scope of the project, for statistical and reporting purposes, as well as for spatial representations. In this framework, the URBASRISKdb structure includes several interrelated data tables, reported elsewhere (Georgescu et al. 2014). Geospatial databases and online tools were used for mapping the buildings in the considered zone, as well as the hazard impacts. As a result, a number of about

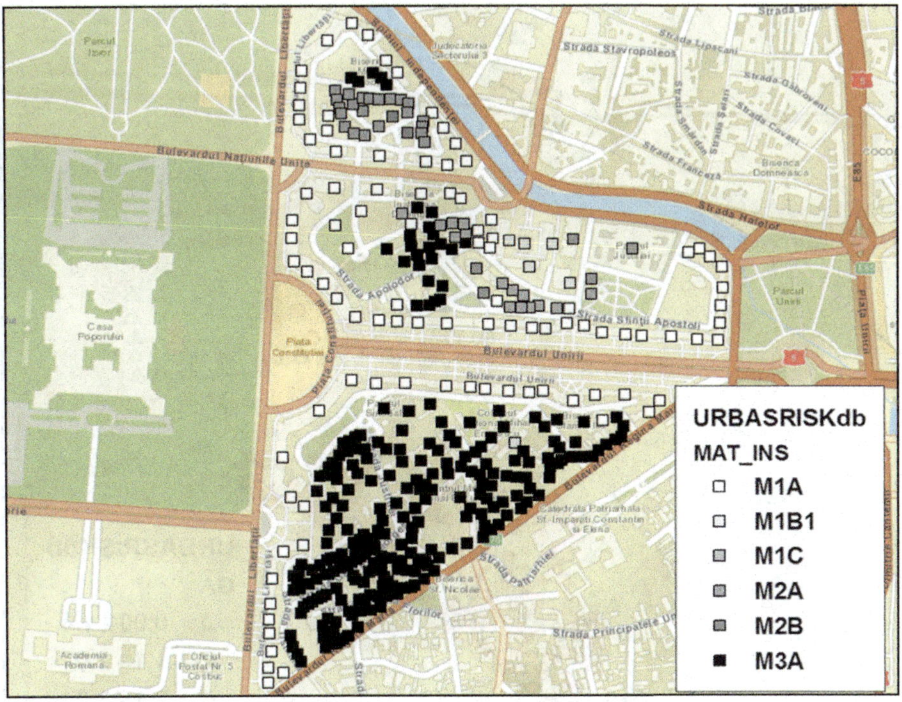

Fig. 2 Mapping of material/structural identifiers for structural vulnerability, with color codes in the study area. Base map: ESRI, World Street Map 2014 (Project URBASRISK, 2014)

400 buildings from the URBASRISKdb geodatabase were identified and mapped, with their associated attributes, using a spatial representation created by using ESRI ArcMap software (Figs. 2 and 3).

It is worth of mentioning that, for the vulnerability analyses, specific fields were also included, for the:

- harmonization with the building classification systems, based on the main construction material and the number of stories, respectively, used by the census;
- harmonization with the vulnerability classes, adapted and recalibrated, based on INCERC data from the 1977 earthquake (Balan et al. 1982; Sandi et al. 2008);
- use of the Mean Damage Degree, GA, based on the histograms computed for the above classes, associated with the Site-Adjusted Damage Degree, GAMA as a proxy indicator (Project URBASRISK, 2014).

In this respect, the basic source, i.e., the ESRI *World Street Map* layer, was verified against satellite, aerial and street views, while field visits also allowed the inspection of buildings to assess their structural vulnerability and, if applicable, the choice of the reduction factors required to obtain the site-modified damage index, GAMA, from the mean damage index, GA.

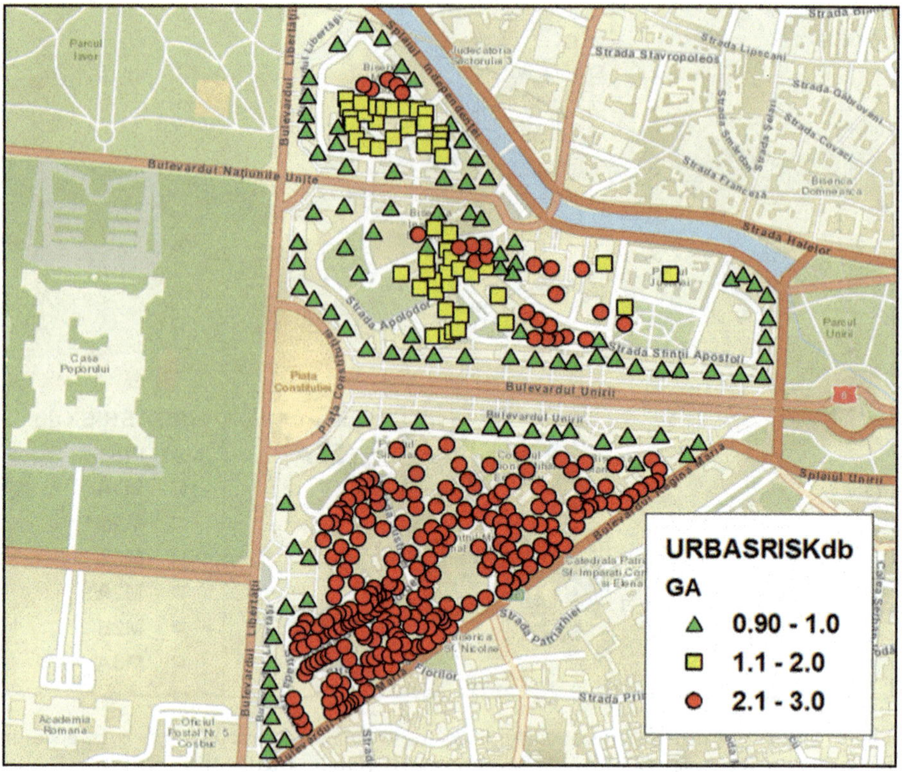

Fig. 3 Mapping of structural vulnerability: Damage Degree GA for ranges of values, with color codes in the study area. ESRI, World Street Map 2014 (Project URBASRISK, 2014)

4 Multi-Hazard Scenarios Results and Impacts That Are Useful for Emergency Preparedness

The multi-hazard scenarios used in the assessment were the following:

- for seismic hazard, earthquakes with intensities I = VIII and I = VIII ½ were considered, with specific vulnerability functions, calibrated and adjusted after the March 4, 1977, Vrancea, Romania, earthquake. When relevant for the exposure, some hypotheses of day and night were used, under an average occupancy ratio per house;
- as local soil hazards, liquefaction and/or settlements were considered; climatic and hydrologic hazards, such as snow, snow-storm, vortex, raising water, and rain were assessed, including accidental flooding from a remote dike;
- for man-made hazards, explosions with chemical release, gas explosions, terrorist attacks on public institutions, and urban fire were evaluated. The terrorist blasting

scenario considered three hypotheses, for two separate sub-zones, with explosive in quantity to be carried by one person, by a compact car, or by a van, respectively.

Earthquake scenarios revealed that the number of buildings with significant damage is around 63 for intensity I = VIII and 165 for intensity I = VIII½, these numbers referring mostly to low-rise structures. The number of heavy injured people was of 29 and 64, while live losses were of 59 and 92, respectively. This resulted in 1809–3061 persons evacuated from the above unsafe buildings. Although this number may indicate the need for a shelter, since the number of people from heavy damaged houses is only of 355 to 677 and houses have gardens and yards nearby, some shelter could be arranged there. The daytime scenario led to exposure and casualties reduced by 50 % in houses. However, an extra exposure of 10,120 people in public institutions around 1000 clients in commercial places, 1,400 in offices and 800 in schools, and 400 in churches was evaluated. Since such buildings are less vulnerable, only 26 light injuries and one hospital admission resulted. The scenario for I = VIII½ indicated the necessity of seismic safety assessment of damaged buildings and possibly a large number of temporary or long-term evacuation of occupants for repairs and structural strengthening. In this case, the already weakened buildings and the heritage buildings would need thorough and long duration works.

In the scenario of climatic and hydrologic hazards, the hypothesis was that of an extreme event of accidental flooding due to Lake Lacul Morii dam/dikes breaking (Drobot et al. 2007). The 14.7 millions cu. m. volume water supply Lake Lacul Morii, located in the Western part of Bucharest, has a concrete dam with a height of 15 m, 7 km earth dikes, and it discharges towards the city Dambovita River Canal that passes near the study area. An early warning can save the community persons, but not the built area. According to the available public data (Drobot et al. 2007), the water cover can be as high as 2.5 m and all 2 story houses as well as first 2 stories of condominium would be under water. Since this scenario is beyond the reaction capacity of the community, and its probability reduced because of permanent monitoring and water control by gates, it was not further evaluated.

The terrorist scenario blasting considered three hypotheses:

- explosive in quantity to be carried by one person (backpack);
- explosive in quantity to be carried by a compact car;
- explosive in quantity to be carried by an urban van.

These scenarios considered some particular locations/institutions that can be a target and it resulted that such impacts would affect very large and densely inhabited areas of irreparable damage, with fragmentation and debris spreading. In the study area, a number of 6 scenarios were considered, with the said three hypotheses for some particular location of an institution as target, in each of two separate sub-zones. The radiuses of impact were evaluated using the tables of IABTI—International Association of Bomb Technicians and Investigators, (https://www.iabti.org/) (Figs. 4 and 5) and in each case the radius is larger as the explosive quantity increases.

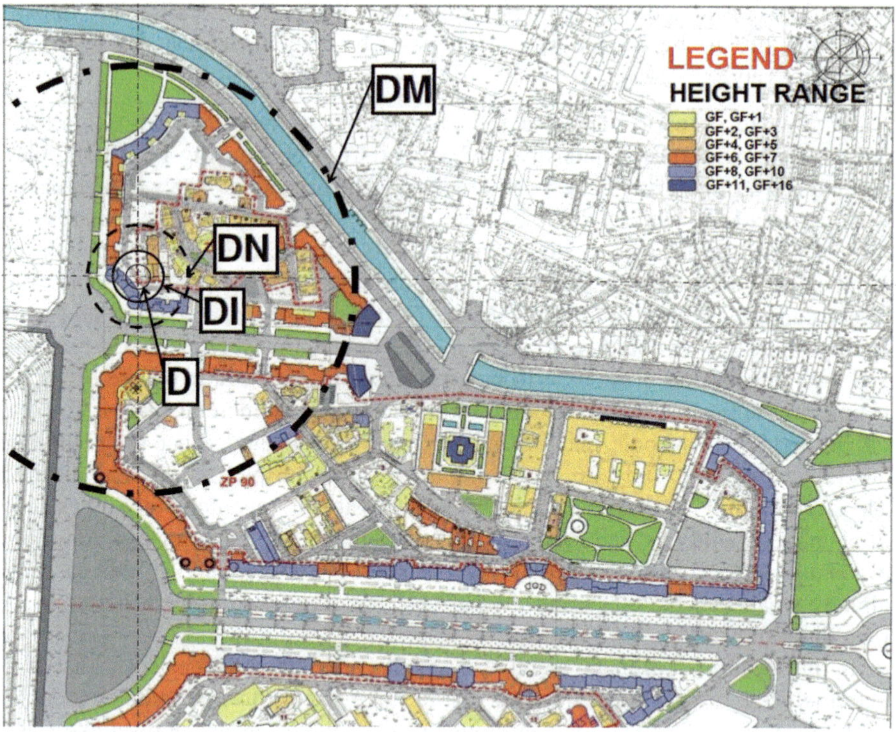

Fig. 4 Radiuses of specific zones in a specific location (Northern zone) within the study area. Position A, Scenario S 4.4, Case 2: explosive in a quantity to be carried by a compact car (Project URBASRISK, 2014)

The legend in Figs. 4 and 5 is as follows (www.iabti.org/):

- D—radius of demolition impact
- DI—irremediable demolition radius
- DN—non-repairable radius
- DM—minor destruction radius
- DF—minimum protection radius against fragments/debris impact.

The results of the scenario of terrorist blasting have shown that, as the explosive quantity increases, the impacts would affect larger areas and would lead to irreparable building damage in densely inhabited areas, with fragmentation and debris spreading and with great potential of injured and casualties. However, these impacts are more or less nominal, since higher buildings could provide a shielding effect.

The scenarios for explosions in other places of Bucharest, having impact on the study area, considered the hazards from various tanks containing gases or chemicals. It is known that in Bucharest such sources can be in thermal power plants, large deposits of toxic gases (ammonia gas NH_3, chlorine gas, and sulfuric acid), or GPL

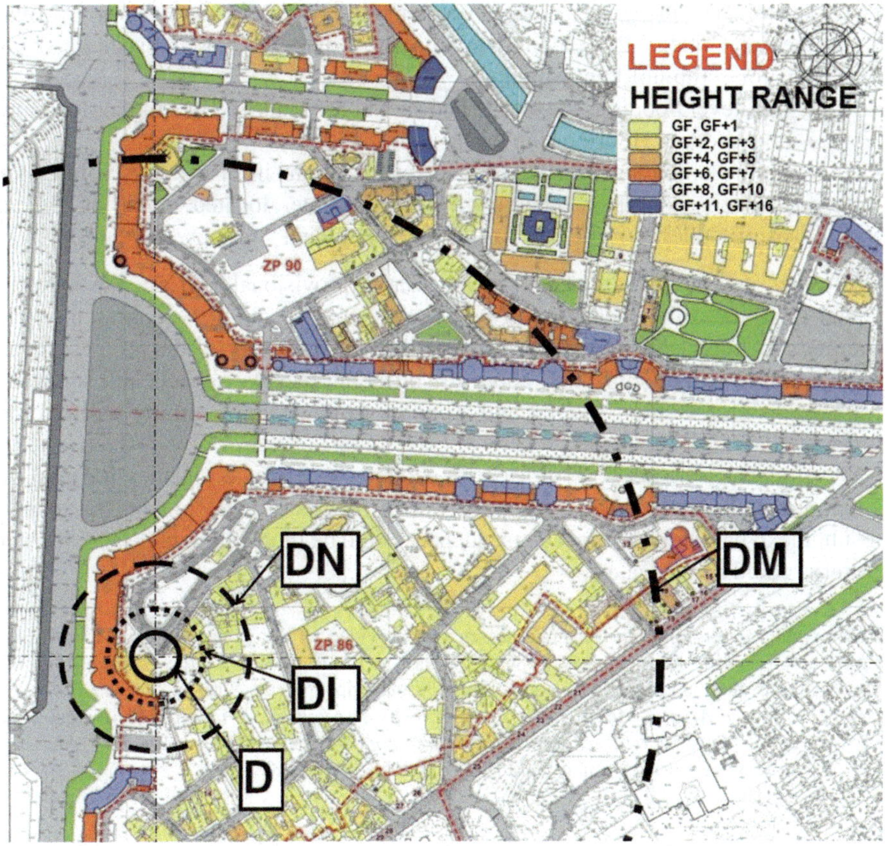

Fig. 5 Radiuses of specific zones in a specific location (Southern zone) within the study area. Position B, Scenario S 4.4, Case 3: explosive in quantity to be carried by an urban van (Project URBASRISK, 2014)

storage tanks. Data provided by Rufat (2009) prove that such plants and storages exist in other areas but their nominal release radiuses cannot reach the study area.

5 Conclusions and Suggestions for Emergency Planning

The multi-hazard assessments and scenarios provided some basic data about the extent of damage and casualties in the studied area. The fingerprint and the dynamic of impacts are different for each hazard, influencing the nature, amount, and dynamic of required logistics flows that are necessary at city scale. According to the results of the above scenarios, the necessity and possibility to create local shelters and security centers are as follows:

- for seismic damage consequences, a center could be created in the nearby "Mihai Eminescu" College, this having a main role for daily feeding and social care, as well as for community information;
- for a high-level flooding, the refuge and shelter area must be in a higher location, near the Academy House or the Parliament;
- for extreme climatic hazards, as well as for urban infrastructure failure, for terrorist scenario and/or other explosions, a center located either in "Mihai Eminescu" College or in Izvor Park is recommended.

Besides the shelter planned in advance as a provisional logistic, the security center is aimed to be a place with permanent activity within the community, for training and knowledge transfer to citizens. If the community of the protected area will be informed and consulted about the different hazards impacts and solutions for post-disaster shelters, the societal attitude of the local community will also be different and more beneficial.

At city scale, the network of security centers could create logistics for disaster situations and a poly nuclear support system in urban development. As a final result, the dynamics of hazards consequences could be more predictable and disaster patterns could be mitigated.

Acknowledgments Funding for this research was provided by the Romanian Ministry of Education and Scientific Research—UEFISCDI Agency in the framework of the National Plan for Research, Development, and Innovation, PNII, Partnership Program, Project "*Urban Blocks in Central Protected Area in Multiple Hazard Approach - Assessment, Mapping and Strategies for Risk Mitigation. Case Study: Bucharest Destructured Zone by Razing Occurring in the Communist Period*," under the Contract Number 53/2012, Project URBASRISK.

References

Balan, St., Cristescu, V., Cornea, I. (coordinators): The Romania Earthquake of March 1977 (in Romanian, with English abstract). Editura Academiei, Bucharest, Romania (1982)

Drobot, R., Amaftiesei, R., Alexandrescu, M.I., Cheveresan, B.: Modelling the effect of a scenario of Lake Lacul Morii breaking (in Romanian - Modelarea efectului unui scenariu de cedare a barajului Lacul Morii). Hidrotehnica **52**(12), 8–14 3 (2007)

Florescu, T., Moscu, C.I. Gociman, C.O., Georgescum M.S., Dragomirm C.S., Craifaleanu, C.S., Georgescu, E.S.: Urban heritage and multi-hazard threats. Case study of seismic vulnerability assessment and mapping in a protected area of Bucharest, Romania. In: Proceedings of the 1st Huixian International Forum on Earthquake Engineering for Young Researchers, Harbin, China, 16–19 August 2014

Georgescu, E.S., Pomonis, A.: The Romanian Earthquake of March 4, 1977 Revisited: new insights into its territorial, economic and social impacts and their bearing on the preparedness for the future. In: Proceedings of the 14th World Conference on Earthquake Engineering, Beijing, China, 12–17 October 2008

Georgescu, E.S., Gociman, C.O., Craifaleanu, I.G., Florescu, T., Moscu, C.I., Georgescu, M.S., Dragomir, C.S.: Urban heritage value and seismic vulnerability mapping: challenges for engineering and architectural assessments. Case study of a protected area in Bucharest, Romania, Paper no. 431, Earthquake Risk Mitigation Policies and Methodologies (EAEE Session), Second European Conference on Earthquake Engineering and Seismology, 2ECEES, Istanbul, 24–29 August 2014. ISBN 978-605-62703-6-9

Gociman, C.O., Florescu, T., Moscu, C.I., Georgescu, E.S.: Decisions on building stock survival and conservation in a multi-hazard environment: cultural and functional identity vs. safety and environment values in protected areas. In: International Conference in Climate Change, Ecology, and Conservation (ICCCECE 2014). The Venetian Macao, Macau, China, 13–14 February 2014a

Gociman, C.O., Moscu, C.I., Georgescu, E.S.: The relation between identity and vulnerability values in carrying out interventions in protected urban areas. In: The 9th International Conference on Urban Regeneration and Sustainability. Sustainable City, Siena, Italy, 23–25 September 2014b

Gociman, C.O., Georgescu, E.S., Florescu, T., Craifaleanu, I.G., Moscu, C.I., Georgescu, M.S.: Concept of a community security centre in the multi-hazard environment of a protected area in Bucharest, Romania. In: Proceedings of the 4th International Conference on Disaster Management and Human Health: Reducing Risk, Improving Outcomes. 20–22 May 2015. Istanbul, Turkey. Proceedings of Disaster Management 2015 - Disaster Management and Human Health Risk IV, 360 pp (Print ISBN: 978-1-84564-926-5; eISBN: 978-1-84564-927-2). Online at the WIT eLibrary in Volume 150 of WIT Transactions on the Built Environment (ISSN: 1746-4498 Digital ISSN: 1743-3509) (2015a). http://library.witpress.com

Gociman, C.O., Georgescu, E.S., Florescu, T., Moscu, C.I., Craifaleanu, I.G., Georgescu, M.S.: Structural methods for multi-hazard risk reduction in protected central areas of cities. In: Proceedings of the 5th International Conference on Building Resilience, Newcastle, Australia, 15–17 July 2015b

IABTI – International Association of Bomb Technicians and Investigators, USA, https://www.iabti.org/

Law No. 50 (1991): Law No. 50/1991 concerning the authorization of construction works, modified by OUG No. 22/2014, Romanian Official Gazette/Monitorul Oficial, Part I, No. 353/14.05.2014 (in Romanian)

Nagurney, A., Nagurney, L.: A mean-variance disaster relief supply chain network model for risk reduction with stochastic link costs, time targets, and demand uncertainty. presentation. In: The 2nd International Conference on Dynamics of Disasters (DOD 2015), Kalamata, Greece, June 29–July 2, 2015; updated July 25, 2015. Electronic copy available at: http://ssrn.com/abstract=2646867

Pardalos, P.M.: Assessing the vulnerability of evacuation plans via critical element detection. AAAS 2013 Annual Meeting, Boston, US, 14–18 February 2013. Abstract retrieved at September 26, 2015 at https://aaas.confex.com/aaas/2013/webprogram/Paper8389.html

Project URBASRISK (2012–2014) Project "Urban Blocks in Central Protected Area in Multiple Hazard Approach - Assessment, Mapping and Strategies for Risk Mitigation. Case Study: Bucharest Destructured Zone by Razing Occurring in the Communist Period" 2012–2015. UAUIM and UEFISCDI Agency. Retrieved on March 28, 2015 from http://www.uauim.ro/cercetare/urbasrisk/en/

Rufat, S.: Estimation relative de la vulnérabilité urbaine à Bucarest. Environnement, ville et société, Université de Lyon. N° 95 (3-2009) (2000). Retrieved at http://mappemonde.mgm.fr/num23/articles/art09301.html

Sandi, H., Pomonis, A., Francis, S., Georgescu, E.S., Mohindra, R., Borcia, I.S.: Development of a nationwide seismic vulnerability estimation system. In: Proceedings of the Symposium Thirty Years from the Romania Earthquake of March 4, 1977, Bucharest, Romania, March 1–3, 2007 (2008). CONSTRUCTII, No. 1/2008, pp. 38–47, Retrieved from http://constructii.incerc2004.ro/Archive/2008-1/2008-1-5.pdf

WCDRR: Sendai framework for disaster risk reduction 2015–2030. A/CONF.224/CRP.1. Third World Conference on Disaster Risk Reduction, 14–18 March 2015, Sendai, Miyagi, Japan. Accessed 28 March 2015, at http://www.wcdrr.org/uploads/Sendai_Framework_for_Disaster_Risk_Reduction_2015-2030.pdf
Zonal Urban Plan – PUZ, Bucharest: Zonal Urban Plan - Protected Areas. University of Architecture and Urban Planning "Ion Mincu" - UAUIM (in Romanian) (2000)

Lean Thinking and UN Field Operations: A Successful Co-existence?

Sulejman Halilagic and Dimitris Folinas

Abstract Lean thinking (LT) and supply-chain management (SCM) are recognized in the case of United Nations (UN) field operations (FO) as the organizational change towards the "One-UN" culture of cooperation which is one of the biggest challenges for the UN. Based on the above, this paper aims to identify the SCM and LT principles and best practices of the business world for the humanitarian operations world and the expected benefits of the application of the proposed principles and best practices. The study concludes that LT and SCM are two languages with the different accents that, when put together in cross functional design, can deliver better services to the field (SCM accent) and at a lower cost to member states (lean accent), leading to the conclusion that the "ideal SCM" will be the "Lean SCM" for the UN.

Keywords Lean thinking • Supply-chain management • United Nations field operations • Humanitarian operations

1 Introduction

Lean thinking (LT) and supply-chain management (SCM) are recognized in the case of the United Nations (UN) field operations (FO) as the organizational change towards the "One-UN" culture of cooperation. The "One-UN" culture is one of the biggest challenges in humanitarian operations (HO) for the UN. Its main objective is to address the current major organizational challenges: first, to provide lower costs of operation to the member states; second, to improve organizational capabilities by strengthening operations to be faster and flexible, and third, to offer better service to beneficiaries by bringing all its operations under one umbrella. This can be achieved

S. Halilagic
Office of Central Support Service, FF Building, Room: 0226, 10017 New York City, NY, USA
e-mail: halilagic@un.org

D. Folinas (✉)
Department of Logistics, TEI-KM, Kanellopoulou 2, 60100 Katerini, Greece
e-mail: dfolinas@gmail.com

I.S. Kotsireas et al. (eds.), *Dynamics of Disasters—Key Concepts, Models, Algorithms, and Insights*, Springer Proceedings in Mathematics & Statistics 185,
DOI 10.1007/978-3-319-43709-5_5

by network consolidation (vertical virtual integration) and the horizontal integration with other field operations, based on the "service provider-customer" relationship where the value chain is orchestrated and integrated by the Department of Field Support (DFS) of the UN within every UN Member State.

The Department of Field Support was established in 2007 in order to provide support functions for current 39 field operations, providing services that include—but are not limit to—the following: information communication technology and related infrastructure, air and ground transportation, medical services, public media, construction and maintenance, general administration (budget, finance, human resource, training, etc.), and a range of other activities that can be only compared with the government public sector (DFS 2013). Therefore, it was imperative for the DFS to improve its SCM performance by translating a number of best practices of commercial and humanitarian sectors as many researchers have argued (Kovács and Spens 2011a, p. 34; Van Wassenhove 2006, p. 475; Balcik et al. 2010, p. 22; Day et al. 2012, p. 22; Hines et al. 2004).

Moreover, it is critical that these best practices be aligned with the strategy of the "One-UN" paradigm and connected with the total supply chain to ensure close links within the humanitarian network. For instance, humanitarian relief must include collaborative procurement actions with suppliers, including best practices in managing such contracts (Balcik et al. 2010, p. 28; Davison and Sebastian 2011, p. 110). Best practices can also be looked at as a knowledge base and a "set of problem-solving management practices" (Hines et al. 2004, p. 1003) where traditional theory will consider that each operation is unique and, therefore, not replicable.

While humanitarian operations are setting a leading edge in the area of agility (Cozzolino et al. 2012; Mason-Jones et al. 2000), there is a valid question of at what cost such effectiveness has been achieved (Womack and Jones 2003; Hines et al. 2004) especially considering that sudden operations represent only 3 % of total humanitarian operations (Kovács and Spens 2011b, p. 7). It is difficult to understand why previous research on humanitarian operations has focused mainly on sudden humanitarian operations while slow-onset/man-made operations (97 % of total HO) have the greatest potential for improvement of humanitarian operations (HO) in terms of efficiency and effectiveness (Tatham 2012, p. 110). As a result, it remains unclear where the lean-agility boundary lies in the range of different humanitarian operations such as wars, civil strife, conflicts, famine, displaced populations, and political disasters, that are characterized as slow-onset/man-made operations (Van Wassenhove 2006, p. 476; Listou 2008, p. 3; Audet 2015).

Furthermore, although previous academic work in the area of humanitarian operations focuses on agility, there is increasingly greater consideration to lean due to the constraints on resources. It is evidenced that, through standardization (Van Wassenhove and Pedraza Martinez 2012, p. 312; Taylor and Pettit 2009), streamlining, and decreasing variability (Hines et al. 2004, p. 998; Mohamed and Berry 1999, p. 112; Salvadó et al. 2015), LT can support the effective application of the "One-UN." In addition, improvement in information and communication technology (Beamon and Balcik 2008, p. 11; Kovács and Spens 2011a) and implementation of the SCM best practices hold great potential towards the lean paradigm (Kovács and Spens 2011b, p. 7; Hines et al. 2004).

As such, the identification of best practices should be used to enhance standardization of work among UN Member States (public, commercial, and military) (Kovács and Spens 2011b, p. 7; Tatham and Worrell 2010). Based on the above, "One UN" means the establishment of an overall standard that defines a common framework and allows synergy between different entities and achieves best practices under the umbrella of the UN Secretariat.

Furthermore, current theory of SCM humanitarian operations does not clearly assess lean thinking applicability to slow-onset/man-made HO (Kovács and Spens 2011b, p. 7; Beamon and Balcik 2008, p. 23; Kovács and Spens 2009, p. 509; Overstreet et al. 2011, p. 115; Tatham 2012, p. 110). For instance, there is no case study related to the UN Department of Field Support (UN DFS) to identify similarities and differences of the SCM concept between peacekeeping and HO (GAO 2008; Van Wassenhove 2006, p. 476) despite the fact that the DFS is the largest and most complex SCM stream under the UN flag (DPKO 2013; Listou 2008; Van Wassenhove 2006, p. 476; Oloruntoba and Kovács 2015).

In such a light, the DFS in the last 3 years has started officially its journey "from silos to lean thinking," "from departmentalization to global operation," "from public to a more commercial business perspective," and "from function to process." Such a transformation can be described by using the "dinosaur's" metaphor (alluding to uniqueness, inflexibility, and large size of an operation) that no longer exists in today's world; making the point that transformation is about coordination that makes the "dinosaur" efficient, fast, lean, and able to survive also in a peacekeeping environment.

The transformation process was based on a theoretical framework consisting of the "Five Lean Principles" theory by Womack and Jones (2003) and the "Learning to evolve" theory by Hines et al. (2004) that examine each level of operation based on observed best practices.

Based on the above, this paper aims to identify how the SCM and lean thinking principles and best practices of the business world can be applied to the humanitarian operations world. The rest of the paper is organized as follows. In the next section a theoretical coverage of the SCM and LT paradigm in humanitarian operations is presented. The identification of SCM best practices and LT principles in UN field operations, as well as their perspective/achievements, is the main objective of the third section. Finally, lessons learned from the above application and further steps are discussed in the conclusions.

2 The SCM and LT Paradigm in Humanitarian Operations

Lean thinking (LT) is not a set of practices (tools) but a philosophy (Schiele and McCue 2011; Bhasin and Burcher 2006, p. 64; Spear and Bowen 1999) aiming to decrease the waste and, at the same time, to increase the value for customers/stakeholders (Womack and Jones 2003, pp. 299–312; Sezen and Erdogan 2009). It is based on universal principles that need to be applied throughout a never-ending journey towards perfection; seen rather as the target than a destination (Schonberger 2008). Furthermore, it has journeyed a long way from the manufacturing floor to the service industry (Suárez-Barraza et al. 2012, p. 201); to overall (strategic) management (Gibson et al. 2005; Womack and Jones 2003) becoming universally recognized as a holistic best practice among many industries (Wood 2004, p. 8). It became widely known in the course of successful implementation that turned around entire industries (Womack and Jones 1996; Womack and Jones 2003), with success coming from roots of process streamlining and a focus on customer value (Tatham and Worrell 2010; Naim and Gosling 2011). However, it has yet to be proven in the humanitarian sector.

2.1 Lean Thinking Principles and Humanitarian Operations

The lean thinking philosophy consists of various management concepts such as business process engineering, total quality management, and systems thinking that can be observed as an LT evolution (Hines et al. 2004, p. 996; Suárez-Barraza et al. 2012, p. 368). As a result, it cannot be applied in piecemeal fashion (Liker and Morgan 2006, p. 5), but, rather, must be applied in an integrated manner that includes the organizational structure, the management system, organizational value, culture, relationships, and the overall supply chain network (Hines et al. 2004; Nagurney et al. 2015). In the UN DFS context such a holistic approach is described as the integration of people, processes, and technology. Furthermore, the lean philosophy can be understood as a total approach that starts as a long-term strategic direction, including product design, process, structure, culture, humans, and technology (Heaslip et al. 2015). As Emiliani and Stec (2004, p. 67) argue, the difficulties of the lean deployment are a lack of a comprehensive approach, direction, planning, and adequate change management. According to Womack and Jones (2003), the key to the implementation of lean is to correctly identify the value and the value stream, make the product flow, let the customer pull value, and pursue perfection as a theoretical base for practical implementation of LT, emphasizing the importance of linking principles internally as well as externally as the way to achieve full benefits. Such long-term thinking requires that lean best practices and principles be put on the way to form synergy, not only internally but also throughout the extended supply chain; bringing it all as one "*Lean enterprise*" similar to the idea of "*One UN*."

The following figure presents the five lean thinking principles for humanitarian operations (Fig. 1).

Specifically, we highlight the following:

- *Specify value*. Womack and Jones (2003, p. 16) define value as "*the critical starting point for Lean Thinking ... that ... can only be defined by the ultimate customer*" and as such must be expressed in the provision of specific services. One of the main problems in defining value from a customer's perspective is the understanding of what the ultimate customer really wants.

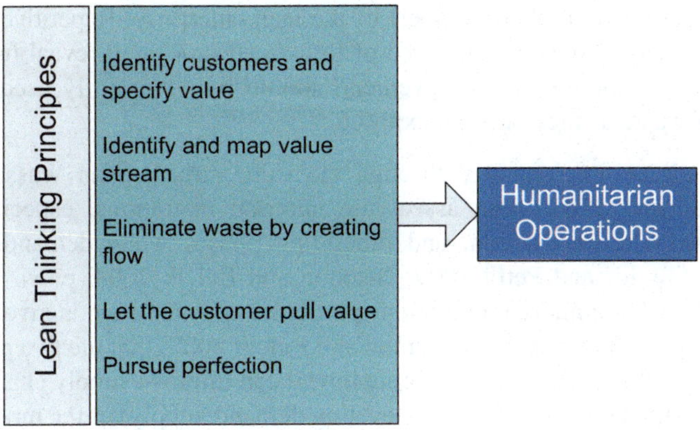

Fig. 1 Lean thinking principles for humanitarian operations

- *Identify the value stream.* While value definition is the starting point of LT, designing a value stream is a critical management task that includes specific actions, problem solving, information management, and physical transformation. Channeling the value stream to remove all non-value-adding steps, in the context of LT, refers to the process of revisiting the established standard, repeatedly, as elimination of one unnecessary step will reveal new opportunities for improvement. According to Womack and Jones (2003, p. 18) "Lean Thinking must go beyond the firm" to include all involved in the value stream where the lean enterprise is seen as a new way of thinking about the organization-to-organization relationship.
- *Flow.* "*The lean alternative is to redefine the work of functions, departments and firms … along the value stream so it is actually their interest to make value flow*" (Womack and Jones 2003, p. 24). This means that the value-adding process is systematically managed through every step at the necessary pace, based on customer demand. This includes insuring that each step is done correctly and with a required quality to avoid any back-flow such as the need for rework or scrap.
- *Pull.* While push delivers often unwanted product sitting idle in the warehouse; pull just makes what is requested by the customer, based on real customer demand where no step in the upstream will take place until the customer signals a need. Such a seamless supply chain connection allows customers to pull what they want, which leads to the elimination of the many types of waste due to: obsolescence, unnecessary inventories, and expensive demand planning systems.
- *Perfection.* Five principles can be recognized as being connected from the perspective that each previous step serves as a necessary input for the next; while, all together and, in repetitive cycles, "*efforts, time, space, cost and mistakes*" are reduced, while offering wanted products (Womack and Jones 2003, p. 26). Such recurring efforts should be continuously made systematically to remove waste throughout the extended supply chain that includes suppliers, contractors;

subcontractors, and other members of the lean enterprise. Repetitively applying principles and increasing the speed of flow (lead time) will reveal further waste that, through continuous improvement, should be persistently used to remove layers of waste as they become exposed.

Lean thinking, implemented through the value stream (Van Wassenhove and Pedraza Martinez 2012), can assist humanitarian operations, especially during the restoration, reconstruction, and rebuilding phases when demands are more predictable (Taylor and Pettit 2009; Beamon and Balcik 2008, p. 8). Taking into account that in humanitarian operations particular consideration is given to limited resources, it can be recognized (Beamon and Balcik 2008) that the five principles of lean can be used to assist in determining how it can improve supply chain visibility, capacity management, and planning (leveling demand/supply) as the most important preconditions for waste elimination (Argollo et al. 2012).

2.2 SCM Best Practices and Humanitarian Operations

Lean thinking represents the paradigm shift beyond applying the best practice at the shop floor to the direction of understanding the organizational environment, the management style, the vision, and strategic direction (Mason-Jones et al. 2000). This way, forward thinking can be seen as the paradigm shift of Womack and Jones (2003) theory that explains "*one best way of achieving lean*" towards extending flexibility and contingency of LT. This includes testing not only the boundaries but also the logic of lean enterprise design. Mason-Jones et al. (2000) find strategic and operational levels of LT at the strategic level dimension that must exist in order to select the right tools at the operational level.

This may especially be the case in the area of continuous improvement culture that is at the center of the lean house that supports the roof (goals/results) (Liker 2004, p. 33). According to Sollish et al. (2011) the transformation towards SCM is impossible without implementing a commercial point of view guided by a supplier/customer-centric approach where integration of projects must be managed from "end to end" as "a single coherent whole." As such, Van Wassenhove and Pedraza Martinez (2012) identify the following best practices as most critical for HO (Fig. 2).

A lean victory cannot be claimed by the adoption of best practices and tools without an overall philosophy and culture that often prove to be difficult not only from the point of long transformation but also from significant changes required in the way people think (Womack and Jones 2003).

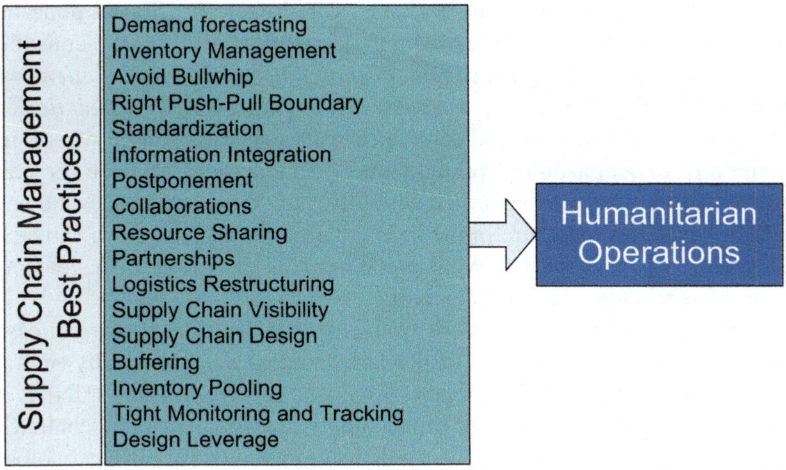

Fig. 2 SCM best practices for humanitarian operations (Source: Van Wassenhove and Pedraza Martinez 2012, p. 312)

3 Applying SCM Best Practices and LT Principles in UN Field Operations

3.1 Best Practices

Based on the above synthesis of the literature there is much evidence that supports the idea of the deployment of lean thinking practices and principles in the UN field operations and especially in slow-onset/man-made operations. However, important differences must be taken into account as the UN operates in an environment that can be described as a combination of the commercial, public, and military types of supply chains. In addition, while commercial business is narrow and deep in its business approach, peacekeeping operations are wide and shallow. Furthermore, peacekeeping field operations use a wide combination of lean thinking practices that consider the world as "one-single-market" with all risks and opportunities (Shenkar and Luo 2004):

- *Consolidation*. Consolidation is the process of merging departments/section/units together in order to eliminate waste, whenever there is overlapping of functions. Good examples are the integration of receiving, warehousing, distribution, and disposal functions under a single management frame. In the second stream of consolidation there is the consolidation of processes within the different levels of operation to place functions on the level that provides the highest benefits. In such a direction, the UN will take the majority of global operational and functional transactions. A good example is the Regional Support Center of Uganda and Entebbe (RSS) that centrally manages processes such as telephone billing, warranty repairs, and entitlement claims, which were previously done in each mission separately (Arif 2010; Maon et al. 2009), as well as, the consolidation projects utilizing "*Lean Six Sigma methodology to identify bottlenecks and opportunities for process improvement and implementation of solutions*" (UN GA 2012b, p. 22). Centralized service increases standardization where, instead of hundreds of people doing the same function spread across a wide geographical area, such functions are performed centrally. In addition, it reduces the need for building large infrastructure in the field; where recent missions in Libya, Syria, and an initial deployment to Mali prove benefits of remote management from points of: lead time, cost, and flexibility.

- *Lean Procurement combined with centralization/consolidation practices.* Procurement in the UN (The UN Global Marketplaces) was historically centralized, providing services to 15 UN Organizations and the World Bank (UNDP 2006). Today is the most important part of the supply management value chain (Sollish et al. 2011, pp. 215–227). This mainly is accommodated by ERP technology that simplifies the procurement process using a "Master Material Catalog" which can be seen as an "Amazon.com" type shopping. In addition, modularization allows a decrease of complexity and a higher level of integration of services and products through project life cycle phases (UNLB 2013, p. 2). Although public procurement "*required to satisfy a broad range of stakeholders*" is subject to greater public scrutiny and accountability (Waterman and McCue 2012), procurement centralization/consolidation brings high managers' expertise (profound market intelligence) together with modularization/standardization (Sollish et al. 2011, p. 215). This is an important step towards improving quality and reducing cost and complexity of FO (UN GA 2012a). Research actually identifies that there is an ongoing review of all system contracts, against a modularization pillar that will be available not only for the Department of Field Support of the UN but also to the "One UN" as a set of standardized projects/solutions. An approach that leverages the global market opportunities, in the concept of Lean, can create capabilities for new missions to be deployed with minimum lead time and can reduce oscillation with peacekeeping operation demand that sometimes cannot be planned in advance. Flexible contracts in volume such as pre-approved system contracts are key capabilities that encourage vendors to allocate their operations closer to a DFS regional center, due to the high volume of demand and the ability to serve larger customer groups in relatively close proximity.
- *Off-shoring.* In business, off-shoring is associated with relocation of products to a low-cost country (Shenkar and Luo 2004). In the Department of Field Support it includes cost of operations but, most importantly, close proximity to customers/missions. In addition, the off-shoring process represents impact in the field towards the goal of becoming more lean, and to improve responsiveness, efficiency, and effectiveness.
- *Use of government incentives for off-shoring.* Decisions on a regional center take into account incentives of the local government (Hoekman and Kostecki 2009, p. 260) such as in the case of Uganda/Entebbe that includes the availability of the international airport capacity for the Department of Field Support of UN

exclusive use and a special arrangement for expediting import/export (free of charge). Furthermore, Kenya, Italy, and Spain have provided sites for free with great logistics infrastructure as well.

- *Increased outsourcing*. Similar to the military concept of the US Army and NATO alliance, the Department of Field Support has increased the use of global commercial capabilities to advance external commercial partners' responsibility in delivering products and services to the "front line" (Christianson 2011, pp. 6–8). Outsourcing has many advantages, especially from the lean point of view, where commercial capabilities set through flexible contracts amortizing sudden changes in demand. Indeed, it can reduce the cost of operation, due to the high level of efficiency in the commercial/business sector. Outsourcing decreases mission footprints from the point of internal resources, and creates a resilient supply chain, based on best features of UN peacekeeping and commercial sectors.
- *Modularization*. Modularization in the Department of Field Support is the methodology of standardization that aims to deliver all elements of the supply chains required in order to deliver a project holistically for field operations. For instance, a water treatment plant requires equipment, service to install, training, daily operation control, and maintenance; therefore, it can be grouped as one project. This approach can be extended to a field camp design and many other services where one product meets the needs of different field operations (DPKO/DFS CPN 2013). Modularization not only reduces complexity through replication of business solutions but it also allows life cycle management, and a focus on total ownership cost that, furthermore, enhances learning, long-term partnerships, and continuous improvements (Kovács and Spens 2011b, p. 7; Nadler and Tushman 1999, p. 55). Although modularization is a valuable option for the field operations, research indicates that it must be carefully implemented in order, not to reduce the international competition required for transparency of public procurement.
- *Use of Information Communication Technology*. The use of ICT can produce many benefits and, from a lean thinking point of view, can reduce a lot of waste due to waiting, overstock, unnecessary steps, etc., as long as there is a "right fit" with business model/strategy (Yasuf et al. 2004). For instance, with the implementation of the ERP system, the visibility of the supply chain has been enhanced, with real demand being linked to each step in the chain and, most importantly, translated to more accurate planning, and action based on real-time data (Christopher 2005, pp. 249–253). This seamless process automates many steps in the supply chain, allowing set performances to be measured, and supporting decision-making at all levels.
- *Other Best Practices*. As the list of best practices collected through this research is almost endless, the following table presents the most critical ones (Table 1).

The list presents a set of best practices in order to support field operations streamlining. This is important in the areas of planning, forecasting, and procurement practices that enable the organization to minimize stock by creating a stable and

Table 1 Best practice observed

Best practice	Description in the context of DFS/FO	Lean perspective/achievements
ABC analysis	Prioritize stock based on higher transaction frequency (Serafimov 2013)	Reduce stock by outsourcing slow-moving stock cards
Avoid bullwhip effect	Improve integration of supply chain management through an integrated planning, reduced lead time and ERP visibility	Stock is not built "Just in Case"
Chain operations reference (SCOR) model	Building a comprehensive picture of the supply chain	Mapping and improving the field operations processes
Collaborations	Avoid duplication through a single orchestration	Waste elimination and elimination of boundaries within and between UN units
Cross-docking	Use of a logistics hub on a port closer to the "centre of gravity" for shipments consolidations	Improving level of services and reducing lead time and cost
Demand forecasting	Engage customers in joint planning	Understanding customer needs and what creates value for them
Design leverage	Implementation of the DFS supply chain management strategy to build a global capacity	Streamlining operations through one single value stream
Development of new products	Developing through internally and externally network, specific products for field operations such as the "Fly away kit"	Applying right design practices that meets specific customer needs in the field operations
Inventory management	Adaptation of the life cycle inventory management system starting from requisition point to disposal. Managing inventory as one single global resource	Waste elimination and increase collaboration between missions. Moreover, the management of transportation means through a consolidated approach
Inventory positioning	More gravity towards regional centers to provide synergy by aggregation of safety stock	Reduction of stock; close proximity of stock to the customers; improve safety of stock by reducing stock in the field
Horizontal integration among field operations	Creating relationship of "service provider"—customer, by sharing products, services, and capacity/capability of the DFS global supply chain system, including the following: collaborative procurement and acquisition planning, piggyback contract arrangements, services sharing, consolidation of logistics operations, join process streamlining, and specialization. Avoidance of competition for same goods and service on the commercial market	Reduction of duplication. Increase interdependency and collaboration among the various field operations. Cost decrease due to economy of scale. Decline of lead time. Better utilizations of capacity through sharing

Best practice	Description in the context of DFS/FO	Lean perspective/achievements
ISO 9001 standard	Standardization of quality control and management. Benchmarking of the best practices. Currently implemented only in SC and missions that have almost a permanent presence	Use of continuous improvement approaches to drive better performance. Focus on root cause of failures
Key performances indicators	Measurement of supply chain management performances and level of implementation. Gaining intelligence about how the supply chain management actually is performing, setting targets, linking performances to staff appraisal process, benchmarking with best in class	Continues improvements, staff motivation, and better management
Life cycle cost management	Total cost of ownership used through assessment of risks and benefits as the main tool to determine cost efficiency of outsourcing (Tchokogué et al. 2015; Nagurney et al. 2011)	Reduction in operational cost. Long-term partnership with commercial sector
Strategic partnership	Close cooperation with main partners and other bilateral actors (DPKO 2013)	Decrease in cost and complexity though use of local solutions
Partnering with institutes	Massachusetts Institute of Technology's (MIT) involvement in the development of the supply chain management design and strategy	Bringing academic knowledge to practice in order to "Make it Right - First Time"
Planning and managing capacity	Forecasting to optimize capacity and use of share resources, especially during opening and closing operations. Establishing closer stakeholder coordination such as the UN Security Council and Military in the field	Maintaining capacity in accordance with the needs of the DFS, call "Rightsizing" (UN GA 2012b, p. 18)
Post-ponement	Equipment is configured in accordance with the real demand	Waste elimination through matching right products configuration with actual demand/environment of the field operations
Resources sharing	More collaborative relations through a matrix organization and a horizontal coordination structure	Increases utilization of resources. Allows resources configuration to be aligned with actual needs
Tight monitoring and tracking	Use of barcode scanning capabilities. Monitoring goods in transit in accordance with INCOTERMS	Improve data quality and visibility from pipeline to the end of life
Vendor-managed inventory	Outsourcing specific supply chains such as medical supply chain to specialized company	Decrease complexity of operations

predictable flow of materials and services; fully integrated, not only within the DFS vertical stream but also through horizontal integration among the field operations family in line of the "The UN" vision.

Our research finds (despite similarity and differences between commercial and public sectors) that the great majority of best practices can be translated to the context of field operations. In such a light, our research finds that field operations have most difficulties in translated coordination and integration elements that in the commercial sector is motivated by profit/incentives; where "Value for Money" needs to be further better understood so that each field operations component fighting their own battles where there is a "One UN" solution/capacity is avoided.

3.2 *Principles*

Womack and Jones (2003) argue that the keys for the implementation of lean are to: correctly identify customers and value, identify and map value stream, eliminate waste by creating flow, and pursue perfection. Specifically, we summarize the keys below.

3.2.1 Identify Customers and Specify Value

Specifying what creates "value for the customer" is the critical starting point of lean thinking deployment (Womack and Jones 1996, p. 570). In the case of the Department of Field Support, customers' (mainly the host countries and UN member states) values are: a short lead time of deployment to secure peacekeeping arrangements, a continuous level of service, and the ability to quickly reconfigure resources based on changes in operation.

Definition of "who is the customer?" (Radnor et al. 2006, p. 62) in the Department of Field Support as well as in the public sector, in general, takes a broadened view of customer's value. From that point, peacekeeping "network support" with the UN Member States must be understood as an important customer value, due to their joint involvement in country rebuilding where often such parallel projects are financed by the same countries (Beamon and Balcik 2008, p. 4, UNHCR 2013). According to the UN strategic plan, the Department of Field Support is moving in a customer-centric direction.

However, such an approach will remain fragmented without extending a customer point of view and achieving one face of the UN in field operations ("One UN") (Williamson et al. 2004, p. 104). Some best practices observed through this research are: (1) modularization of products in accordance with customer needs, (2) establishment of new operations/mission within the specified time-frame, (3) establishment of more consolidated acquisition plans that start with customer inputs based on needs/quality required, (4) development of a customer-oriented culture, and (5) country team integration and collaboration (UN OCHA 2006).

3.2.2 Identify and Map Value Stream

Value stream mapping in the case of the DFS takes a global picture to link each field operation stream through a SCM holistic approach. On the contrary, the local suboptimization can negatively affect the overall performance of the global supply chain. This approach brings flexibility, resilience, and speed required to optimally prioritize resources based on the needs. But it also requires a project integration that must seamlessly ensure that products and services are delivered as a value stream coherent project (Dobrzykowski et al. 2012, p. 570). Some best practices observed through the research include: (1) focus on synergies across the DFS, (2) regularly reviewing the appropriateness of the supply chain structure and infrastructure, especially at the operational level in regards to the mission life cycle changes, (3) increased emphasis on long-term partnerships with suppliers, (4) more focus to coordinate and monitor the performance of system contracts through all levels of the SCM, (5) strategic sourcing review at the strategic, regional, and local levels, (6) contracts with more flexibility in regards to contract amendments and changes, (7) translation of processes to diagram flow and policies, and finally, (8) a broad involvement and coordination of all parts linked in the value chain.

3.2.3 Eliminate Waste by Creating Flow

Waste elimination is achieved through the creation of a continuous flow of products and information that can level both the demand and supply side (Day et al. 2012, p. 23), from the opening to the closing of one mission. Aggregation of supply and demand at the regional and strategic level not only removes the major sources of waste but also allows standardization to be applied across all field operations processes.

In the concept of the DFS, the elimination of waste can improve demand planning and external capacity (flexible contract, outsourcing) that can surge during peaks of demand (Gibson et al. 2005, p. 22). Enhanced information visibility and coordination at different levels within the supply chain bring high potential for further elimination of waste. Research also indicates the importance of the transformation of the strategic stock (SDS 2013) in a rotation manner; thereby, allowing other DFS operations to perform at a very lean level, especially when stock is close to customers' demands (Naylor et al. 1999, p. 112). Some best practices include the following: (1) reduction of stock with a higher turnover rate, (2) establishment of strategic demand planning, (3) focus on best use of resources to the level that does not negatively affect responsiveness and effectiveness, (4) simplification of procurement requirements in order that needs can be easily met by the marketplace, (5) resource coordination and reallocation through a global SCM foundation, (6) enhancement of the information management system to a more integrated supply chain, (7) extended use of electronic tendering process to reduce lead time, and (8) significant reduction of strategic stock.

3.2.4 Let the Customer Pull Value

The Department of Field Support with the implementation of many projects has increased the capacity to optimally respond to demand, moving from a push to a more pull system. The lean concept can ensure that goods are delivered on time but only replenished when absolutely necessary (Liker 2004). In the case of peacekeeping under uncertainty, additional capacity such as strategic stock (Brazier 2009; Nagurney et al. 2011) and flexible arrangements with suppliers and governments is still required to cope with changes in demand (Christopher 2005, pp. 254–258; Tatham and Worrell 2010). Further, a decoupling point is used for goods with a long lead time (Mason-Jones et al. 2000, p. 4065) such as firefighting trucks or armored vehicles that are pushed to closer demand locations and then pulled from that point. Although DFS is increasing the shift to a pull system, due to risk of not having critical goods on hand, there is still a necessity for a relatively high level of contingency (Hines et al. 2004, p. 994). Moreover, the existence of extra capacity can smooth imbalances between demand and supply (Christopher 2005, pp. 237–243) and avoid expedite/contingency procurement actions that come with a high cost.

Using a vendor managed inventory approach and direct delivery from the factory/vendor site to the final destination, the lead time can be reduced up to 70 days. Here DFS has made significant progress in integrating external manufacturing capabilities to avoid satisfying demand for stock. Research indicates that new missions are excellent examples of the SCM and LT approaches in delivering goods from the producers directly to final destination in accordance with the schedule deployment plans. Proposed best practices are: (1) better management of demand, (2) creation of a simplified and a more streamlined SCM process, trigger by "iNEED" internet customer interface, (3) satisfying demand through direct delivery from manufacturing locations, (4) establishment of robust ICT systems and infrastructure that can translate customer demands to the most appropriate action to the various levels of the global supply chain, (5) focus on the right trade-offs to balance responsiveness and efficiency that continue to meet customers' needs without operational interruption, and (6) increased stock aggregation, supported with intermission transfers.

3.2.5 Pursue Perfection

Considering that DFS is in the early phase of the establishment of the lean approach, it is difficult to find evidence of lean perfection in peacekeeping operations; nevertheless, significant improvements have been made in such a direction. What establishes the best value towards perfection (business excellence) (Oakland 2003, pp. 225–227) is the increasing involvement in lean thinking (Hines et al. 2004) that continuously desires to make things better (Holguin-Veras et al. 2013).

TR News, July–August (2013). This includes, at first, the benchmarking with the other UN Member States. This, according to Hines et al. (2004, p. 1001), embraces the "organizational learning" paradigm. Specifically, continuous improvement builds a "Learning Organization" that can capture new opportunities and achieves sustainability in the long run, providing elements of flexibility and contingency to lean philosophy. Reform projects are great examples of putting LT into practice that even now are making significant improvements in reducing the steps, amount of time, resource level, and information needed to serve the customer (Womack and Jones 2003; Hines et al. 2004, p. 1003).

Some best practices observed through this research are the following: (1) implementation and upgrade of the SCM practices and processes to remain better or with best in class, (2) establishment of the global supply chain center of excellence (McOwan 2013), (3) development of a global inventory performance management, (4) recognition of efficiency achievements as the methods to strengthen the implementation of the SCM and lean thinking and build a lean culture, (5) establishment of partnerships with certificated supply chain agencies, universities, and institutes (McOwan 2013), (6) benchmarking SCM performances against similar organizations, and (7) mapping of processes and establishment of standards so as to support continuous improvements.

4 Conclusions

Lean thinking and SCM are two languages with different accents that, when put together in cross functional design, can deliver better services to the field (SCM accent) and lower cost to member states (lean accent), leading to the conclusion that the ideal SCM will be the "Lean SCM." Lean thinking in the Department of Field Support can be understood as the moving mind shift from vertical to horizontal process-oriented thinking. This means that lean thinking in the DFS case is mostly a strategic guiding philosophy that builds culture throughout the entire supply chain.

In this paper, the authors argue that humanitarian operations, especially in the slow-onset/man-made segment, have a huge potential to benefit from exposure to lean principles/practices, where the Department of Field Support has just made pioneering steps in inheriting commercial best practices. In that direction, a higher level of coordination and cooperation within field operations is required as the "One UN" strategic direction can work only when all streams act together in a synchronized way, especially taking advantage of a large SCM integrator such as the Department of Field Support. This suggests that the current trend of horizontal integration will become more important with the adoption of SCM best practices among field operations, where it can be identified as the right step in such a direction is closer coordination and integration of SCM by the UN Office for Coordination of Humanitarian Affairs and/or country team. In general, our research confirmed that lean SCM can be implemented even in the most challenging environment; however, not with the same results as in lean manufacturing, due to the need for addressing contingency requirements through a very robust, flexible, and resilient SCM and some limitations such as infrastructure, political factors, and lack of skilled staff.

SCM blended with lean thinking is acknowledged as a critical strategic philosophy to reduce response time and a smooth level of flow that takes advantage through the field operations global network consolidation; supporting the idea that lean thinking is applicable in a dynamic environment when applied from a strategic point of view (Fawcett et al. 2014; Gligor and Holcomb 2012).

While integration is still to be implemented to its full capacity, many barriers can be removed by bringing best practices and translating them into the exact circumstances of each organization that forms a Unity Enterprise ("One UN"). A lean enterprise, obviously, through consolidation of business as one global operation can provide additional synergy, especially, where oscillations in demand can be observed by size, consolidation, and cooperation within organizations.

Lean thinking has a huge potential in field operations not only to flatten organizational structures and streamline processes but also to create a modern/virtual organization. Such an organization takes the advantage of coordination where motivation can be achieved through lean thinking that focused on people as the main resource. Research findings are mainly applicable to slow-onset humanitarian operations that can be compared with dynamic global business, which increasingly operates under a different kind of business/social/political uncertainty and supply chain interdependency (Qiang and Nagurney 2012).

Further empirical inquiry is needed to determine how the research findings can be used to enhance coordination and cooperation within UN global supply chain operations. The main objective is to clearly determine how competitive advantages of each partner can be best used for benefits of the UN SCM ("One UN"). Further research on slow-onset type of operations is also required to capture organizations smaller in size and with more fragmented field operations, such as in nongovernmental organizations (NGO).

Acknowledgments The authors acknowledge with gratitude, the University of Liverpool, UK, Faculty of Humanity and Social Sciences that provided the technical inputs to this research.

References

Argollo, S.R., Vânia, B.G., Albergaria, R.: Supply chains in humanitarian operations: cases and analysis. Proc. Soc. Behav. Sci. **54**(n/a), 598–607 (2012)

Arif, N.: Internal Memo [Comet]. Property management performance report for the period 01 July 2009 to 30 June 2010, United Nations DFS (13/09/2010)

Audet, F.: From disaster relief to development assistance: why simple solutions don't work. Int. J. Can. J. Glob. Policy Anal. **70**(1), 110–118 (2015)

Balcik, B., Beamon, M., Krejci, C., Muramatsu, M., Ramirez, M.: Coordination in humanitarian relief chains: practices, challenges and opportunities. Int. J. Prod. Econ. **07**(126), 22–34 (2010)

Beamon, B.M., Balcik, B.: Performance measurement in humanitarian relief chains. Int. J. Public Sect. Manage. **21**(1), 4–25 (2008)

Bhasin, S., Burcher, P.: Lean viewed as a philosophy. J. Manuf. Technol. Manage. **17**(1), 56–72 (2006)

Brazier, D.: Heavy Lifting for United Nations Peacekeeping: Strategic Deployment Stocks. Chartered Institute of Logistics & Transport (UK), 11 (2009)
Christianson, C.V.: Challenge and change in supply chain management: pointed questions and blunt answers. Def. AT&L **40**(4), 4–11 (2011)
Christopher, M.: Logistics and Supply Chain Management, 3rd edn. FT Prentice-Hall, London (2005)
Cozzolino, A., Rossi, S., Conforti, A.: Agile and lean principles in the humanitarian supply chain: the case of the United Nations World Food Programme. J. Humanit. Logist. Supply Chain Manage. **2**(1), 16–33 (2012)
Davison, B., Sebastian, R.J.: A detailed analysis of the relationship between contract administration problems and contract types. J. Pub. Procurement **11**(1), 108–125 (2011)
Day, J.M., Melnyk, S.A., Larson, P.D., Davis, E.W., Whybark, D.C.: Humanitarian and disaster relief supply chains: a matter of life and death. J. Supply Chain Manage. **48**(2), 21–36 (2012)
DFS: Department of Field Support [Online]. Available from: http://www.un.org/en/peacekeeping/about/dfs/ (2013). Accessed 19 Dec 2013
Dobrzykowski, D.D., Hong, P.C., Park, J.S.: Building procurement capability for firm performance: a service-dominant logic view. Benchmarking **19**(4), 567–584 (2012)
DPKO: Department of Peacekeeping Operations [Online]. Available from: http://www.un.org/en/peacekeeping/about/dpko/ (2013). Accessed 13 Jan 2013
DPKO/DFS CPN: DPKO/DFS Communities of Practice Networks Peace operation, Policy and Best Practices Sharing knowledge & Empowering Teams [Online]. Available from: http://cop.dfs.un.org/ (2013). Accessed 22 Dec 2014
Emiliani, M.L., Stec, D.J.: Using value-stream maps to improve leadership. Leadersh. Organ. Dev. J. **25**(8), 622–645 (2004)
Fawcett, S., Fawcett, A., Watson, B., Magnan, G.: Peeking inside the black box: toward an understanding of supply chain collaboration dynamics. J. Supply Chain Manage. **48**(1), 44–72 (2014)
GAO: Challenges Obtaining Needed Resources could Limit further Large Deployments and should be Addressed in U.S. Reports to Congress [Online]. Available from: http://www.gao.gov/assets/290/284560.pdf (2008). Accessed 24 July 2014
Gibson, B.J., Mentzer, J.T., Cook, R.L.: Supply chain management: the pursuit of a consensus definition. J. Bus. Logist. **26**(2), 17–25 (2005)
Gligor, D.M., Holcomb, M.C.: Understanding the role of logistics capabilities in achieving supply chain agility: a systematic literature review. Supply Chain Manage. Int. J. **17**(4), 438–453 (2012)
Heaslip, G., Kovács, G., Haavisto, I.: Supply chain innovation: lessons from humanitarian supply chains. In: Stentoft, J., Paulraj, A., Vastag, G. (eds.) Research in the Decision Sciences for Global Supply Chain Network Innovations, pp. 9–26. Pearson Press, New York (2015)
Hines, P., Holwe, M., Rich, N.: Learning to evolve - a review of contemporary LT. Int. J. Oper. Prod. Manage. **24**(9–10), 994–1011 (2004)
Hoekman, B.M., Kostecki, M.M.: The Political Economy of the World Trading System from GATT to WTO, 3rd edn. Oxford University Press, Oxford (2009)
Holguin-Veras, J., Jaller, M., Wachtendorf, T.: Improving postdisaster humanitarian logistics. TR News 4–10 (2013)
Kovács, G., Spens, K.M.: Identifying challenges in humanitarian logistics. Int. J. Phys. Distrib. Logist. Manage. **39**(6), 506–528 (2009)
Kovács, G., Spens, K.M.: Trends and developments in humanitarian logistics – a gap analysis. Int. J. Phys. Distrib. Logist. Manage. **41**(1), 32–45 (2011a)
Kovács, G., Spens, K.M.: Humanitarian logistics and supply chain management: the start of a new journal. J. Humanitarian Logist. Supply Chain Manage. **1**(1), 5–14 (2011b)
Liker, J.K.: The Toyota Way: 14 Management Principles from the World's Greatest Manufacturer. McGraw-Hill, New York (2004)
Liker, J.K., Morgan, J.M.: The Toyota way in services: the case of lean product development. Acad. Manage. Persp. **20**(2), 5–20 (2006)

Listou, R.: Non-commercial supply chains and the challenge of demand uncertainties and political processes. In: 7th International Meeting for Research in Logistics, pp. 1–17. AVIGNON, 24–26 September 2008

Maon, F., Lindgreen, A., Vanhamme, J.: Developing supply chains in disaster relief operations through cross-sector socially oriented collaborations: a theoretical model. Supply Chain Manage. Int. J. **14**(2), 149–64 (2009)

Mason-Jones, R., Naylor, B., Towill, D.R.: Lean, agile or leagile? Matching your supply chain to the marketplace. Int. J. Prod. Res. **38**(17), 4061–70 (2000)

McOwan, S. (mcowan@un.org), 07 March 2013. Fw: Supply Chain Strategy as at 8 March. Email to P. Smith (devaulxdechambord@un.org)

Mohamed, M.N., Berry, D.: Leagility: integrating the lean and agile manufacturing paradigms in the total supply chain. Int. J. Prod. Econ. **62**(1), 107–118 (1999)

Nadler, D., Tushman, M.: The organization of the future: strategic imperatives and core competencies for the 21st century. Organ. Dyn. **28**(1), 45–60 (1999)

Nagurney, A., Masoumi, A., Yu, M.: An integrated disaster relief supply chain network model with time targets and demand uncertainty. In: Nijkamp, P., Rose, A., Kourtit, K. (eds.) Regional Science Matters: Studies Dedicated to Walter Isard, pp. 287–318. Springer International Publishing Switzerland (2015)

Nagurney, A., Yu, M., Qiang, Q.: Supply chain network design for critical needs with outsourcing. Pap. Reg. Sci. **90**, 123–142 (2011)

Naim, M., Gosling, J.: On leanness, agility and leagile supply chains. Int. J. Prod. Econ. **131**(1), 342–354 (2011)

Naylor, B.J., Naim, M.M., Berry, D.: Leagility: integrating the lean and agile manufacturing paradigms in the total supply chain. Int. J. Prod. Econ. **62**(1), 107–118 (1999)

Oakland, J.S.: TQM: Text with Cases, 3rd edn. Butterworth-Heinemann, Oxford (2003)

Oloruntoba, R., Kovács, G.: A commentary on agility in humanitarian aid supply chains. Supply Chain Manage. Int. J. **20**(6), 708–716 (2015)

Overstreet, R.E., Hall, D., Hanna, J.B., Rainer Jr., R.K.: Research in humanitarian logistics. J. Humanitarian Logist. Supply Chain Manage. **1**(2), 114–131 (2011)

Qiang, Q., Nagurney, A.: A bi-criteria indicator to assess supply chain network performance for critical needs under capacity and demand disruptions. Transp. Res. A **46**(5), 801–812 (2012)

Radnor, Z., Walley, P., Stephens, A., Bucci, G.: Evaluation of the lean approach to business management and its use in the public sector [Online]. Available from: http://www.scotland.gov.uk/Resource/Doc/129627/0030899.pdf (2006). Accessed 10 Dec 2014

Salvadó, L., Lauras, M., Comes, T., Van de Walle, B.: Towards more relevant research on Humanitarian disaster management coordination. In: Palen, Bscher, Comes, Hughes (eds.) Proceedings of the ISCRAM 2015 Conference, Kristiansand, 24–27 May 2015. Short Paper–Researching Crisis: Methodologies (2015)

Schiele, J.J., McCue, C.P.: LT and its implications for public procurement: moving forward with assessment and implementation. J. Pub. Procurement **11**(2), 206–239 (2011)

Schonberger, R.: Best Practices in Lean Six Sigma Process Improvement [electronic book] a Deeper Look/Richard Schonberger. Wiley, Hoboken, c2008 [Online] (2008)

SDS: UNLB official page [Online]. Available from: http://www.unlb.org/sds.asp (2013). Accessed 13 Jan 2014

Serafimov, K.: 'ABC Analysis of Expendables', Property Management Conference. United Nations Property Management Unit, COE&PMSS/SSS/LSD/DFS, New York (2013)

Sezen, B., Erdogan, S.: Lean philosophy in strategic supply chain management and value creating. J. Glob. Stratg. Manag. **5**, 68–72 (2009)

Shenkar, O., Luo, Y.: International Business, 4th edn. Wiley, New York (2004)

Sollish, F., Semanik, J., Morris, P.W.G., Pinto, J.K.: Planning and Administering Project Contracts and Procurement. Laureate Education, Inc., custom ed. Wiley, Hoboken (2011)

Spear, S., Bowen, H.K.: Decoding the DNA of the Toyota Production System. Harv. Bus. Rev. **77**(5), 96–106 (1999)

Suárez-Barraza, M.F., Smith, T., Dahlgaard-Park, S.: Lean service: a literature analysis and classification. Total Q. Manage. Bus. Excell. **23**(3), 359–380 (2012)
Tatham, P., Worrell, D.: LT in an uncertain environment: the implications for UK defence acquisition. Int. J. Def. Acquis. Manage. **3**(n/a), 1–22 (2010)
Tatham, P.: Some reflections on the breadth and depth of the field of humanitarian logistics and supply chain management. J. Humanitarian Logist. Supply Chain Manage. **2**(2), 108–111 (2012)
Taylor, D., Pettit, S.: A consideration of the relevance of lean supply chain concepts for humanitarian aid provision. Int. J. Serv. Technol. Manage. **12**(4), 430–444 (2009)
Tchokogué, A., Nollet, J., Fortin, J.: Outsourcing Canadian armed forces logistics in a foreign theatre. Can. J. Adm. Sci. **32**(2), 113–127 (2015)
UN GA: Department of Field Support Tells Budget Committee it Managed to 'Do Better With Less', Owing to Global Strategy, Steadfast Commitment of Blue Helmets, GA/AB/4028 [Online]. Available from: http://www.un.org/News/Press/docs/2012/gaab4028.doc.htm (2012a). Accessed 22 Dec 2014
UN GA: Third annual progress report on the implementation of the global field support strategy, A/67/633 [Online]. Available from: http://www.un.org/ga/search/view_doc.asp?symbol=A/67/633 (2012b). Accessed 22 Dec 2014
UN OCHA: UNDAC Handbook [Online]. Available from: https://docs.unocha.org/sites/dms/Documents/UNDAC%20Handbook-dec2006.pdf (2006). Accessed 15 Dec 2014
UNDP: United Nations System General Business Guide, 20th Edition [Online]. Available from: https://www.ungm.org/Publications/Documents/gbg_master.pdf (2006). Accessed 10 Dec 2014
UNHCR: Handbook for emergencies Edition [Online]. Available from: http://www.sicht.ucv.ve:8080/biblioteca/Cd/e/publicaciones/Handbook/index.htm (2013). Accessed 22 Dec 2014
UNLB: Strategic planning and work plan guidelines 2012-13 [Online]. Available from: http://www.pms-qms.unlb.org/Logistics%20Services%20Strategic%20Work%20Plan%202012-13,pdf.pdf (2013). Accessed 15 Dec 2014
Van Wassenhove, L.N., Pedraza Martinez, A.J.: Using OR to adapt supply chain management best practices to humanitarian logistics. Int. Trans. Oper. Res. **19**(1), 307–322 (2012)
Van Wassenhove, L.N.: Humanitarian aid logistics: supply chain management in high gear. J. Oper. Res. Soc. **5**(n/a), 475–489 (2006)
Waterman, J., McCue, C.: LT within public sector purchasing department: the case of the U.K. public service. J. Pub. Procurement **12**(4), 505–527 (2012)
Williamson, D., Cooke, P., Jenkins, W., Moreton, K.M.: Strategic Management and Business Analysis. Butterworth-Heinemann, Amsterdam (2004)
Womack, J.P., Jones, D.T.: Lean Thinking. Free Press, New York (2003)
Womack, J.P., Jones, D.T.: Beyond Toyota: how to root out waste and pursue perfection. Harv. Bus. Rev. **74**(5), 140–158 (1996)
Wood, N.: LT: what it is and what it isn't, management services. Bus. Source Premier **48**(2), 8–10 (2004)
Yasuf, Y., Gunasekaran, A., Abthorpe, M.: Enterprise information systems project implementation: a case study of ERP in Rolls-Royce. Int. J. Prod. Econ. **87**(3), 251–266 (2004)

Collaborative Incident Planning and the Common Operational Picture

Georgios Marios Karagiannis and Costas E. Synolakis

Abstract Disasters are notorious for extending across multiple jurisdictions, both geographical and functional, and the modern disaster response operational environment is fraught with a multitude of agencies with different mandates and objectives. The complexity and unpredictability of interactions between various actors contribute to the "fog" and "friction" of what constitutes a crisis, similar to the fog and friction of war. Therefore, although situational awareness is an absolute necessity in disaster response, it is impossible to achieve without effective coordination and communication.

Here, we focus on the common operational picture in disaster response, with a view to bridging the gap between its technological and operational components. We use a typical incident planning outline to highlight how software solutions developed at the disaster preparedness phase can reduce the uncertainty during disaster response and streamline the operational planning process. We identify the capabilities and categories of existing applications, and we correlate the capabilities with the stages of the incident planning process to highlight how software supports disaster response coordination. Finally, we discuss the gaps between existing products and modern operational needs and suggest avenues for further research and product development.

Keywords Common operational picture • Incident planning • Risk modeling • Operations management

G.M. Karagiannis (✉)
Technical University of Crete, School of Environmental Engineering, Chania, Greece
e-mail: gmkaragiannis@gmail.com

C.E. Synolakis
University of Southern California, Viterbi School of Engineering, Los Angeles, CA, USA
e-mail: costas@usc.edu

I.S. Kotsireas et al. (eds.), *Dynamics of Disasters—Key Concepts, Models, Algorithms, and Insights*, Springer Proceedings in Mathematics & Statistics 185,
DOI 10.1007/978-3-319-43709-5_6

1 Introduction

Disasters create overwhelming demands to affected communities and pose unique response problems. Effective disaster response is not only a function of the magnitude of the hazard and its effects, but also of complexity. Early disaster response research highlighted that disasters cannot be adequately managed merely by mobilizing more resources, and that scaling up procedures for responding to daily routine emergencies may be inadequate for major incidents (Auf der Heide 1989). Furthermore, disasters are notorious for extending across multiple jurisdictions, both geographical and functional, and the modern disaster response operational environment is fraught with a multitude of agencies with different mandates (Geist et al., 2006; Salmon et al. 2011). The complexity and unpredictability of interactions between various actors contribute to the "fog" and "friction" of what constitutes a crisis, similar to the fog and friction of war discussed by Carl von Clausewitz (Howard and Paret 2008).

Therefore, although situational awareness is an absolute necessity in disaster response, it is impossible to achieve without effective coordination and communication. The importance of situational awareness in high uncertainty environments has been well-known to military leaders since early war history. While on campaign, Roman Emperor and stoic philosopher Marcus Aurelius wrote (Stephens 2012):

> Everything we hear is an opinion, not a fact. Everything we see is a perspective, not the truth.

In addition to the uncertainty inherent in disaster response operations, the sheer number of organizations involved in the field and the existence of multiple decision-making centers makes coordination and communication an arduous task. In 2005, the challenges during the response to Hurricane Katrina were partly attributed to the existence of three separate operational commands and the involvement of more than 500 organizations. Poor coordination and lack of situational awareness were two of the major issues of concern during the entire response. Some of the coordination and communication problems subsided when United States Coast Guard Rear Admiral Thad Allen was appointed as both Principal Federal Official and Federal Coordinating Officer (Moynihan 2009). He subsequently described the situation as "a weapon of mass destruction without a criminal dimension", thus highlighting the need for framing the issue from the outset (Lagadec 2015).

This dire need for coordination between the myriad of responding organizations and for situational awareness can be addressed almost exclusively through joint planning. Multi-agency coordination systems facilitate collaborative planning and assist all involved jurisdictions and responding agencies to achieve situational awareness (Kapucu et al. 2010). They need to be established before the disaster and regularly maintained through training and exercising. Emergency operations centers (EOCs) are a major component of multi-agency coordination systems (Karagiannis et al. 2014). They are critical decision-making hubs and provide a link between the strategic level and the resources in the field (McEntire 2006).

Here, we focus on the common operational picture in disaster response. Although the term "common operational picture" was originally coined by the military (U.S. Department of Defense 2015; Mittu and Segaria 2000), it has been increasingly used by emergency management practitioners to denote a single identical display of relevant operational information, such as disaster affected areas; position and status of critical infrastructure and critical buildings; and position and types of own resources, shared by more than one agencies or jurisdictions. Modern communications technology and a common operational picture (shared electronically) allow collaboration between multiple decision-making centers from distant locations, increase information sharing, and improve decision-makers' visualization of the situation. Furthermore, the common operational picture allows planning staff to take advantage of tactical units' input and knowledge of the situation and develop better courses of action faster (US Army 2010).

The notion of a common operational picture is well-known among experienced practitioners (Huder 2012). However, although several technical solutions have been developed to build and share a common operational picture, its actual operational use remains a matter of debate in professional cycles (Alexander 2002). Similarly, most academic discussions of the common operational picture focus on the technological component rather on the operational aspects thereof (Topper and Lagadec 2013; Warren Mills et al. 2008; Wolbers and Boersma 2013). This chapter endeavors to contribute to the discussion on the operational component of the common operational picture, by identifying the capabilities of the technological component and how they are put to the test in the field.

This chapter is structured in four parts. First, we present an overview of the incident planning process and how the common operational picture can improve collaborative incident planning in the time-constrained post-disaster environment. Then, we establish a typology of the capabilities of common operational picture solutions and identify major software categories, based on their utility for incident planning. Next, we discuss potential developments towards more integrated common operational picture solutions and, finally, we summarize our findings.

2 Incident Planning and the Common Operational Picture

Uncertainty, complex demands and the need to plan in a time-constrained environment are the hallmarks of disaster response operations. Emergency managers mitigate uncertainty, complexity, and time-constraints by planning ahead. Emergency operations plans (EOPs) are developed before the disaster strikes and describe protective actions, assign responsibilities, identify resources, and address coordination and communication (Perry 2007). They are typically based on planning assumptions, or hypotheses about the intensity of the hazard, the affected areas, the projected demands, and available resources. Yet, disasters tend to disprove planning assumptions (Karagiannis et al. 2013), which brings about the need for incident planning to address disaster-generated demands as they occur.

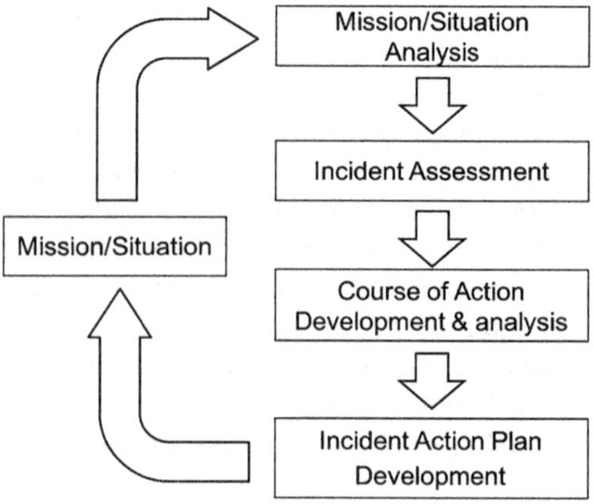

Fig. 1 Incident planning process

2.1 *Incident Planning*

Incident planning is "planning associated with an actual or potential incident, likely under emergency conditions, that involves developing procedures for responding to actual or projected effects" (Federal Emergency Management Agency 2012). It typically includes an analytical part and a planning (or synthesis) part. The analytical part encompasses the collection and organization of critical information that enables disaster managers to identify the needs generated by the disaster. The synthesis part includes the identification of an appropriate course of action and the development of a corresponding plan of action for a determined operational period. The process is circular, with a new operational planning process beginning as each developed plan of action is approved and executed.

More specifically, incident planning includes four phases (Fig. 1):

- Situation/mission analysis
- Incident assessment
- Course of action development and analysis
- Incident action plan development

Table 1 illustrates the incident planning process in further detail. Although the process is described in a linear fashion, it is in reality iterative and planners revisit several steps in a back-and-forth fashion as more information on the situation becomes available (Karagiannis and Synolakis 2015a).

Typically, this kind of incident planning process is implemented by command staff at an emergency operations center or an incident command post, that is, the location from which the primary command, control, coordination, and communication functions are executed for a specific incident. Shared authority implies that the command staff for a given incident likely comes from various agencies with functional or geographical jurisdiction. This heterogeneity of the planning group complicates incident planning and coordination, unless working relationships and trust have been developed long before the disaster strikes (Moynihan 2009).

Table 1 Incident planning process steps (Karagiannis et al. 2015b)

Step	Description
	Situation/mission analysis
Area of operations	Terrain, affected population, weather, incident date and time
Essential elements of information	Incident type and location, lifesaving needs, status of critical infrastructure and facilities, imminent hazards and/or threats, potential for cascading events, actions already taken
	Incident assessment
Needs assessment	Identification of agent-generated and response-generated demands
Resource evaluation	Analyze the status of resources, and determine the extent to which they can cover the needs generated by the disaster and to what extent they can support the objectives and tasks required to accomplish them. Identify required vs. available, en route and ordered resources
Identification of objectives	Objectives provide guidance and direction for developing and selecting a course of action. They are based on realistic expectations of what can be accomplished with the available resources. Objectives describe the desired end state and express the intent of decision-makers
	Course of action development and analysis
Course of action development	A course of action is an option that will accomplish or contribute to the accomplishment of one or more objectives, and from which a detailed plan is developed. One or more alternative courses of action are developed, depending on staff experience and the time available. Courses of action should include specific tasks, the organization or subordinate unit the task is assigned to, the location and time of the action, as well as the available and required resources
Course of action analysis	Courses of action are analyzed with a view to developing an incident action plan. When multiple courses of action are developed, they are eventually compared against set criteria to select the best alternative. When a single course of action is contemplated, the analysis serves to modify the course of action based on the war-game results
	Incident action plan development
Course of action approval	Regardless of whether one or more courses of action have been developed and analyzed, the final call is made by a senior member of the staff with decision-making authority. Planners submit the results of their analysis to the appropriate decision-maker who is required to select, approve, modify, or disapprove courses of action
Incident action plan	The incident action plan may be written or oral, depending on the complexity of the situation and the incident tempo. It includes an overview of the situation; a statement of the mission, a concept of operations, an outline of the structure and specific assignments, logistics support and command, control, coordination, communication, and information

Coordination becomes even harder in larger incidents, where multiple decision-making centers are activated, as the demands exceed the capabilities of individual organizations. This decentralization of the planning process combined with the requirement for rapid development of an incident action plan brings about the need for collaborative planning to identify and achieve shared objectives.

2.2 Common Operational Picture: Collaborative Incident Planning in a Time-Constrained Environment

Collaborative incident planning improves coordination in the post-disaster time-constrained environment by allowing decision-makers to jointly explore capabilities and develop better courses of action faster. Modern information and communication systems and a common operational picture can facilitate joint planning by providing planners and decision-makers with a single identical display of the necessary information. Information requirements may be different during each phase of the incident action planning process, as illustrated in the following paragraphs.

Situation Analysis Situational awareness is one of major challenges in disaster response. For example, the US National Response Plan, developed by the Department of Homeland Security, points out that disaster response activities may have to begin without a complete assessment of critical needs. In addition, the response to Hurricane Katrina is a well-documented example of the failure to establish situational awareness (Harrald 2006). Information about the effects of a disaster and the operational environment comes from various sources. A common operational picture can streamline this process by providing a centralized platform for all responding organizations to input information on the effects of the disaster, based on geographical and/or functional jurisdiction. In addition, modern information and communications technologies (ICT) help planning staff take into consideration first-hand accounts of the situation from field units. The added value of the common operational picture originates from the combination of information provided by various sources. For example, a local Department of Health Services may provide information about the number of available beds in local hospitals, while the Fire Department may advise the estimated number of injured victims in the affected area. It is only through the combination of both pieces of information that decision-makers can get an accurate picture of the status of health services in the affected area.

Incident Assessment In the aftermath of a disaster, numerous agencies assess critical response needs, usually based on their functional or geographical jurisdiction. The lack of coordination between agencies at this stage leads to gaps in the provision of critical services and limited efficiency and effectiveness as some needs are identified and addressed by more than one organization, while others may go unnoticed in the early stages. Furthermore, the lack of situational awareness may lead to agencies setting conflicting objectives or planning to address the same objectives, and therefore wasting precious resources. A common operational picture expedites the identification of needs and helps identify synergies between the myriad of responding organizations by providing a centralized platform linking decision-making centers in distant locations. In addition, it allows organizations to plan joint operations, ultimately sharing resources, and maximizing effectiveness.

Course of Action Development and Analysis Once common objectives have been decided upon, organizations can use a common operational picture to jointly develop and war-game one or more courses of action. Here, a single identical display of teams, equipment, facilities, affected areas, and critical infrastructure helps planners from distant locations to visualize the area of operations, identify elements, or capabilities to achieve objectives and analyze courses of action faster.

Incident Action Plan Development At the incident action plan development stage, the common operational picture allows emergency operations centers at distant locations to share a jointly developed incident action plan. As incident planning goes from situation analysis to the development of the incident action plan, the common operational picture transforms from a planning to a communication tool.

A common operational picture can therefore increase the efficiency all four phases of the incident planning process. Given the wide range and diversity of the agencies and organizations involved in disaster response, a combination of planning, training, and modern information and communications systems is required to maintain a common operational picture throughout the incident planning process.

2.3 Common Operational Picture and Emergency Planning

The implementation of a common operational picture requires, by definition, modern information and communications technology (ICT). Consequently, the common operational picture encompasses both a technological and an operational component. The technological component includes the ICT technology assets and systems that connect distant emergency operations centers and decision-making focal points. The operational component is comprised of the non-technological activities that build and maintain the emergency response system, such as collaborative emergency planning, joint training, and disaster exercises.

Several parts of both components can be developed before a disaster strikes. Although technically not a part of the incident planning process, emergency planning can greatly enhance situation analysis and incident assessment. Several elements of these two phases can actually be addressed before a disaster occurs. For example, information about the area of operation, populated areas, critical infrastructure assets, and resources (both existing and available through mutual aid agreements) can be gathered while the emergency operations plan is developed and displayed on maps.

Maps are highly useful in incident planning because they help decision-makers visualize the area of operations and provide a common reference system for planning and communication (Alexander 2002). A lot of information can be displayed on maps created during the preparedness stage. In addition, emergency planning typically includes the development of disaster scenarios as the baseline for planning. Although disasters never happen according to plan, emergency planning is fundamentally an educational activity (FEMA 2010). In this sense, the purpose of developing disaster scenarios is not to foresee every possible development of the situation. Scenarios are rather intended to help disaster planners estimate post-disaster needs, determine response objectives, identify the resources necessary to achieve them, and establish an appropriate organization before the disaster strikes.

Besides its more or less tangible result, which is the development of appropriate emergency operations plans, disaster preparedness also helps build trust and willingness to share information among the various stakeholders, central to coordination on the volatile and dynamic post-disaster environment (Janssen et al. 2010). In other words, emergency planning builds a network of people and stakeholders that forges the concept of coordination into effective disaster response. However, the high uncertainty and fast operational tempo of emergencies need to be counter-balanced by efficient technical information management and communications capabilities. The advent of modern information and communications technology has increased real-time information sharing capabilities and has spawned several solutions to facilitate joint and parallel planning in the aftermath of a disaster.

3 Common Operational Picture Capabilities and Solutions

Several ICT-based solutions have been developed experimentally or have been commercialized to support decision-making and joint planning in emergencies by establishing the technological component of common operational picture capabilities. Despite the vast array of software available, most available solutions encompass similar or comparable functions.

Our approach has been based on the use of the incident planning process to investigate how software solutions reduce uncertainties during disaster response. First, we reviewed available information on common operational picture solutions available commercially or through research projects, such as user's guides, technical documents, and advertising documentation. In addition, we observed how these applications are employed during several disaster exercises. Then, we used the incident planning process as a guide for identifying the capabilities of applications. Specifically, we war-gamed a number of all disaster scenarios to highlight the information needs of emergency managers during each phase of the incident planning process and matched these needs to the functions of available applications to develop a typology of the available capabilities. Finally, the known common operational picture solutions were assigned to categories according to the nature of services they provide.

This section discusses common operational picture software capabilities and categories. The first part establishes a typology of the key functions of common operational picture software, while the second presents the major categories of software solutions.

3.1 A Typology of COP Software Functions and Capabilities

As the locus of incident planning shifts from analyzing the situation to identifying one or more courses of action and then building an incident action plan, the common operational picture needs to accommodate different types of information. During situation analysis, emergency managers need both static (e.g., topographic features of the area of operations) and dynamic (e.g., disaster effects) information. Yet, in either case, information is mostly descriptive of the situation and the level of certainty is relatively high. During the incident assessment and incident planning stages, information management becomes more dynamic as the agent- and response-generated demands accumulate, the status of available and on-call resources changes and emergency managers develop one or more courses of action to address set objectives. The level of certainty also decreases at this stage, as emergency managers need to make several assumptions about the future situation.

Given the wide range of information types that needs to be included in a common operational picture and the multi-disciplinary nature of disaster management, the capabilities of existing solutions naturally span across a wide range of science and engineering disciplines. Although not all software incorporate all capabilities, most of the ones commercially available today or developed through research have one or more of the functions described in the following paragraphs.

Hazard Modeling Hazard modeling capabilities include numerical models of the hazard-generating natural process that yield the consequences of the hazard. Inputs include specific data on the hazard-generating process, while outputs include quantifiable parameters of the effects of the hazard. For example, tsunami hazard models require input on the dislocation of the ocean floor and the bathymetry and topography of the undersea and onshore environment, while they yield tsunami arrival times and run-up (Synolakis and Bernard 2004).

Hazard Mapping Hazard maps are used to represent critical characteristics of one or more hazards over a given territory (Alexander 2002). For example, a flood hazard map could depict the inundated area for the 100-year flood, while an industrial accident hazard map could highlight the affected areas for various accident scenarios. Maps are especially useful for hazards that can be spatially defined, e.g., technological accidents, landslides, and floods (Karagiannis 2012). The advent of geographical information systems (GIS) has streamlined hazard mapping capabilities and enabled hazard modeling software to project hazard-affected areas as a shape file on a GIS base map.

Vulnerability and Risk Mapping Vulnerability maps illustrate the spatial distribution of elements at risk (Alexander 2002). They are generally used in conjunction with hazard maps. While hazard maps delineate the area that will likely be affected from a hazard of a given intensity level, vulnerability maps are part of the base map. Once the hazard map is overlaid on the base map, elements at risk are identified either visually or through appropriate database queries. Risk maps combine hazard and vulnerability maps. They illustrate the level of risk from the combination of the probability of occurrence and severity of single or multiple hazard scenarios. Although risk maps are often developed together with hazard and vulnerability maps, they reflect a combined probabilistic and deterministic approach and, therefore, they are predominantly used to support decision-making regarding prevention measures. Yet, incident planning essentially requires a deterministic estimation of the effects of the hazard, hence the utility of risk mapping in disaster response is somewhat limited.

Incident Assessment Incident assessment capabilities include interfaces that allow users to log information about the effects of hazards, the needs generated by the disaster, and activities performed by responding organizations. These logs can present information in a variety of formats, including chronological, geographical, or incident-based. Chronological formats present information about incidents in the order that they occurred or the information was logged into the system, essentially building a "timeline" of the emergency. Geographical formats display incident information based on geographical clusters, which may or may not coincide with administrative boundaries. Incident-based formats describe incidents in groups according to major events. In all cases, modern incident assessment functions can be linked to resource management capabilities and enable users to assign tasks to specific resources and monitor, either manually or automatically, the status of completion of these tasks.

Two-Way Communications Two-way communications capabilities integrate horizontal and vertical operational communications using diverse communications technologies. The use of several communication systems reduces dependency on any one asset and lowers the risk of communications breakdown in the aftermath of disasters. The capabilities of two-way communications systems have steadily increased over the years, and modern systems can provide for digital and/or encrypted communications, and can be linked via GPS systems with resource management functions to help track individual resources in real time.

Resource Management In their earliest form, these functions were essentially databases of individual resources (people, teams, equipment, facilities, and vehicles), built to store basic information, such as resource name, function, technical capabilities, contact details, and location. Later versions allowed the user to group individual resources to create task forces (i.e., combination of different resources with a specific mission) or strike teams (i.e., a set number of resources of the same type). An improvement over these functions was the capability of manually inputting resource status and assigning resources to specific tasks, thus linking resource management with incident assessment. Modern versions use GPS locators to track resources in real time.

Automated Response Checklists Automated response checklists are used to ensure consistency and completeness in carrying out critical tasks. Two types are generally used. The first includes digitized versions of checklists included in standard operating procedures, field operations guides, or even emergency operations plans. They are essentially an electronic version of resources that already exist on paper. The second checklist type is more dynamic, as it allows the user to log the completion of each task included in the checklist. More advanced versions of these systems include the time that each task was accomplished in the incident log and generate alerts when certain tasks have not been completed on a set timeframe.

Alert Notifications This capability sends automated or semi-automated messages to selected respondents, including emergency operations center staff, specialized units, or elected officials. These messages are used to activate emergency response arrangements, staff emergency operations centers, issue warning orders for key resources and transmit critical information to elected officials. Pagers were among the earliest versions of this capability, but have rapidly been replaced by automated text messages and other solutions. Notwithstanding information security concerns, some disaster management organizations are also considering social media as surrogate alert notification applications (Karagiannis and Synolakis 2015b).

Social Media Monitoring In the aftermath of disasters, the public turns to social media for up-to-date and unfiltered information on the hazard event and to check the whereabouts of family and friends. In addition, crowdsourcing applications have been used to gather information from social media users in the affected area after major disasters. For example, in the aftermath of the 2011 earthquake in Haiti and the 2010 Deepwater Horizon oil spill, university volunteers used reports from social media users to rapidly develop online maps of the incidents (Gao et al. 2011). Social media monitoring capabilities are used sporadically by emergency operations centers to enhance situational awareness by complementing the information received through official channels. However, the use of social media as an active crisis communications tool remains controversial (Lindsay 2011; Karagiannis and Synolakis 2014).

Multi-Channel Communication Multi-channel communication capabilities are predominantly used to disseminate crisis information to the public, for example, broadcasting warnings, news, and situation updates. They encompass a diversity of communications media, including television and radio ads, pre-recorded telephone messages, text messages, pagers, e-mail, and social media, with a view to reaching as wide an audience as possible.

Despite the diversity of capabilities of common operational picture solutions, the type of information that needs to be shared and the nature of the activities performed during each stage of the incident planning process are different. Therefore, not every function is required for every stage of the process. Table 2 illustrates which capabilities are mostly required during each stage of the incident planning process.

Table 2 Common operational picture software capabilities required to support the incident planning process at various stages

Function/capability	Situation/mission analysis	Incident assessment	Incident planning	Incident action plan development
Hazard modeling	X			
Hazard mapping	X			
Vulnerability and risk mapping	X	X		
Incident assessment		X		
Two-way communications	X	X		
Resource management		X	X	
Automated response checklists	X			X
Alert notifications		X		X
Monitoring social media	X	X		
Multi-channel communication				X

3.2 Common Operational Picture Software Categories

Several software solutions have been developed by research institutions or have been commercialized by software companies to support decision-making and incident planning in emergencies. Each solution encompasses one or more of the functions delineated above. The following paragraphs describe the capabilities of the most common software solutions.

3.2.1 Hazard Modeling Software

Hazard modeling software includes computer codes that run hazard-specific models. They provide running estimates of hazard intensity and help identify the areas that are likely to be affected should a hazard occur. Their use is essentially based on computing time and required input of an appropriate likely event and its magnitude or duration. Software solutions with long computing times or requiring excessive or difficult-to-obtain data are predominantly used to identify assumptions for the development of emergency operations plans. Examples include a number of codes using computational fluid dynamics to simulate hazardous materials releases. On the other hand, software that requires little computing time and easily obtainable input data is more effective during disaster response. Areal Location Of Hazardous Atmospheres (ALOHA), a hazardous materials release modeling software developed by the US Environmental Protection Agency and the US National Oceanic and Atmospheric Administration (Available via http://www2.epa.gov/cameo/aloha-software) is one example. In other cases, part of the required input data can be collected and set up in advance, therefore sharply reducing computing times. Examples include the FARSITE wildland fire growth simulation software (Available via http://www.firelab.org/project/farsite) and the Method Of Splitting Tsunami (MOST—Available via the National Oceanic and Atmospheric Administration's Pacific Marine Environment Laboratory at http://nctr.pmel.noaa.gov/model.html) (Titov and Synolakis 1998). The former uses spatial information on fuel, weather, and topography to simulate the spread of wildland fires, while the latter uses spatial bathymetric, topographic, and seismic data to simulate tsunami generation, propagation, and run-up. In either case, the GIS baseline map can be developed in advance to speed the simulation process and help provide almost real-time results.

In addition to computing time, operator training is also an important issue regarding the use of hazard modeling software. The models run by software codes are hazard-specific and require expert knowledge on the natural phenomena simulated and the inner workings of the model. Yet, emergency operations center staffers have typically only generalist-level knowledge on a number of hazards. One solution is to train emergency managers and EOC staff members to use the software, provide that they software interface is simple enough and that non-specialist personnel can be trained in a reasonable timeframe. For example, the duration of a standard ALOHA user training course is about 2 days (EPA and NOAA 1999).

If the model is too complicated to teach over a short timeframe, technical experts can be assigned to the emergency operations center to support decision-makers by running simulations to help assess the effects of the hazard or war-game courses of action (Karagiannis and Synolakis 2014). Finally, modern information and communications technology allows experts to support decision-making without being physically present at the emergency operations center. For example, several Tsunami Watch Centers are using software such as MOST to run tsunami simulations with a view to providing early warning to civil defense agencies (Kanoglu et al. 2015). In this case, the experts run the simulation from their office or facility and send the results to the authorities having jurisdiction. The advantage of this solution is that scientists work from their own facility, where they can more easily get support from peers, access documentation, or use more powerful and faster computers.

3.2.2 Hazard, Vulnerability, and Risk Maps

Hazard, vulnerability, and risk maps are disaster prevention and hazard mitigation tools. They are developed before a disaster strikes and are primarily intended to support disaster risk assessments and the establishment of risk management priorities (Karagiannis 2012). Modern GIS maps can provide a wide range of geospatially referenced information, including the extent of the hazard-affected area and the elements at risk, such as critical buildings and infrastructure in the hazard zone.

In addition to disaster prevention, hazard and vulnerability maps are also useful in disaster response, as they provide estimations about the impact of selected hazards under a number of pre-designated, high-risk scenarios. Although disasters do not follow risk assessment scenarios, hazard and vulnerability maps can support incident planning assumptions about the impact of hazards (Billa et al. 2006; Bouamrane et al. 2012). They are particularly useful during the early stages of a disaster, when information is scarce and uncertainty about the situation is at its highest. By providing information about risk assessment scenarios, hazard and vulnerability maps help emergency managers develop better courses of action faster.

3.2.3 Emergency Operations Center Software

Emergency Operations Center (EOC) software solutions typically incorporate several information management capabilities. At their simplest form, EOC solutions include incident assessment and resource management options. A mapping interface may be provided to indicate the location of incidents and position of resources. One example is the Sahana Eden software, an open-source platform designed for disaster response (Available via http://sahanafoundation.org/products/eden/). Sahana Eden is highly configurable and easy to modify to fit different situations. It is being used by the City of New York Office of Emergency Management, by the International Federation of Red Cross and Red Crescent Societies and other agencies around the world (Sahana Software Foundation 2015).

In addition, a number of software solutions are available commercially. Some agencies, especially in Europe, use custom-made software, designed to meet their needs. Several solutions have also been developed by academic institutions of research purposes (for example, Assilzadeh and Gao 2010). These software solutions usually incorporate additional functions, such as two-way communications, automated response checklists, and alert notifications.

EOC software typically seeks to address the coordination aspect of emergency management, and support the situation analysis and incident assessment phases of incident planning. These applications operate in stand-alone or network mode. The stand-alone configuration only allows one user to input information, which is usually enough for simple incidents with minor communication needs. In this case, a single user needs to centralize information from a few responding agencies. However, disasters and emergencies involve by nature many organizations and cross jurisdictional and geographical boundaries. As multiple decision-making centers are involved, it is both ineffective and impractical to centralize information management to a single data entry point. In addition, resources are assigned to the emergency, but remain under the control of their parent organizations. In these situations, EOC software in network configuration allows multiple users from various decision-making centers to contribute information on the hazard-affected area, the demands and constraints generated by the emergency, as well as the status and capabilities of resources. In turn, this single identical display can be used by all responding organizations as a reference for joint planning from distant locations. In other words, by establishing and maintaining a common operational picture, organizations can develop better courses of action faster, and improve decision-making.

3.2.4 Crisis Communication Software

Crisis communication applications generally focus on sharing information through multiple channels, such as text messages, social media, e-mails, and mobile telephone applications. They were primarily developed for business continuity and were initially intended to provide private companies with multi-channel communications capabilities to support two-way messaging with employees, clients, and other stakeholders during a crisis. Nowadays, these first-generation crisis communication tools have been readily replaced by various instant messaging, chat rooms, and other social media applications, for example, Facebook, Twitter, and Google+, as several businesses use social media as a primary crisis communication tool (Karagiannis and Synolakis 2015b; Lindsay 2011).

Later versions of crisis communication applications can be operated on handheld devices, such as smartphones and tablet PCs. They have expanded capabilities, including social media monitoring and automated alert notifications. Automated alert notifications are essentially an expansion of the original capabilities of crisis communication software, oftentimes combined with the capability of sending messages to multiple recipients. On the other hand, social media monitoring aims not at sending, but at gathering information to support corporate crisis response.

3.2.5 Integrated Software

Integrated software is an augmented version of common operational picture applications, including multiple capabilities, such as cartography, alert notifications, automated response checklists, resource management, incident assessment, and two-way communications. Cartography is usually provided in a GIS-based format. Although several applications provide users the option of using web mapping services as an alternative, GIS formats are usually preferred because of configurability and the risk of security breaches. In addition to baseline cartography and the possibility of mapping events and resources, some integrated applications either include or can be linked with hazard modeling software, thereby allowing the user to make on-the-spot projections of the hazard-affected area and hazard impact.

These solutions are arguably the most comprehensive member of the common operational picture software family. They provide the widest range of capabilities, but with three caveats. First, given the wide range of capabilities they provide, user training is highly recommended by most commercial providers of integrated software. Second, the acquisition budget also needs to include the cost for one or more workstations, communications infrastructure (e.g., high-speed internet access), and potentially additional space in the EOC. Third, they mark the transition to an ICT-based operations management paradigm, which usually comes with strategic-level changes in emergency plans, but also required a mature multi-agency coordination system to operate.

4 Discussion: The Way Forward

Common operational picture software applications provide a wide range of capabilities to emergency managers and help streamline the incident planning process. As the technological component of the common operational picture is essentially based on information and communications technology, new capabilities will undoubtedly be developed and existing ones will keep improving. However, common operational picture software cannot substitute for prior network building, emergency planning, and the development of interpersonal relations and trust, all of which are key requirements for coordination in disaster response.

Operational vs. Technological Component The importance of information and communications technology (ICT) lies in facilitating coordination rather than the information it processes. Therefore, prior planning and network-building are required to forge the coordination mechanism that is needed in the volatile post-disaster environment. In the absence of prior planning, common operational picture technical solutions cannot be implemented, because people are not familiar with the concept of working together. For example, the command, control, communications, and integration (C4I) system designed to coordinate Athens' security during the 2004 Olympic Games is often referred to as a crucial failure. Although the Greek government invested in 100 new command centers, it was the central software, which was intended to centralize information from all sensors, that failed to work (Samatas 2007). In addition, a series of administrative deficiencies, arguably indicative of an immature public administration and homeland security system, also contributed to the failure.

In another case, a relatively straightforward EOC application was used to augment a local multi-agency coordination system during a tsunami disaster exercise. Here, the multi-agency coordination system was steered by a decision-making group and an information management unit (encompassing situation awareness and resources management responsibilities), both of which operated from the jurisdiction's EOC. The EOC application was developed specifically for this exercise and incorporated four capabilities: it delineated the hazard-affected area on a GoogleMaps® background, it pointed out the location of incidents and resources, and it supported the development of an incident log. Moreover, it had a multi-user interface, allowing multiple users from distant locations to log into the system and input incident log entries and data about incidents and resources.

However, the lack of prior planning and training made it difficult for emergency managers and decision-makers to use the software (Karagiannis et al. 2014). The jurisdiction did not have a tsunami emergency operations plan and the multi-agency coordination structure was being implemented for the first time. Networking was limited at a few official meetings per year and decision-makers were arguably not used at working together in an emergency setting. This slowed down decision-making and hampered the entire incident planning process. Furthermore, neither decision-makers nor key staff had training on use of the EOC application, which had been developed without input from operational stakeholders. It was also the first time that a common operational picture was established in an operational setting. The net result was a very poor utilization of the EOC application. Actually, although the EOC application was available throughout the simulated incident, decision-makers only ever used the "paper" version of hazard maps.

Yet perhaps the most flagrant example of the importance of the operational component of the common operational picture is the European refugee crisis. The recent armed conflicts in Iraq, Syria, and Libya have exacerbated the humanitarian situation in the Middle East and North Africa. The prolonged conflict and insecurity has led to migratory movements throughout the region, resulting in an upsurge of the number of irregular refugees arriving to Europe through the Mediterranean Sea in the summer of 2015. The Office of the United Nations High Commissioner for Refugees (UNHCR) estimates that more than one million people arrived in Greece and Italy in 2015, almost twice the number of 2014 (UNHCR 2016). This spawned a major crisis, which started with closures of borders between several European-Union Member-States in the summer of 2015. After heated political debates, the European Union only took action after the numbers of migrants had risen off the chart, highlighting Europe's lack of readiness (Barnard 2015).

However, the European Commission had established in 2013 a new Emergency Response Coordination Center, which also operated as the hub of its Integrated Political Crisis Response arrangements. This state-of-the-art emergency operations center is equipped with the best information management systems and arguably represents the epitome of the technological component of the common operational picture. Nevertheless, in this case, information and communication technology could not counteract the lack of preparedness neither accelerate decision-making.

All three cases point to the need for emergency response systems to focus on the operational component of the common operational picture. Poorly prepared organizations often regard the technological component of the common operational picture as the panacea to all their problems. However, early research has highlighted that it is the network built long before a disaster that carries the weight of coordination and decision-making in the aftermath. Technological solutions can improve and augment the network, but cannot create it from the outset. In addition, technological solutions rely on communications systems that may not be available in the early phases of disaster response. Many organizations that have successfully employed common operations picture solutions frequently train their key staff to operate without the software by switching to a non-ICT system in regular intervals. Therefore, disaster preparedness, including planning, training, and exercising, should take precedence over high-tech solutions.

The above is not to belittle the role of the technological component. Common operational picture applications have proven their worth during disaster response. The only limitation is that any information management software is as good as the system it helps coordinate. A common operational picture solution cannot create a disaster response system, but it can augment and improve an existing one.

Disaster and Humanitarian Logistics Logistics is a critical function of the emergency response system. It is part of response-generated demands, which are frequently neglected or underestimated in emergency planning and during disaster response (McEntire 2006). Equipment shortages, interoperability issues, and supply chain failures are among the most common failures in disaster response operations (Karagiannis 2010). By helping develop better courses of action faster, the common operational picture can streamline disaster and humanitarian logistics.

Hazard modeling software and risk maps can help identify the extent of disaster affected areas and the intensity of the hazard, and inform supply chain management decisions. For example, the International Federation of Red Cross and Red Crescent Societies used a weather forecasting software to enhance flood disaster preparedness in West African countries. Based on the hazard simulations, they prepositioned relief supplies closer to the areas that were expected to be most severely hit, therefore reducing disaster relief delivery times from weeks to days (Braman et al. 2013). EOC and integrated software can correlate required resources with response objectives and tasks, therefore helping to identify logistical requirements and reduce lead times. Even more important, these tools allow stakeholders from distant decision-making centers to communicate their capabilities in relation to resource requirements and "pledge" resources to the response. One example is the European Commission's Common Emergency Communications and Information System, a secure online platform for countries to request international assistance in times of disaster. Member-States wishing to assist can use the system to declare their intent to provide resources according to the demands identified by the affected country.

However, it is the operational component of the common operational picture that can really make the difference in disaster logistics. This is especially true in humanitarian settings, where the multitude of actors, each with different mandates, cultures, and operating procedures, can make coordination nearly impossible. Duplication of effort and gaps in relief coverage are frequent concerns in humanitarian settings, which have been attributed to poor coordination and communication between responding organizations (UN/OCHA 2013). Maintaining a common operational picture can help identify needs, gaps, and resource requirements (Van Wassenhove 2006), and optimize the relief supply chain to address the fog and friction of disaster response operations (Qiang and Nagurney 2012).

Training Besides the strategic advantages it provides, the common operational picture can also be used as a training instrument. First, the operational component includes several educational activities, such as emergency planning and disaster exercises. Here, the common operational picture aims at identifying disaster response needs and the courses of action to address them. And second, the software solutions which make up the technological component can also be used to enhance training, essentially during disaster exercises. For example, Karagiannis and Synolakis (2014) have pointed out that hazard models can also support the design, conduct, and evaluation of disaster exercises. Besides providing a credible and accurate foundation for the development of the exercise scenario, these numerical simulations help exercise designers present the scenario to players more effectively, therefore players' exercise experience and fostered participation.

Limitations and Further Research The goal of this chapter has been to identify the capabilities and categories of common operational picture solutions and correlate them with the incident planning process, in an effort to bridge the gap between the technological and operational components of the common operational picture. However, the framework we have presented is somewhat abstract. A comprehensive and detailed treatment of the common operational picture would go well beyond the scope of a single chapter. So, although we have attempted to identify the major capabilities that information and communications technologies contribute to the incident planning process, future research is still needed to specify further how a common operational picture can streamline the incident planning process and identify future research goals.

As the capabilities of ICT systems improve at a fast pace, common operational picture software will become more affordable and information management more dynamic. For example, current resource management capabilities either allow users to input information on the status and position of resources manually, or at best can track resources using GPS locators. Future systems may be able to use sensors to monitor additional information on assets, for example, the level of water and foam concentrate remaining in the tank of a fire engine, therefore improving decision-makers' visualization of the situation in the field. As more information becomes available in emergency operations centers, common operational picture applications will likely become more useful in the development and analysis of courses of action, by involving disciplines such as operations research and systems engineering.

Operations research has often been used to aid decision-making regarding complex aspects of emergency response, such as disaster relief supply chain optimization (Nagurney and Nagurney 2015) and evacuation flow modeling (Vogiatzis et al. 2012). However, current capabilities generally remain at the emergency planning level and have yet to be expanded to incident planning. As sensor-based automated information collection and management improves, future research could develop capabilities for building mathematical models of courses of action in real-time, therefore speeding course of action analysis.

In addition to operations research, functional modeling has been used to model emergency plans with a view to analyzing their performance. Functional models provide the platform for the assessment of the risk of failure of the functions of an emergency plan. As the capability of building mathematical models of functions and activities of emergency plans becomes available, further research could use systems modeling approaches to identify potential critical points in courses of action before they occur, and streamline course of action analysis and decision-making (Karagiannis et al. 2013).

5 Conclusion

The goal of this chapter has been to identify the capabilities and categories of common operational picture solutions and correlate them with the incident planning process, in an effort to bridge the gap between the technological and operational components of the common operational picture. We have used the incident planning process as a guide to identify the information requirements of emergency managers during disaster response operations. This chapter establishes a typology of the current capabilities and identifies the major categories of existing software. Several of these capabilities can be developed before a disaster strikes. However, the technological component is but one half of the common operational picture. These solutions cannot create a disaster response system, but they can augment and improve an existing one. Therefore, the operational component of the common operational picture, that encompasses the activities that build and maintain the emergency response system, should be given priority before major investments in information and communications technologies are considered. Finally, as the capabilities of ICT systems improve at a fast pace, common operational picture software will become more affordable, information management will become more dynamic and common operational picture applications will likely become more useful in the development and analysis of courses of action.

Acknowledgements This research was partially supported by the project ASTARTE (Assessment, STrategy And Risk Reduction for Tsunamis in Europe) FP7-ENV2013 6.4-3, Grant 603839 to the Technical University of Crete and the University of Southern California.

References

Alexander, D.E.: Principles of Emergency Planning and Management. Terra Publishing, Hertfordshire (2002)

Assilzadeh, H., Gao, Y.: Designation of an interactive oil spill management system. Disaster Prev. Manag. **19**(2), 233–242 (2010)

Auf der Heide, E.: Disaster Response: Principles of Preparation and Coordination. Mosby, St. Louis (1989)

Barnard, A.: Exodus casts spotlight on Europe's lack of readiness. The International New York Times, 5–6 September (2015)

Billa, L., Shattri, M., Mahmud, A.R., Ghazali, A.H.: Comprehensive planning and the role of SDSS in flood disaster management in Malaysia. Disaster Prev. Manag. **15**(2), 233–240 (2006)

Bouamrane, K., Hamdadou, D., Yachba, K., Guenachi, K.: Towards a decision support system, application: itinerary road modification for road transport of hazardous materials. Int. J. Decis. Sci. Risk Manag. **4**(3/4), 175–196 (2012)

Braman, L.M., van Aalst, M.K., Mason, S.J., Suarez, P., Ait-Chellouche, Y., Tall, A.: Climate forecasts in disaster management: red cross flood operations in West Africa, 2008. Disasters **37**(1), 144–164 (2013)

Federal Emergency Management Agency: Incident Action Planning Guide. Federal Emergency Management Agency, Washington (2012)

Federal Emergency Management Agency: Developing and Maintaining Emergency Operations Plans: Comprehensive Preparedness Guide (CPG) 101, Version 2.0. Federal Emergency Management Agency, Washington (2010)
Gao, H., Barbier, G., Goolsby, R.: Harnessing the crowdsourcing power of social media for disaster relief. IEEE Intell. Syst. **26**(3), 10–14 (2011)
Geist, E.L., Titov, V.V., Synolakis, C.E.: Tsunami: wave of change. Sci. Am. **294**, 56–63 (2006)
Harrald, J.R.: Agility and discipline: critical success factors for disaster response. Ann. Am. Acad. Polit. Soc. Sci. **604**(1), 256–272 (2006)
Howard, M., Paret, P.: Carl Von Clausewitz - On War. Oxford University Press, Oxford (2008)
Huder, R.C.: Disaster Operations and Decision Making. Wiley, Hoboken (2012)
Janssen, M., Lee, J.K., Bharosa, N., Cresswell, A.: Advances in multi-agency disaster management: key elements in disaster research. Inf. Syst. Front. **12**(1), 1–7 (2010)
Kanoglu, U., Titov, V., Bernard, E., Synolakis, C.E.: Tsunamis: bridging science, engineering and society. Philos. Trans. R Soc. A **373**, 20140369 (2015). doi:10.1098/rsta.2014.0369
Kapucu, N., Arslan, T., Demiroz, F.: Collaborative emergency management and national emergency management network. Disaster Prev. Manag. **19**(2), 452–468 (2010)
Karagiannis, G.M., Synolakis, C.E.: Perception or reality? A common operational picture for enhancing situational awareness and decision-making in disasters. Paper presented at the 2nd International Conference on Dynamics of Disasters, Kalamata, Greece, 29 June–2 July 2015 (2015a)
Karagiannis, G.M., Synolakis, C.E.: Using social media to manage disaster volunteers in Greece. Paper presented at the 2015 International Crisis and Risk Communication Conference – Accountability, Metrics and Critique, University of Central Florida, Orlando, 2–4 March 2015 (2015b)
Karagiannis, G.M., Synolakis, C.E.: Using risk analysis and hazard numerical simulation models to support the design, conduct and evaluation of disaster exercises. In: Stal, M., Sigrist, D., Ammann, W. (eds.) Extended Abstracts of the 5th International Disaster and Risk Conference 2014, Davos, Switzerland, 24–28 August 2014
Karagiannis, G.M., Saini, K.S., Synolakis, C.E.: Lessons from the first European Union tsunami simulation exercise. In: Knezic, S., Chhetri, M.P. (eds.) Proceedings of the International Emergency Management Society USA Conference. Global Response for Capacity Building of Disaster Preparedness, Hattiesburg, MI, July 2014. The International Emergency Management Society, Brussels, Belgium (2014)
Karagiannis, G.M., Piatyszek, E., Flaus, J.M.: Model-driven and risk-based performance analysis of industrial emergency plans. J. Conting. Cris. Manag. **21**(2), 96–114 (2013)
Karagiannis, G.M.: Model Local/Regional Disaster Prevention Policy Plan. General Secretariat for Civil Protection, Athens, Greece (2012)
Karagiannis, G.M.: Méthodologie pour l'analyse de la robustesse des plans de secours industriels. Doctoral Dissertation, Ecole Nationale Supérieure des Mines de Saint-Etienne (2010)
Lagadec, P.: Framing the issue – Hurricane Katrina: a weapon of mass destruction without criminal dimension. In: Patrick Lagadec – Training Videos. Available via http://www.patricklagadec.net/fr/ (2015). Accessed 15 Jul 2015
Lindsay, B.R.: Social Media and Disasters: Current Uses, Future Options, and Policy Considerations. Congressional Research Service, Washington. Available via http://www.fas.org/sgp/crs/homesec/R41987.pdf (2011). Accessed 27 Feb 2013
McEntire, D.A.: Disaster Response and Recovery: Strategies and Tactics for Resilience. Wiley, New York (2006)
Mittu, R., Segaria, F.: Common Operational Picture (COP) and Common Tactical Picture (CTP) Management via a Consistent Networked Information Stream (CNIS). Naval Research Laboratory, Washington (2000)
Moynihan, D.P.: The network governance of crisis response: case studies of incident command systems. J. Pub. Admin. Res. Theor. **19**, 895–915 (2009)

Nagurney, A., Nagurney, L.: A mean-variance disaster relief supply chain network model for risk reduction with stochastic link costs, time targets, and demand uncertainty. Paper presented at the 2nd International Conference on Dynamics of Disasters, Kalamata, Greece, 29 June–2 July 2015

Office for the Coordination of Humanitarian Affairs: United Nations Disaster Assessment and Coordination Field Handbook, 6th edn. United Nations, Geneva (2013)

Office of the United Nations High Commissioner for Refugees: Refugees/Migrants Emergency Response – Mediterranean. http://data.unhcr.org/mediterranean/regional.php (2016). Accessed 11 Jan 2016

Perry, R.W.: Emergency Planning. Wiley, Hoboken (2007)

Qiang, P., Nagurney, A.: A bi-criteria indicator to assess supply chain network performance for critical needs under capacity and demand disruptions. Transp. Res. Part A Policy Pract. **46**(5), 801–812 (2012)

Sahana Software Foundation: Deployments. http://sahanafoundation.org/about-us/deployments/ (2015). Accessed 07 Aug 2015

Salmon, P., Stanton, N., Jenkins, D., Walker, G.: Coordination during multi-agency emergency response: issues and solutions. Disaster Prev. Manag. **20**(2), 140–158 (2011)

Samatas, M.: Security and surveillance in the Athens 2004 Olympics: some lessons from a troubled story. Int. Crim. Just. Rev. **17**(3), 220–238 (2007)

Stephens, W.O.: Marcus Aurelius: A Guide for the Perplexed. Continuum, London (2012)

Synolakis, C.E., Bernard, E.N.: Tsunami science before and after boxing day 2004. Philos. Trans. R. Soc. **364**(1845), 2231–2265 (2006)

Titov, V.V., Synolakis, C.E.: Numerical modeling of tidal wave runup. J. Waterway Port Coast. Ocean Eng. **124**(4), 157–171 (1998)

Topper, B., Lagadec, P.: Fractal crises – a new path for crisis theory and management. J. Conting. Cris. Manag. **21**(1), 4–16 (2013)

U.S. Army: The Operations Process – FM5-0. Department of the Army, Washington (2010)

U.S. Department of Defense: Dictionary of Military and Associated Terms – Joint Publication 1-02. Available via DTIC. http://www.dtic.mil/doctrine/new_pubs/jp1_02.pdf (2015). Accessed 25 Feb 2015

U.S. Environmental Protection Agency, National Oceanic & Atmospheric Administration: Areal Location Of Hazardous Atmospheres (ALOHA) Instructor Manual. EPA, Washington, DC & NOAA, Seattle, WA (1999)

Van Wassenhove, L.N.: Humanitarian aid logistics: supply chain management in high gear. J. Oper. Res. Soc. **57**(5), 475–489 (2006)

Vogiatzis, C., Walteros, J.L., Pardalos, P.M.: Evacuation through clustering techniques. In: Goldengorin, B., Kalyagin, V.A., Pardalos, P.A. (eds.) Models, Algorithms, and Technologies for Network Analysis. Springer Proceedings in Mathematics & Statistics, vol. 32. Springer, Heidelberg (2012)

Warren Mills, J., Curtis, A., Pine, J.C., Kennedy, B., Jones, F., Ramani, R., Bausch, D.: The clearinghouse concept: a model for geospatial data centralization and dissemination in a disaster. Disasters **32**(3), 467–479 (2008)

Wolbers, J., Boersma, K.: The common operational picture as collective sensemaking. J. Conting. Cris. Manag. **21**(4), 186–199 (2013)

Metaheuristic Optimization for Logistics in Natural Disasters

Thomai Korkou, Dimitris Souravlias, Konstantinos Parsopoulos, and Konstantina Skouri

Abstract Logistics in natural disasters or emergencies involve highly complicated optimization problems with diverse characteristics. The contribution of the present paper is twofold. First, it introduces a multi-period model aiming to minimize the shortages of different relief products in a number of affected areas. The relief products are transported via multiple modes of transportation from dispatch centers to these areas, while adhering to traffic restrictions. A test suite of benchmark problems with diverse characteristics is generated from the proposed model and solved to optimality with CPLEX. The test suite is used for benchmarking a number of established metaheuristics. Necessary modifications are introduced in the algorithms, in order to fit the special requirements of the specific problem type. The algorithms' performance is assessed in terms of solution accuracy with respect to the optimal solutions. Comparisons among the employed metaheuristics offer valuable insight regarding their ability to tackle humanitarian logistics problems.

Keywords Humanitarian logistics • Metaheuristics • Algorithm portfolios

1 Introduction

Humanitarian Logistics (HL) has attracted increasing interest due to the exponential surge in natural and man-made disasters (Özdamar et al., 2004). Ranging from earthquakes to tsunamis, natural disasters have produced startling devastation with major death tolls and economical consequences worldwide. Recent examples of severe natural disasters include the Nepal earthquake in 2015, Japan earthquake and tsunami in 2011, Haiti earthquake in 2010, Myanmar cyclone Nargis in 2008, and

Th. Korkou • D. Souravlias • K. Parsopoulos (✉)
Department of Computer Science and Engineering, University of Ioannina, GR-45110 Ioannina, Greece
e-mail: thkorkou@cs.uoi.gr; dsouravl@cs.uoi.gr; kostasp@cs.uoi.gr

K. Skouri
Department of Mathematics, University of Ioannina, GR-45110 Ioannina, Greece
e-mail: kskouri@uoi.gr

I.S. Kotsireas et al. (eds.), *Dynamics of Disasters—Key Concepts, Models, Algorithms, and Insights*, Springer Proceedings in Mathematics & Statistics 185,
DOI 10.1007/978-3-319-43709-5_7

Pakistan earthquake in 2005. All these events caused thousands of deaths and left numerous people homeless, needing emergent assistance. The long lasted, slowly progressing recovery efforts forced thousands of wounded people to continue living in refugee camps that were set up immediately after the disaster.

HL plays a crucial role in addressing disaster relief operations problems. Quoting from (Thomas and Kopczak, 2005), HL is responsible for "*planning, implementing and controlling the efficient, cost-effective flow and storage of goods and materials, as well as related information, from point of origin to point of consumption for the purpose of alleviating the suffering of vulnerable people*." It is widely perceived that HL constitutes a powerful tool, capable of making the difference between success and failure in managing disaster relief operations (Cozzolino et al., 2012; Van Wassenhove, 2006).

Moreover, according to the *Pan American Health Organization* (PAHO), HL focuses on the procurement, transportation, storage, and distribution of relief supplies (PAHO, 2001). Procurement ensures the availability of the demanded resources, while transportation is responsible for the transfer of the latter to the wounded people in the affected areas. Transportation plans for the goods must take into account the available infrastructure, which is frequently damaged from the disaster. Storage protects the relief commodities until their delivery to the points of consumption. Finally, distribution is responsible for the aid delivery to beneficiaries, avoiding problems triggered by external factors (Balcik et al., 2008).

Although HL is significant to prevent from consequences on people's health or life loss, the relevant literature is limited when compared to commercial logistics. In the latter, the main goal is typically the cost reduction (Van Wassenhove, 2006). HL promotes different priorities than cost, thereby introducing new aspects to the underlying problems. Various aspects of HL have been investigated in several studies (Diaz et al., 2013; Galindo and Batta, 2013). Transportation and routing were studied by Barbarosoglu and Arda (2004), Han et al. (2011), Huang et al. (2013), Yi and Özdamar (2007), Yi and Kumar (2007), and Yuan and Wang (2009). Specifically, Barbarosoglu and Arda (2004) proposed a two-stage stochastic programming model to plan the transportation of vital relief resources to the affected areas, taking into account the variations in demand, supply, and route capacity. Han et al. (2011) proposed a model based on Lagrangian relaxation to address a problem of delivering relief commodities.

Huang et al. (2013) studied a continuous approximation approach by utilizing aggregated instance data to explore appropriate routes for aid response teams. Yi and Özdamar (2007) addressed a dynamic coordination model for evacuation and support in disaster response situations. Yi and Kumar (2007) employed a meta-heuristic algorithm, namely ant colony optimization to explore optimal solutions for wounded people with respect to speed delivery. Yuan and Wang (2009) presented two mathematical models for path-selection, taking into account the travel speed on each route as well as hectic situations that frequently follow disasters.

Supply chain and procurement were investigated by Balcik and Beamon (2008), Clark and Culkin (2013), Falasca and Zobel (2011), Peng and Chen (2011), as well as by Tatham and Kovacs (2010), and Taylor and Pettit (2009). More specifically,

Balcik and Beamon (2008) presented a model aiming to define the optimal number of distribution centers and the quantity of supplies. Clark and Culkin (2013) proposed a mathematical transshipment multi-commodity supply chain flow model to satisfy demand for affected people. Falasca and Zobel (2011) explored the need for models that support procurement through a two-stage decision model. Peng and Chen (2011) studied how transportation and information delays affect disaster relief operations. Tatham and Kovacs (2010) suggested a model of swift trust aiming to enhance disaster relief operations. Finally, Taylor and Pettit (2009) explored how lean logistics techniques such as value chain analysis perform in HL.

In addition, a number of research works are devoted to the study of distribution and supply location (Chang et al., 2007; Sheu et al., 2005; Van Hentenryck et al., 2010; Vitoriano et al., 2011; Zhang et al., 2012). Specifically, Chang et al. (2007) addressed a flood emergency logistics problem by presenting two stochastic programming models that were solved by using a sample average approximation scheme. Sheu et al. (2005) solved a disaster relief distribution problem by employing fuzzy clustering techniques and fuzzy linear programming. Van Hentenryck et al. (2010) proposed a hybrid optimization algorithm to solve the single commodity allocation problem. Vitoriano et al. (2011) solved a distribution model, taking into account performance measures such as equity of the distribution and security of routes. Zhang et al. (2012) proposed a heuristic approach based on linear programming and network optimization to solve a multiple-resource and multiple-depot disaster relief problem.

The complexity of HL optimization problems requires efficient solvers that can produce satisfactory solutions within strict time constraints. *Metaheuristics* have been recognized as valuable optimization tools for this purpose. The term *metaheuristics* mostly refers to nature-inspired algorithms with stochastic components (Glover, 1986). Such algorithms are able to offer (sub-)optimal solutions to difficult optimization problems within reasonable amount of time. However, this comes at the cost of dubious optimality of the detected solution. The dynamic of metaheuristics is governed by two major properties, namely *exploration* and *exploitation* (Blum and Roli, 2003). The first one is the ability to perform diverse search without neglecting regions of the search space. The latter is the ability to conduct more refined search in the neighborhood of already detected candidate solutions. Global optimality of the solutions is highly related to the appropriate balance of these two components.

Metaheuristics have gained increasing popularity in academia and industry due to their successful application in solving complex real-world problems. This can be attributed to their efficiency in decision making, simplicity, noise tolerance, and easy implementation (Liu and Ye, 2014). There is a significant amount of research studying the performance of metaheuristics in various problems in logistics, while recently several works appeared also in the growing area of HL (Yan and Shih, 2012; Yi and Kumar, 2007; Zheng et al., 2014).

Recently, Liu and Ye (2014) studied a multi-period problem, taking into account limited supply and transportation capacity that aims to minimize losses caused by (i) the mismatch between supplies and demand and (ii) the transportation time due to logistics processes. In the present study, we consider a similar model where the objective is the minimization of losses caused by the mismatch between supply and demand of relief resources in the affected areas, while taking into account the already existing quantities (if any) and the importance of the different resources. Furthermore, apart from constraints related to the number, volume, and load capacity of vehicles, we consider also road capacity constraints. The latter is the source of bottleneck in supply chain due to the increase of relief vehicles and possible decrease in transportation capacity, defined by the authorities (Besiou et al., 2011).

A test suite of benchmark problems is produced for the proposed model and they are solved to optimality with the commercial CPLEX solver. In addition, a number of established metaheuristics, namely differential evolution (DE) (Price et al., 2005) and particle swarm optimization (PSO) (Parsopoulos and Vrahatis, 2010), are considered and their efficiency is studied on the test suite. In order to fit the special requirements of the test problems, appropriate modifications are made in their basic operations. Moreover, we consider an enhanced DE (eDE) variant (Mohamed, 2014), which is shown to produce significant performance improvement. All the aforementioned algorithms are also utilized in a parallel algorithm portfolio (AP) framework (Gomes and Selman, 1997, 2001; Huberman et al., 1997; Peng et al., 2010; Tang et al., 2014), according to a recently proposed scheme (Souravlias et al., 2014).

The rest of the paper is organized as follows: in Sect. 2 the mathematical formulation of the proposed model is given. The employed algorithms are comprehensively discussed in Sect. 3, while Sect. 4 contains details on the experimental setting and presentation of the obtained results. Sect. 5 concludes the paper.

2 Problem Formulation

In our model, we consider a set J of affected areas (AAs) and a set I of dispatch centers (DCs). Relief resources (commodities) are transported from DCs to AAs through a number of vehicles of different type and mode. In our case, ground and aerial vehicles of two sizes (big and small) are considered. We henceforth denote as C the set of commodities, M the set of transportation modes, and O_m the set of vehicles of mode $m \in M$. The planning horizon is finite and denoted as T. The complete notation used in our model is presented in Table 1.

The main optimization goal lies in specifying the optimal delivered quantities s^t_{cijm} per commodity $c \in C$ from DC i to AA j, using vehicles of transportation mode m, for each time period t. Moreover, we need to specify the optimal number v^t_{cijmo} of type o, mode m vehicles that are used to transport the commodities at each time period t. All decision variables assume integer values. The corresponding minimization problem is defined as follows:

$$\min \sum_{t\in T} \sum_{j\in J} \sum_{c\in C} b_{cj} \left(d^t_{cj} - \sum_{i\in I} \sum_{m\in M} s^t_{cijm} - L^{t-1}_{cj} \right)^2 , \qquad (1)$$

Table 1 Notation used in the proposed model

Model parameters	*Description*
T	Planning horizon
I	Set of dispatch centers (DCs)
J	Set of affected areas (AAs)
C	Set of commodities
M	Set of transportation modes
m	Index denoting the transportation mode (ground, air)
O_m	Set of vehicle types of transportation mode m
o	Index denoting the vehicle type (big vehicle, small vehicle)
b_{cj}	Importance weight of commodity c in AA j
w_c	Unit weight of commodity c
$volume_c$	Unit volume of commodity c
cap_{mo}	Capacity of type o, mode m vehicle
vol_{mo}	Volume capacity of type o, mode m vehicle
d^t_{cj}	Demand for commodity c in AA j at time period t
k^t_{ijm}	Traffic restriction for mode m vehicles from DC i to AA j at time t
v^t_{imo}	Number of type o, mode m vehicles at DC i at time t
L^t_{cj}	Inventory level of commodity c in AA j at time t
Decision variables	*Description*
s^t_{cijm}	Delivered quantity of commodity c from DC i to AA j through transportation mode m at time t
v^t_{cijmo}	Number of type o, mode m vehicles used at period t to transport commodity c from DC i to AA j

where b_{cj} is a scalar weight of importance of commodity c at AA j. The model is subject to the following constraints:

$$L^0_{cj} = Y_{cj}, \qquad \forall c \in C,\ \forall j \in J, \tag{2}$$

$$L^t_{cj} = \sum_{i \in I} \sum_{m \in M} s^t_{cijm} - d^t_{cj} + L^{t-1}_{cj}, \qquad \forall t \in T,\ \forall c \in C,\ \forall j \in J, \tag{3}$$

$$\sum_{c \in C} \sum_{j \in J} s^t_{cijm}\, w_c \leqslant \sum_{o \in O_m} v^t_{imo}\, cap_{mo}, \qquad \forall t \in T,\ \forall i \in I,\ \forall m \in M, \tag{4}$$

$$\sum_{c \in C} \sum_{j \in J} s^t_{cijm}\, vol_c \leqslant \sum_{o \in O_m} v^t_{imo}\, vol_{mo}, \qquad \forall t \in T,\ \forall i \in I,\ \forall m \in M, \tag{5}$$

$$s^t_{cijm} \leqslant \min\left\{ \frac{\sum_{o \in O_m} v^t_{cijmo}\, cap_{mo}}{w_c}, \frac{\sum_{o \in O_m} v^t_{cijmo}\, vol_{mo}}{volume_c} \right\}, \tag{6}$$

$$\sum_{c \in C} \sum_{o \in O_m} v^t_{cijmo} \leqslant k^t_{ijm}, \qquad \forall t \in T,\ \forall i \in I,\ \forall m \in M,\ \forall j \in J, \tag{7}$$

$$\sum_{c \in C} \sum_{j \in J} v^t_{cijmo} \leqslant v^t_{imo}, \qquad \forall\, t \in T,\ \forall\, i \in I,\ \forall\, m \in M,\ \forall\, o \in O, \tag{8}$$

$$s^t_{cijm}, v^t_{cijmo}, L^t_{cj} \in \mathbb{N}, \quad \text{for all } t, c, i, j, m, o, \tag{9}$$

$$v^t_{imo}, d^t_{cj}, vol_{mo}, cap_{mo}, volume_c, w_c \in \mathbb{N}^+, \quad \text{for all } t, c, i, j, m, o, \tag{10}$$

$$\sum_{c \in C} b_{cj} = 1, \quad b_{cj} \in [0, 1], \qquad \forall j \in J. \tag{11}$$

Equation (2) accounts for the initial inventory level of commodity c that pre-exists at DC j. Equation (3) determines the inventory balance, which takes into account the demand of the commodity c and the replenishment quantity. Equations (4) and (5) refer to capacity and volume constraints, respectively. Equation (6) defines upper limits of the delivered quantity s^t_{cijm}, which is useful for bounding the decision variables. Equation (7) stands for traffic flow restrictions expected in natural disasters, e.g., roads that are partially damaged or destroyed, thereby reducing traffic capacity. Equation (8) ensures that the number of vehicles transporting the commodities in a particular AA does not exceed the total number of vehicles. Equations (9)–(11) define the appropriate domains of the decision variables and problem parameters.

The squared error in Eq. (1) can be replaced by the absolute error if metaheuristics are used. Nevertheless, the quadratic form is selected in order to render the problem solvable by CPLEX. Note that the objective function is also convex since it constitutes the sum of convex functions.

3 Employed Algorithms

In the following paragraphs, we briefly present the main features of the employed metaheuristics. For presentation purposes, we assume that the considered minimization problem is defined in the general form,

$$\min_{\mathbf{x} \in X \subset \mathbb{R}^n} f(\mathbf{x}), \tag{12}$$

where X is the (real-valued) search space. The only requirement on the objective function is the availability of $f(\mathbf{x})$ at any $\mathbf{x} \in X$. Appropriate modifications of the algorithms to handle the integer search spaces of the considered HL model are given later.

3.1 Differential Evolution

Differential Evolution (DE) was introduced by Storn and Price (1997). Being a population-based optimization algorithm, it proceeds by iteratively improving a population of candidate solutions,

$$S = \{\mathbf{x}_1, \mathbf{x}_2, \ldots, \mathbf{x}_N\},$$

consisting of N search points, called the *individuals*. The population is randomly and uniformly initialized over the search space. DE applies biologically inspired operators, namely *mutation*, *crossover*, and *selection*, on each individual in order to produce new candidate solutions by combining existing ones.

At each iteration (also called *generation*) g of the algorithm, a mutant vector $\mathbf{v}_i$ is generated for each individual $\mathbf{x}_i$, $i = 1, 2, \ldots, N$. This vector is produced by combining existing individuals from the population according to various *mutation operators*. The following are among the most popular ones:

$$\text{DE1}: \quad \mathbf{v}_i^{(g+1)} = \mathbf{x}_{\text{best}}^{(g)} + F\left(\mathbf{x}_{r1}^{(g)} - \mathbf{x}_{r2}^{(g)}\right), \tag{13}$$

$$\text{DE2}: \quad \mathbf{v}_i^{(g+1)} = \mathbf{x}_{r1}^{(g)} + F\left(\mathbf{x}_{r2}^{(g)} - \mathbf{x}_{r3}^{(g)}\right), \tag{14}$$

$$\text{DE3}: \quad \mathbf{v}_i^{(g+1)} = \mathbf{x}_i^{(g)} + F\left(\mathbf{x}_{\text{best}}^{(g)} - \mathbf{x}_i^{(g)}\right) + F\left(\mathbf{x}_{r1}^{(g)} - \mathbf{x}_{r2}^{(g)}\right), \tag{15}$$

$$\text{DE4}: \quad \mathbf{v}_i^{(g+1)} = \mathbf{x}_{\text{best}}^{(g)} + F\left(\mathbf{x}_{r1}^{(g)} - \mathbf{x}_{r2}^{(g)}\right) + F\left(\mathbf{x}_{r3}^{(g)} - \mathbf{x}_{r4}^{(g)}\right), \tag{16}$$

$$\text{DE5}: \quad \mathbf{v}_i^{(g+1)} = \mathbf{x}_{r1}^{(g)} + F\left(\mathbf{x}_{r2}^{(g)} - \mathbf{x}_{r3}^{(g)}\right) + F\left(\mathbf{x}_{r4}^{(g)} - \mathbf{x}_{r5}^{(g)}\right), \tag{17}$$

where $\mathbf{x}_{\text{best}}^{(g)}$ denotes the individual with the lowest objective value in the population at iteration g. The randomly selected indices $r_j \in \{1, 2, \ldots, N\} \setminus \{i\}, j = 1, 2, \ldots, 5,$ are taken to be mutually different. The user-defined parameter $F \in [0, 2]$ is called the *scale factor* and defines the size of the steps taken towards the search directions defined by the differences between existing individuals.

After generating the mutant vectors, crossover takes place. A *trial vector*, $\mathbf{u}_i = (u_{i1}, u_{i2}, \ldots, u_{in})^\top$, is generated for each individual $\mathbf{x}_i^{(g)}$, as follows:

$$u_{ij}^{(g+1)} = \begin{cases} v_{ij}^{(g+1)}, & \text{if } R_j \leqslant CR \text{ or } j = RI(i), \\ x_{ij}^{(g)}, & \text{otherwise,} \end{cases} \tag{18}$$

where $j = 1, 2, \ldots, n$; R_j is the j-th evaluation of a uniform random number generator in the range [0, 1]; $CR \in [0, 1]$ is a user-defined *crossover rate*; and $RI(i)$ is a randomly selected index in $\{1, 2, \ldots, n\}$.

Eventually, the selection operator is applied, where the trial vectors compete against their original individuals. If the trial vector achieved a better objective value, it replaces the original individual in the population, as follows:

$$\mathbf{x}_i^{(g+1)} = \begin{cases} \mathbf{u}_i^{(g+1)}, & \text{if } f\left(\mathbf{u}_i^{(g+1)}\right) < f\left(\mathbf{x}_i^{(g)}\right), \\ \mathbf{x}_i^{(g)}, & \text{otherwise.} \end{cases} \tag{19}$$

The parameters F, CR, and N must be carefully selected due to their impact on DE's performance. A comprehensive presentation of DE-related research can be found in Price et al. (2005).

3.2 Enhanced Differential Evolution

Recently, Mohamed (2014) proposed an enhanced DE (eDE) variant. It defines an alternative mutation scheme, while crossover is based on probabilistic selection between the new and the DE2 scheme of Eq. (14). Moreover, the algorithm is enhanced by using restart to alleviate local minima.

Putting it formally, eDE is based on the mutation scheme,

$$\mathbf{w}_i^{(g+1)} = \mathbf{x}_{r1}^{(g)} + F_1\left(\mathbf{x}_{\text{best}}^{(g)} - \mathbf{x}_{r1}^{(g)}\right) + F_2\left(\mathbf{x}_{r1}^{(g)} - \mathbf{x}_{\text{worst}}^{(g)}\right), \tag{20}$$

where $\mathbf{x}_{r1}^{(g)}$ is a randomly selected individual; $F_1, F_2 \in [0, 2]$ are scalar parameters called the *differential weights*; and $\mathbf{x}_{\text{best}}^{(g)}$, $\mathbf{x}_{\text{worst}}^{(g)}$, denote the best and worst individuals at iteration g, respectively. The trial vector is given as follows:

$$u_{ij}^{(g+1)} = \begin{cases} w_{ij}^{(g+1)}, & \text{if } \left(R_j \leqslant CR \text{ or } j = RI(i)\right) \text{ and } R \geqslant \left(1 - \frac{g}{g_{\max}}\right), \\ v_{ij}^{(g+1)}, & \text{if } \left(R_j \leqslant CR \text{ or } j = RI(i)\right) \text{ and } R < \left(1 - \frac{g}{g_{\max}}\right), \\ x_{ij}^{(g)}, & \text{otherwise,} \end{cases} \tag{21}$$

where $g_{\max}$ is the total number of iterations and R is a uniform random number generator in the range [0,1]. The rest of the parameters are identical to the standard DE. Also, note that v_{ij} is the j-th component of the mutation vector $\mathbf{v}_i$ produced through Eq. (14).

A restart mechanism is also incorporated in eDE to avoid premature convergence. The restart mechanism is applied on each individual except for the best one, which is kept unaltered. In our case, we adopt restarts from mild perturbations $\mathbf{x}_i'$ of current individuals $\mathbf{x}_i$, as follows:

$$x_{ij}' = x_{ij} \pm 1. \tag{22}$$

According to this scheme, the probability of plunging into local minima is drastically decreased and the local search capability is enhanced through the perturbation of individuals into their imminent neighborhood. The sign "+" or "−" in Eq. (22) is randomly selected with equal probability for each j. The bias is selected equal to 1 since it constitutes the smallest step size in integer search spaces as the ones in the proposed model.

3.3 Particle Swarm Optimization

Particle swarm optimization (PSO) was introduced by Kennedy and Eberhart (1995). Similarly to DE it is a population-based algorithm, although with special emphasis in cooperation. PSO does not have any direct selection operator. Instead, a population (called a *swarm*) of candidate solutions (called the *particles*) probes the search space. Each particle retains in memory the best position it has ever visited. This position, along with information shared with the rest of the swarm, is used to bias the move of the particles.

Let $S = \{\mathbf{x}_1, \mathbf{x}_2, \ldots, \mathbf{x}_N\}$ be a swarm of N particles, each one being an n-dimensional vector in the search space, $\mathbf{x}_i = (x_{i1}, x_{i2}, \ldots, x_{in}) \in X$, $i = 1, 2, \ldots, N$. The particle moves by adding to its current position an adaptable bias vector, called the *velocity*,

$$\mathbf{v}_i = (v_{i1}, v_{i2}, \ldots, v_{in})^\top,$$

while it has also a personal memory, called the *best position*,

$$\mathbf{p}_i = (p_{i1}, p_{i2}, \ldots, p_{in})^\top \in X.$$

Apart from its own best position, each particle assumes a *neighborhood* in the form of a set of particle indices. The particle exchanges information with other particles from its neighborhood by adopting their best findings. In the case where the particle's neighborhood is the entire swarm, the best position in the neighborhood is referred as *global best* particle, and the resulting algorithm is denoted as *gbest* PSO. On the other hand, when smaller neighborhoods are used the algorithm is denoted as *local best* PSO (*lbest* PSO) (Parsopoulos and Vrahatis, 2010). The size of the neighborhood has crucial impact on the dissemination of information in the swarm and, hence, on convergence speed.

In literature, various neighborhood topologies have been proposed. Most commonly used are the star, Von Neumann, and the ring topology. According to the ring, all particles are assumed to lie on a ring with respect to their indices. Then, for a given particle, its neighborhood is defined by its immediate neighbors in the ring. The considered number of neighbors per neighborhood is called the neighborhood's *radius*. Thus, a ring neighborhood of radius r for a particular particle $\mathbf{x}_i$ is defined as follows:

$$NB_i^r = \{i - r, i - r + 1, \ldots, i, \ldots, i + r - 1, i + r\}. \tag{23}$$

Let best_i be the index of the best position found so far by any individual in the neighborhood NB_i^r of $\mathbf{x}_i$, i.e.,

$$\text{best}_i = \arg \min_{j \in NB_i^r} f(\mathbf{p}_j). \tag{24}$$

Also, let g denote the iteration counter. Then, the swarm is updated according to,

$$v_{ij}^{(g+1)} = \chi \left[v_{ij}^{(g)} + c_1 R_1 \left(p_{ij}^{(g)} - x_{ij}^{(g)} \right) + c_2 R_2 \left(p_{\text{best}_i j}^{(g)} - x_{ij}^{(g)} \right) \right], \tag{25}$$

$$x_{ij}^{(g+1)} = x_{ij}^{(g)} + v_{ij}^{(g+1)}, \tag{26}$$

where χ is the *constriction coefficient* parameter; c_1, c_2, are positive acceleration parameters; and R_1, R_2, are random variables, uniformly distributed in the range $[0, 1]$. The constriction coefficient promotes convergence by reducing the magnitude of the velocities. For the reader's convenience, we mention the typical values $\chi = 0.729$, $c_1 = c_2 = 2.05$, which are widely accepted as the default parameter set on the basis of the PSO's stability analysis due to Clerc and Kennedy (2002).

The best positions of the particles are updated at each iteration according to,

$$\mathbf{p}_i^{(g+1)} = \begin{cases} \mathbf{x}_i^{(g+1)}, & \text{if } f\left(\mathbf{x}_i^{(g+1)}\right) < f\left(\mathbf{p}_i^{(g)}\right), \\ \mathbf{p}_i^{(g)}, & \text{otherwise.} \end{cases} \tag{27}$$

A compendium of PSO-related research can be found in Parsopoulos and Vrahatis (2010).

3.4 Algorithm Portfolios

The term *Algorithm Portfolio* (AP) refers to a framework where different algorithms (*heterogeneous AP*) or different copies of the same algorithm (*homogeneous AP*) are combined in a single algorithmic scheme (Huberman et al., 1997). APs constitute a modern approach for solving challenging optimization problems. Their computational efficiency against common metaheuristics has led to a constantly increasing research production (Gomes and Selman, 1997; Huberman et al., 1997; Peng et al., 2010; Souravlias et al., 2014, 2015; Tang et al., 2014).

The performance of APs depends on the selection of appropriate constituent algorithms. The constituent algorithms can either interact or run independently. In the present work, the AP framework proposed by Souravlias et al. (2015) is used to define interactive algorithmic schemes consisting of the metaheuristics described in the previous sections. The AP's algorithms interact with each other and employ a typical *master-slave* parallelization model. Each metaheuristic runs on one of M

slave nodes for a pre-specified budget of total running time. The budget is divided into *execution time*, T_{exec}, i.e., time used by the algorithm for its own execution, and *investment time*, T_{inv}, i.e., time used to buy elite solutions from the other algorithms.

Slave nodes can communicate via a *master node*, which is responsible for two basic operations. Firstly, it maintains an archive of the M elite solutions detected by the algorithms. Secondly, it assigns *prices* to the elite solutions whenever solution trading takes place. For this purpose, the solutions are sorted in descending order with respect to their objective values. Then, the price of each one is determined according to its corresponding position ρ_i in the ranking, i.e.,

$$C_i = \frac{\rho_i \times BC}{M}, \qquad (28)$$

where $BC = \beta\, T_{\text{inv}}$ is a fixed *base cost* and β is a constant that takes values in [0, 1] as suggested by Souravlias et al. (2014).

When an algorithm (slave node) fails to improve its solution for some predefined amount of time, it requests from the master node to buy an elite solution from the archive. For the buyer algorithm, this comes at the cost of a fraction of its investment time, which is immediately credited to the seller algorithm that offered the purchased solution. The master node proposes to the slave node its archived elite solutions that are better than its own current best solution. Then, the buyer algorithm gets the one that maximizes the *Return on Investment* (ROI) index, among the solutions it can afford. The ROI index comes from trading theory and in our case is defined as follows:

$$ROI_j = \frac{f - f_j}{C_j}, \quad j \in \{1, 2, \ldots, M\}, \qquad (29)$$

where f denotes the objective value of the buyer's best solution, f_j denotes the objective value of the seller's elite solution, and C_j is the assigned price (Souravlias et al., 2014). The paid investment time from the buyer algorithm is then added to the total execution time of the seller algorithm. The purchased solution replaces the worst solution in the population of the buyer. In case of no affordable solution, the buyer algorithm simply restarts, retaining only its best solution.

Apparently, better algorithms of the AP sell solutions more frequently and, consequently, gain additional execution time. It is worth mentioning that the total execution time assigned to the AP remains constant, since time portions are only dynamically transferred from inferior to the most promising constituent algorithms. This is a significant property in modern high-performance platforms where usage and execution time have a significant cost. Also, the final distribution of execution time of the constituent algorithms offers useful insight regarding the best-performing one for the problem at hand (Souravlias et al., 2014, 2015).

3.5 Further Applicability Issues

Two main issues need to be addressed prior to the application of the presented metaheuristics on the problem of Sect. 2. The first one is related to the discrete nature of the search space, while the second one refers to constraint handling.

Regarding the first issue, simple rounding to the nearest integer is used. Specifically, the algorithms are applied on the corresponding real search space and, for the function evaluation, the vectors are rounded to the nearest integer ones. In DE and eDE, the rounded vectors are also retained in the population. In PSO, rounded vectors replace best positions solely. Rounding is a common approach successfully applied in similar problems (Piperagkas et al., 2012; Parsopoulos et al., 2015).

The constraint handling problem is tackled with the widely used *penalty function* approach, combined with a set of preference rules between feasible and infeasible solutions:

1. Between two infeasible solutions, the one that violates fewer constraints is selected.
2. Between a feasible and an infeasible solution, the feasible one is preferred.
3. Between two feasible solutions, the one with the lowest objective value is preferred.

These rules have been previously used with PSO and DE (Parsopoulos et al., 2015). The employed penalty function has a simple form,

$$P(\mathbf{x}) = f(\mathbf{x}) + \sum_{i \in VC(\mathbf{x})} |V(i)|, \tag{30}$$

where $f(\mathbf{x})$ is the actual objective value of $\mathbf{x}$; $V(i)$ is the violation magnitude of the i-th constraint; and $VC(\mathbf{x})$ is the set of constraints violated by $\mathbf{x}$. Note that the penalty for a violated constraint depends on the magnitude of violation. Apparently, in absence of violated constraints the penalty function is equal to the original objective function.

4 Performance Assessment

In this section we expose the experimental settings and the obtained results from the application of the described algorithms on the test suite produced for the proposed HL model.

Table 2 Capacity and volume information for vehicle types I (small) and II (big)

	Transportation mode			
	Ground		Air	
	I	II	I	II
Load capacity (ton)	3	10	4	9
Load volume (m^3)	20	44	35	75

Table 3 Commodities information

	Water	Medicines	Food
Importance weight	0.35	0.35	0.30
Unit weight (kg)	650	20	200
Unit volume (m^3)	1.44	0.125	0.60

Table 4 Number of vehicles per DC

	Transportation mode			
	Ground		Air	
	I	II	I	II
DC_1	4	5	1	1
DC_2	4	5	1	1

4.1 Experimental Setting

The main goal in our model is the minimization of losses caused by the mismatch between supply and demand, as well as the determination of the optimal number of vehicles for the transportation of relief resources to the stricken areas. In our experiments we considered three life-essential commodities, namely *water*, *medicines*, and *food*. Among them, the first two were assumed to have slightly higher importance weights than the third one.

Moreover, we assumed the existence of two DCs responsible to supply two AAs, and two modes of transportation, ground and aerial, using trucks and helicopters, respectively. For each transportation mode, two vehicle types were considered, namely small and big vehicles, henceforth denoted as type I and II, respectively. Tables 2 and 3 report all relevant information regarding vehicles and commodities, respectively. Note that, motivated by the Kefalonia island earthquake in 2014, the reported data are based on real-world values (e.g., palettes of water bottles, typical transportation boxes for medication, etc). Also, Table 4 reports the number of available vehicles per DC.

In the context of the proposed model, a test suite of 10 benchmark problems with diverse characteristics was generated. The test problems are henceforth denoted as P1-P10. The problems were initially solved to optimality with the CPLEX solver. Subsequently, extensive experiments were conducted with the following algorithms: PSO, DE, eDE, as well as APs consisting of PSO+DE, PSO+eDE, DE+DE, DE+eDE, eDE+eDE, and PSO+DE+eDE.

In a preprocessing phase, all five mutation operators of Eqs. (13)–(17) were considered for DE and eDE, along with all combinations of their parameters $F \in [0, 2]$ and $CR \in [0, 1]$, discretized with step size 0.05. Various population sizes were also investigated. Preliminary experiments provided clear evidence that DE2 with,

$$F = F_1 = F_2 = 0.4, \qquad CR = 0.05,$$

was the most promising setting. The PSO algorithm was considered in its lbest model with ring topology of radius $r = 1$, and the default parameter set,

$$\chi = 0.729, \qquad c_1 = c_2 = 2.05.$$

The population size for all algorithms was set to $N = 150$, which was identified as a promising value for our 144-dimensional optimization problem. The boundaries for the decision variables were the ones imposed by the given data (for the vehicles) and the constraints given in Sect. 2 (for the delivered quantities).

In order to statistically validate each algorithm, 30 independent experiments were performed per problem instance. The experiments were conducted on Intel i7 servers with 8GB RAM. The running time for each experiment was set to 10 min in order to be comparable with the time needed by CPLEX to provide accurate lower bounds for the solutions. Note that, on average CPLEX required around 10 min to find good solution approximations, although their optimality guarantee required even hours. The algorithms were run and analyzed also for 5 and 20 min in order to investigate their sensitivity with respect to running time. The time-variation of the solution error is illustrated in Fig. 1.

For each algorithm and experiment, the best solution $\mathbf{x}^*_{\text{alg}}$ and its value f^*_{alg} were recorded, along with the *solution error* from the global minimum detected by CPLEX, i.e.,

$$\textit{solution error} = f^*_{\text{alg}} - f^*_{\text{cplex}}.$$

The plain solution error is utilized instead of the relative error, because the optimal objective value of some problems was zero. Average values of solution error over the 30 experiments, along with standard deviation, minimum, and maximum values, were also recorded for performance comparison purposes.

4.2 Results and Discussion

A summary of all recorded results is reported in Table 5, where the best-performing approach is boldfaced. Also, the results are graphically illustrated to facilitate visual comparisons. The average solution error from the global minimum is presented in

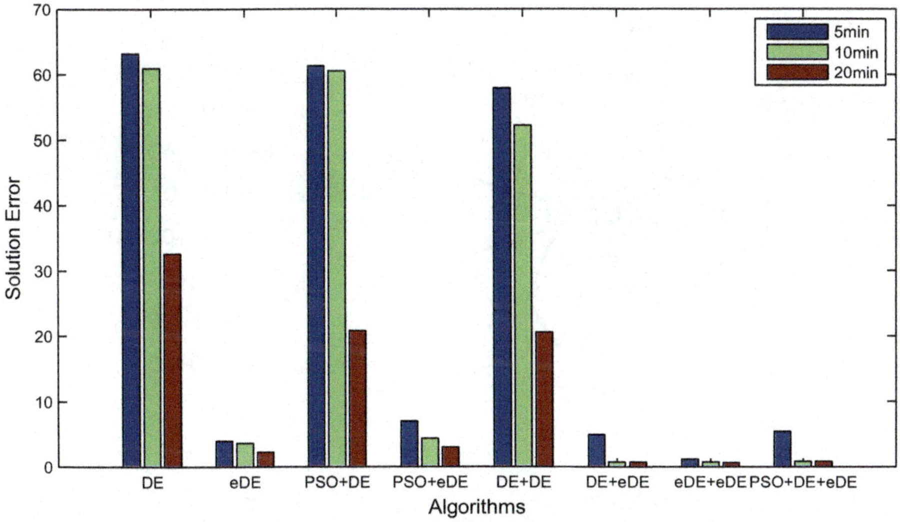

Fig. 1 Solution error averaged over all problem instances per algorithm and running time

Table 5 Mean, standard deviation, minimum, and maximum solution error values for all algorithms, averaged over all problems. Best values are boldfaced. The "+" symbol denotes AP approach constituting of the corresponding algorithms

Algorithm	Mean	St.D.	Min	Max
PSO	513.80	235.85	197.00	2442.20
DE	63.31	40.45	26.97	160.21
eDE	3.54	3.42	0.29	11.80
PSO+DE	52.28	31.11	27.01	129.42
PSO+eDE	4.14	3.99	0.16	13.77
DE+DE	59.65	55.36	21.15	193.81
DE+eDE	0.76	0.91	0.00	2.91
eDE+eDE	**0.75**	**0.85**	**0.00**	**2.27**
PSO+DE+eDE	0.84	1.18	0.00	3.74

the upper part of Fig. 2, per problem and algorithm. In the lower part of Fig. 2, the central region around the origin is zoomed, exposing the corresponding curves of the most competitive algorithms. Similarly, in the upper and lower part of Fig. 3, we illustrate the averaged standard deviation per problem and algorithm. Note that in all figures we excluded the results of PSO due to scaling reasons.

Furthermore, we also recorded the success rate per algorithm, i.e., the percentage of experiments where it succeeded to reach the optimal solution within the available execution time. Figure 4 presents the resulting success rates per problem instance

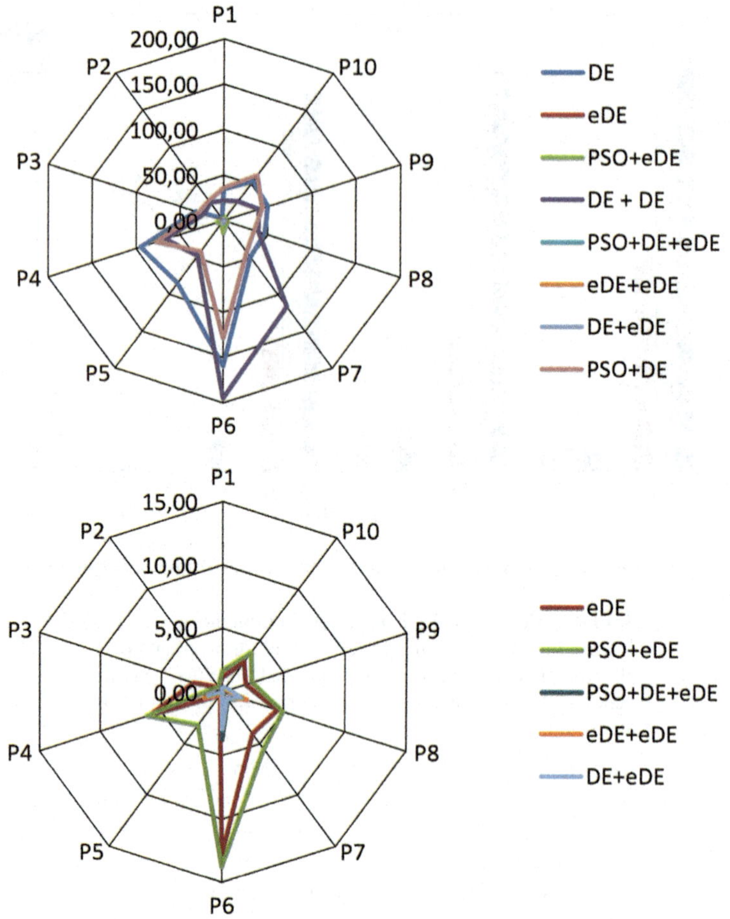

Fig. 2 Averaged solution error per algorithm and problem (*upper part*) and zoom in center area (*lower part*)

for the most promising algorithms. Finally, the boxplots of Fig. 5 illustrate the distribution of the obtained solution error values in all experiments.

The reported results offer interesting conclusions. Firstly, we can easily see that the homogeneous AP approach eDE+eDE as well as the heterogeneous PSO+DE+eDE outperformed the rest of the algorithms, yielding higher success rates. Also, these two approaches exhibited almost equivalent performance. However, in problems P6-P8, which were proved to be the most difficult ones with respect to the success rates of the algorithms, the eDE+eDE approach dominated in terms of efficiency.

In order to quantitatively study this behavior, we further analyzed the solution purchases between the algorithms of the AP approaches. The analysis verified that, especially for the aforementioned problems, the number of purchases between the

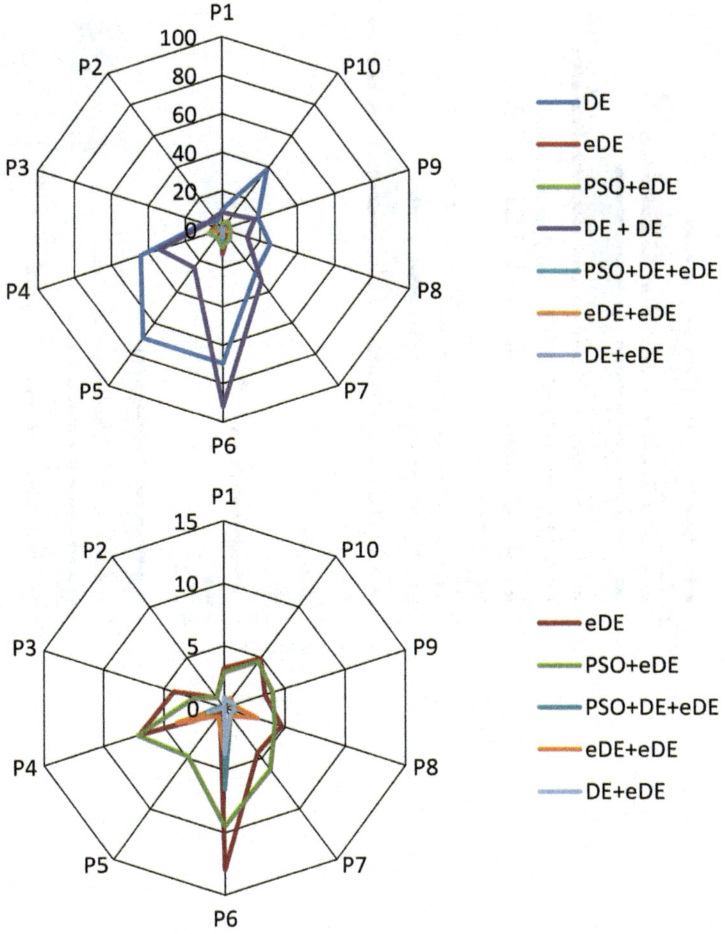

Fig. 3 Standard deviation of the solution error per algorithm and problem (*upper part*) and zoom in center area (*lower part*)

algorithms was remarkably high. This leads to the conclusion that, due to the complexity of these problems, the constituent algorithms of the AP experienced severe difficulties in reaching the optimal solution. Therefore, they were more prone to exchange information in order to improve their performance.

Also, in the case of PSO+DE+eDE, the assigned execution time per algorithm was shorter than that of each eDE instance in eDE+eDE, because in the first case the total time of the AP is divided by 3, while in the latter one it is divided in 2 equal parts. Since PSO was proved to be less efficient than eDE, the assigned time in PSO+DE+eDE was consequently proved to be insufficient.

Regarding the standalone algorithms, eDE was clearly the dominant one, exhibiting undoubtful advantages against the rest. This can also explain the superiority of the eDE-based AP approaches. Obviously, the special probabilistic operator of eDE as well as the restart mechanism with mild perturbations (see Sect. 3.2)

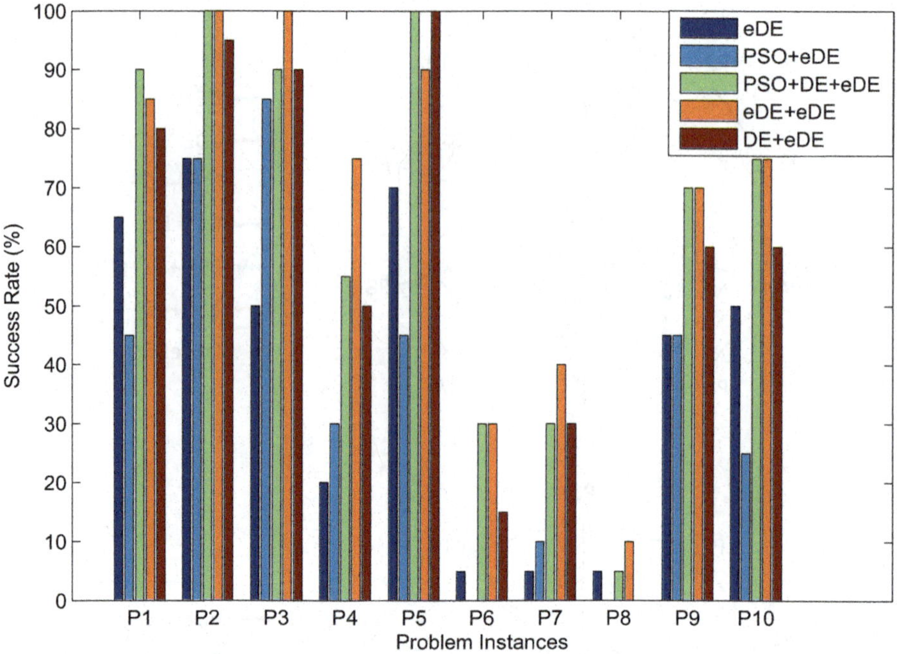

Fig. 4 Success rates of the most promising algorithms per problem

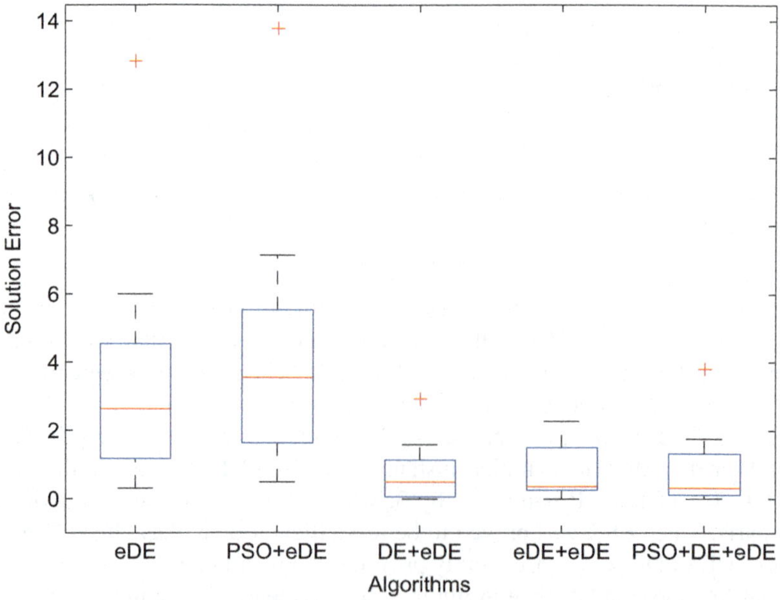

Fig. 5 Solution error distribution of the most promising algorithms for all test problems

were beneficial for the algorithm. Experimental evidence suggested that this can be attributed to the alleviation of search stagnation caused by the rounding of the real-valued vectors to their nearest integers. Moreover, this can be related also to the domination of the DE2 operator, which offers the necessary diversity to avoid stagnation. These properties were also identified in Souravlias et al. (2014).

Although there is a clear advantage of some algorithms against the rest, there are marginal differences among the most promising approaches. In order to investigate whether the observed differences were the outcome of random fluctuations, we conducted statistical significance tests among the most competitive algorithms. Specifically, pairwise comparisons of the algorithms were conducted using the Wilcoxon ranksum tests at 95 % confidence level for all test problems. Whenever an algorithm was statistically superior to another, we counted it as *win* of the algorithm. On the other hand, if it was statistically inferior, we counted it as a *loss*. The lack of statistical significance was counted as a *draw* for both algorithms.

The results concerning wins/losses/draws are presented in Table 6 and the corresponding graphical illustration is given in Fig. 6 for all problem instances. The superiority of DE+eDE, eDE+eDE, and PSO+DE+eDE was anew confirmed. In almost all comparisons, these approaches were prevalent against the rest. Yet, most of the comparisons among them resulted in draws, despite the marginal differences reported in Table 6. Especially for DE+eDE and eDE+eDE, no losses were reported. Thus, our initial assumption regarding the superiority of eDE-based approaches was corroborated by the statistical evidence, placing these AP approaches in a salient position among the most promising solvers.

5 Conclusions

The contribution of the present work was twofold. On one hand, we introduced a model that aims at minimizing the losses caused by the mismatch between supply and demand, while concurrently determining the number of different types of vehicles used to transport relief commodities from dispatch centers to stricken areas. A number of test problems with diverse characteristics was generated for the proposed model and solved to optimality using CPLEX.

Table 6 Wins/losses/draws of row versus column algorithms for all problem instances

	eDE	PSO+eDE	DE+eDE	eDE+eDE	PSO+DE+eDE
eDE	–	1 / 1 / 8	0 / 5 / 5	0 / 5 / 5	0 / 8 / 2
PSO+eDE		–	0 / 7 / 3	0 / 9 / 1	1 / 7 / 2
DE+eDE			–	0 / 0 / 10	0 / 0 / 10
eDE+eDE				–	0 / 0 / 10
PSO+DE+eDE					–

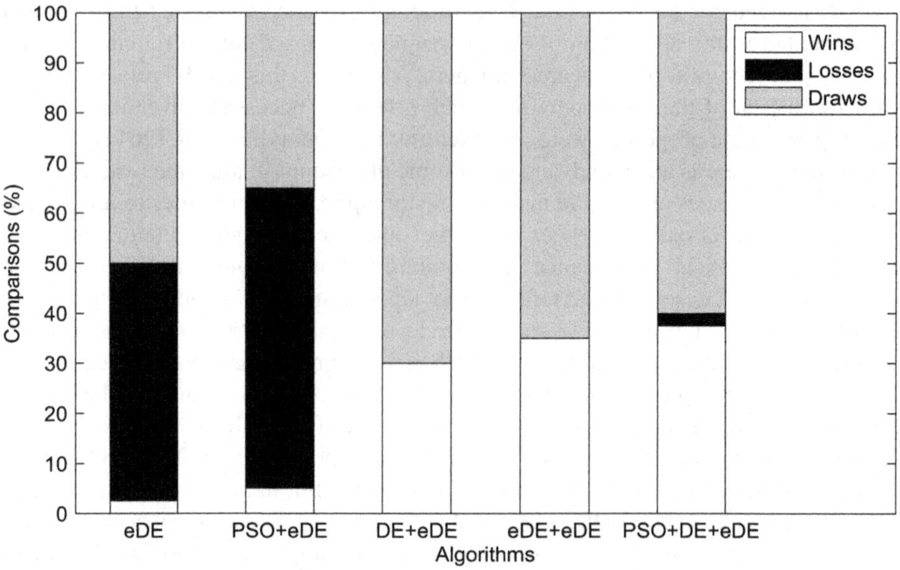

Fig. 6 Results of the pairwise statistical comparisons among the most competitive algorithms for all test problems

On the other hand, prevalent modern metaheuristics were studied in solving Humanitarian Logistics problems. Our approach was based on DE, eDE, PSO, and heterogeneous/homogeneous APs consisting of combinations of these algorithms. Proper modifications and refinements were introduced to tackle the special requirements of the test problems.

From the extracted results, we concluded that APs based on eDE offer remarkable performance efficiency and solution quality. Also, it became evident that APs can offer crucial insight in gathering information regarding the most appropriate metaheuristic for the problem at hand.

Future work will extend the test suite, aiming at an abundant set of test problems with a multitude of different characteristics and peculiarities. Also, the study of APs will be enriched by employing larger and diverse collections of metaheuristics, in order to efficiently deal with problems of higher complexity.

References

Balcik, B., Beamon, B.M.: Facility location in humanitarian relief. Int. J. Log. Res. Appl. **11**(2), 101–121 (2008)

Balcik, B., Beamon, B.M., Smilowitz, K.: Last mile distribution in humanitarian relief. J. Intell. Transp. Syst. **12**(2), 51–63 (2008)

Barbarosoglu, G., Arda, Y.: A two-stage stochastic programming framework for transportation planning in disaster response. J. Oper. Res. Soc. **55**(1), 43–53 (2004)

Besiou, M., Stapleton, O., Van Wassenhove, L.N.: System dynamics for humanitarian operations. J. Humanitarian Logist. Supply Chain Manag. **1**(1), 78–103 (2011)

Blum, C., Roli, A.: Metaheuristics in combinatorial optimization: Overview and conceptual comparison. ACM Comput. Surv. **35**(3), 268–308 (2003)

Chang, M.S., Tseng, Y.L., Chen, J.W.: A scenario planning approach for the flood emergency logistics preparation problem under uncertainty. Transp. Res. E Logist. Transp. Rev. **43**(6), 737–754 (2007)

Clark, A., Culkin, B.: A network transshipment model for planning humanitarian relief operations after a natural disaster. In: Decision Aid Models for Disaster Management and Emergencies, Atlantis Computational Intelligence Systems, vol. 7, pp. 233–257. Atlantis Press, Paris (2013)

Clerc, M., Kennedy, J.: The particle swarm-explosion, stability, and convergence in a multidimensional complex space. IEEE Trans. Evol. Comput. **6**(1), 58–73 (2002)

Cozzolino, A., Rossi, S., Conforti, A.: Agile and lean principles in the humanitarian supply chain. J. Humanitarian Logist. Supply Chain Manag. **2**(1), 16–33 (2012)

Diaz, R., Behr, J., Toba, A.L., Giles, B., Ng, M., Longo, F., Nicoletti, L.: Humanitarian/emergency logistics models: A state of the art overview. In: Proceedings of the 2013 Summer Computer Simulation Conference (SCSC13), pp. 24:1–24:8. Society for Modeling & Simulation International, Vista (2013)

Falasca, M., Zobel, C.W.: A two-stage procurement model for humanitarian relief supply chains. J. Humanitarian Logist. Supply Chain Manag. **1**(2), 151–169 (2011)

Galindo, G., Batta, R.: Review of recent developments in OR/MS research in disaster operations management. Eur. J. Oper. Res. **230**(2), 201–211 (2013)

Glover, F.: Future paths for integer programming and links to artificial intelligence. Comput. Oper. Res. **13**(5), 533–549 (1986)

Gomes, C.P., Selman, B.: Algorithm portfolio design: Theory vs. practice. In: Proceedings of the 13th Conference on Uncertainty in Artificial Intelligence (UAI'97), pp. 190–197. Morgan Kaufmann Publishers, San Francisco (1997)

Gomes, C.P., Selman, B.: Algorithm portfolios. Artif. Intell. **126**(1-2), 43–62 (2001)

Han, Y., Guan, X., Shi, L.: Optimization based method for supply location selection and routing in large-scale emergency material delivery. IEEE Trans. Autom. Sci. Eng. **8**(4), 683–693 (2011)

Huang, M., Smilowitz, K., Balcik, B.: A continuous approximation approach for assessment routing in disaster relief. Transp. Res. B Methodol. **50**, 20–41 (2013)

Huberman, B.A., Lukose, R.M., Hogg, T.: An economics approach to hard computational problems. Science **27**, 51–53 (1997)

Kennedy, J., Eberhart, R.: Particle swarm optimization. In: Proceedings on IEEE International Conference on Neural Networks, vol. 4, pp. 1942–1948 (1995)

Liu, N., Ye, Y.: Humanitarian logistics planning for natural disaster response with Bayesian information updates. J. Ind. Manag. Optim. **10**(3), 665–689 (2014)

Lourenço, H.: Logistics management: An opportunity for metaheuristics. In: Metaheuristics Optimization via Memory and Evolution, vol. 30, pp. 329–356. Springer, New York (2005)

Mohamed, A.W.: RDEL: Restart differential evolution algorithm with local search mutation for global numerical optimization. Egypt. Inform. J. **15**(3), 175–188 (2014)

Özdamar, L., Ekinci, E., Kucukyazici, B.: Emergency logistics planning in natural disasters. Ann. Oper. Res. **129**(1-4), 217–245 (2004)

PAHO: Humanitarian Supply Management and Logistics in the Health Sector. Pan American Health Organization, Washington D.C. (2001)

Parsopoulos, K.E., Konstantaras, I., Skouri, K.: Metaheuristic optimization for the single-item dynamic lot sizing problem with returns and remanufacturing. Comput. Ind. Eng. **83**, 307–315 (2015)

Parsopoulos, K.E., Vrahatis, M.N.: Particle Swarm Optimization and Intelligence: Advances and Applications. Information Science Publishing (IGI Global), Hershey (2010)

Peng, F., Tang, K., Chen, G., Yao, X.: Population-based algorithm portfolios for numerical optimization. IEEE Trans. Evol. Comput. **14**(5), 782–800 (2010)

Peng, M., Chen, H.: System dynamics analysis for the impact of dynamic transport and information delay to disaster relief supplies. In: 2011 International Conference on Management Science and Engineering (ICMSE 2011), pp. 93–98 (2011)

Piperagkas, G.S., Konstantaras, I., Skouri, K., Parsopoulos, K.E.: Solving the stochastic dynamic lot-sizing problem through nature-inspired heuristics. Comput. Oper. Res. **39**(7), 1555–1565 (2012)

Price, K., Storn, R.M., Lampinen, J.A.: Differential Evolution: A Practical Approach to Global Optimization. Springer, New York (2005)

Sheu, J., Chen, Y., Lan, L.: A novel model for quick response to disaster relief distribution. In: Proceedings of the Eastern Asia Society for Transportation Studies, vol. 5, pp. 2454–2462 (2005)

Souravlias, D., Parsopoulos, K.E., Alba, E.: Parallel algorithm portfolio with market trading-based time allocation. In: International Conference on Operations Research 2014 (OR2014). Aachen, Germany (2014)

Souravlias, D., Parsopoulos, K.E., Kotsireas, I.S.: Circulant weighing matrices: A demanding challenge for parallel optimization metaheuristics. Optim. Lett. **10**(6), 1303–1314 (2015)

Storn, R., Price, K.: Differential evolution - a simple and efficient heuristic for global optimization over continuous spaces. J. Global Optim. **11**(4), 341–359 (1997)

Tang, K., Peng, F., Chen, G., Yao, X.: Population-based algorithm portfolios with automated constituent algorithms selection. Inform. Sci. **279**, 94–104 (2014)

Tatham, P., Kovacs, G.: The application of swift trust to humanitarian logistics. Int. J. Prod. Econ. **126**(1), 35–45 (2010)

Taylor, D., Pettit, S.: A consideration of the relevance of lean supply chain concepts for humanitarian aid provision. Int. J. Serv. Technol. Manag. **12**(4), 430–444 (2009)

Thomas, A., Kopczak, L.: From Logistics to Supply Chain Management - The Path Forward to the Humanitarian Sector. Fritz Institute, San Francisco (2005)

Van Hentenryck, P., Bent, R., Coffrin, C.: Strategic planning for disaster recovery with stochastic last mile distribution. In: Integration of AI and OR Techniques in Constraint Programming for Combinatorial Optimization Problems, Lecture Notes in Computer Science, vol. 6140, pp. 318–333. Springer, Berlin-Heidelberg (2010)

Van Wassenhove, L.N.: Humanitarian aid logistics: Supply chain management in high gear. J. Oper. Res. Soc. **57**(5), 475–489 (2006)

Vitoriano, B., Ortuño, M., Tirado, G., Montero, J.: A multi-criteria optimization model for humanitarian aid distribution. J. Glob. Optim. **51**(2), 189–208 (2011)

Yan, S., Shih, Y.L.: An ant colony system-based hybrid algorithm for an emergency roadway repair time-space network flow problem. Transportmetrica **8**(5), 361–386 (2012)

Yi, W., Kumar, A.: Ant colony optimization for disaster relief operations. Transp. Res. E Logist. Transp. Rev. **43**(6), 660–672 (2007)

Yi, W., Özdamar, L.: A dynamic logistics coordination model for evacuation and support in disaster response activities. Eur. J. Oper. Res. **179**(3), 1177–1193 (2007)

Yuan, Y., Wang, D.: Path selection model and algorithm for emergency logistics management. Comput. Ind. Eng. **56**(3), 1081–1094 (2009)

Zhang, J.H., Li, J., Liu, Z.P.: Multiple-resource and multiple-depot emergency response problem considering secondary disasters. Expert Syst. Appl. **39**(12), 11,066–11,071 (2012)

Zheng, Y.J., Ling, H.F., Xue, J.Y., Chen, S.Y.: Population classification in fire evacuation: A multiobjective particle swarm optimization approach. IEEE Trans. Evol. Comput. **18**(1), 70–81 (2014)

Tsunami of the Meteoric Origin

Andrey Kozelkov and Efim Pelinovsky

Abstract Approaches to modeling a tsunami of meteoric origin are discussed. A brief overview of the asteroid and meteorite danger to the Earth is given. Formulas assessing the parameters of the tsunami caused by an asteroid entering the water are derived. The results of the numerical simulation of the effect of the angle of entry of the body into water on the characteristics of the resulting waves in the near field are given. The model based on the Navier–Stokes equations for multiphase flows with a free surface is used in calculations. The dimensions of perturbation are studied and the regularities of changes in the parameters of the source are discovered.

Keywords Natural hazards • Asteroid damage • Tsunami • Physical models

1 Introduction

At the turn of the twentieth and twenty-first centuries there was a substantial reassessment of the threat of falling of small solar system bodies to the Earth. This was due to the accumulation of serious fundamental knowledge of physical and dynamic evolution of the solar system as a whole. As the systematization of knowledge grew, the subject of the interaction of the planet with large celestial bodies has started to attract more and more attention. Currently, this topic is paid special attention to, and the fall of the meteorite in the Chelyabinsk region (Russia) on February 15, 2013, further fueled this interest.

A. Kozelkov
Russian Federal Nuclear Center, All-Russian Research Institute of Experimental Physics, Sarov, Russia

Alekseev Nizhny Novgorod State Technical University, Nizhny Novgorod, Russia

E. Pelinovsky (✉)
Alekseev Nizhny Novgorod State Technical University, Nizhny Novgorod, Russia

Institute of Applied Physics, Nizhny Novgorod, Russia

National Research University - Higher School of Economics, Moscow, Russia
e-mail: pelinovsky@gmail.com

I.S. Kotsireas et al. (eds.), *Dynamics of Disasters—Key Concepts, Models, Algorithms, and Insights*, Springer Proceedings in Mathematics & Statistics 185,
DOI 10.1007/978-3-319-43709-5_8

Increasingly, information about the rapprochement of celestial bodies in size from a few tens of meters to kilometers with the Earth began to appear. Many of these objects fly in close proximity to the Earth at a distance comparable with the distance to the Moon. In this regard, forecasts of asteroids dangerously approaching the Earth began to be worked out for the coming decades and even centuries, and in science the term symbolizing the threat—the asteroid–comet hazard (ACH) is firmly entrenched.

The collision with the Earth of large celestial bodies with a diameter of a few kilometers is quite a rare event, but the clashes with small and medium-sized bodies repeatedly occurred in the past. Our planet has repeatedly collided with celestial bodies, and the number of proved impact craters as far back as 1978 was more than a hundred with the expected redoubling every 5.8 years. To date, the number of craters is about two hundred (Kozelkov et al. 2015a; Earth Impact Database), although, most likely, this figure can easily double due to the inconsistency of some data (Chapman 2004).

Over the past 15 years, humanity has seriously come up to the study of ACH by performing a comprehensive search and analysis of potentially hazardous near-Earth objects (NEO), various programs of studying the latter led to the discovery of about ten thousand near-Earth asteroids (Near-Earth Object Program). Most of these asteroids are capable of crossing the Earth orbit (Earth-crossing asteroids), but only a small part of all of these objects is systematically studied and cataloged.

The so-called long-period comets (long-period comets) can approach a potentially hazardous orbit of the collision with the Earth just two months prior to its discovery (Jack et al. 1999). Currently, more than 800 asteroids of a larger diameter (Fig. 1) whose orbits can intersect with the Earth are discovered. Every year 2–3 spans of bodies of 100–1000 m in diameter at a distance of 0.5–3 mln km from the Earth are recorded, and often such a span is detected after the closest approach.

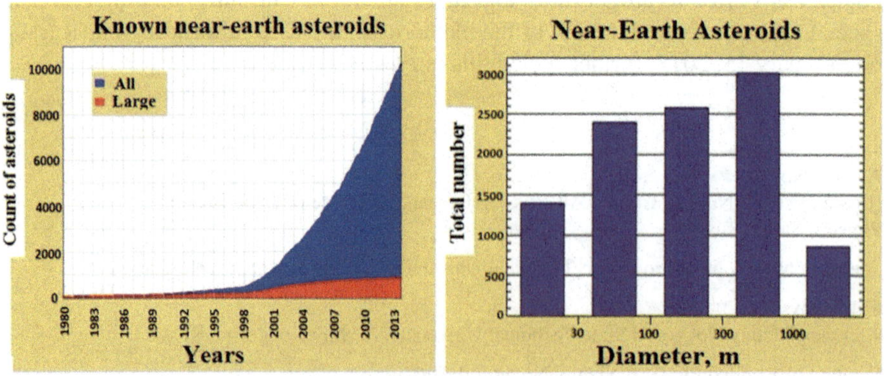

Fig. 1 Statistics of detection of near-Earth celestial bodies (*left*) and their diameter distribution (*right*) (Near-Earth Object Program)

In general, in the range of 0.2 AU (astronomical units) from the Earth rotate more than 7000 comets and asteroids that can potentially collide with our planet (Near-Earth Object Program; Shustov and Rykhlova 2010), with 90 % of them having the diameter of more than a kilometer (Fig. 1), only 7 % of which are cataloged (Jack et al. 1999). Moreover, according to available data, the Earth's orbit is crossed by a few dozen active comets and hundreds of extinct comets with core diameter up to 1 km.

It is assumed that such comets have left their mark on the Earth surface in the form of craters, the amount of which exceeds 20 % of the total. Collisions with asteroids are seen as the most probable, as periodic comets are responsible only for 0.4 % of dangerous approaches (Shustov and Rykhlova 2010). Active comets account for most of the approaches; however, the number of extinct comets which by their supervisory properties do not differ from asteroids can be considerably larger.

Humanity is not prepared to meet such "space aliens" and this is shown in Shustov and Rykhlova (2010), where is given the example of the discovery of the "C/1983 H1" comet, two weeks later flying around the Earth at a distance of only 0.0312 AU. Now the utmost attention is focused on the asteroid "Apophis" with a diameter of 300 m discovered in 2004. In 2029 "Apophis" will be flying at a distance very close to the Earth—only 40,000 km, and after its return in 2036; it has more than a non-zero probability of collision with our planet.

The main danger on a global scale is represented by bodies of more than 1 km in diameter, while the major continental or regional destruction can be caused by bodies of a much smaller diameter (Hills and Goda 1998). The first documented catastrophe on a regional scale is the Tunguska meteorite, after the fall of which, the forest was knocked over an area of 2000 km^2 (Fig. 2). The consequences of collisions of different-sized asteroids with the Earth are discussed in Shustov and Rykhlova (2010) and Adushkin and Nemchinov (2008).

Fig. 2 Forest fall at the crash site of the Tunguska meteorite (*left*—right after the fall, *right*—at the moment)

Explosions of large meteorites can significantly affect the climate and the structure of the Earth's crust not only in the mainland, but also globally. The fall of an asteroid into the ocean will trigger the release of millions of tons of water vapor into the atmosphere, and will inevitably cause a huge tsunami, the effects of the collapse of which on land are difficult to predict. All this makes the study of the problem of falling asteroids by far the most urgent problem of modern science.

Naturally, the bodies of a smaller size have a higher probability of collision with the Earth than large or gigantic asteroids. The approximate probability of falling asteroids of various sizes and their characteristics are presented in Hills and Goda (1998) and Paine (1999). ACH for the existing civilization involves the study of all celestial bodies with dimensions less than the threshold of a global catastrophe (average diameter of 1–2 km) and more than the threshold of protection by the Earth's atmosphere (average diameter of 50–100 m) (Morrison 2005).

The bodies of a smaller diameter will significantly slow at collapse at the height as a result of exposure to atmospheric friction and internal voltage shift. Regarding the height, the shock wave will only insignificantly damage the Earth's surface. Upon the contact with the Earth the main role is played by the mass, speed, and angle of the meteorite trajectory; the composition, density, and shape of a meteorite are of secondary importance. Meteorite classification according to mass, velocity, and trajectory angle are presented in Shustov and Rykhlova (2010).

The estimation of the probability of collisions with meteorites is based on the analysis of known craters on the Earth and determination of the time of their occurrence by isotopic methods (Benest and Froeschele 1998). As estimated, every 100 million years on the Earth are formed from 500 to 1300 craters of 1 km or more in diameter, that is, the Earth collides with huge meteorites every 100–200 thousand years on average. Collisions with large asteroids on a scale of several generations of humankind occur infrequently, but in its history, the planet of Earth has collided with several thousands of meteorites with the diameter of 1 km and dozens of meteorites with the diameter of 10 km. Clashes with meteorites are random and unpredictable, and a contact with a meteorite of any size can occur at any time. The brief history of meteorite falls and possible problems caused by asteroid collisions with the planet Earth is described in Chapman (2004), Shustov and Rykhlova (2010), Adushkin and Nemchinov (2008), and Bobrowsky and Rickman (2007).

ACH is recognized as one of the most important of the existing environmental problems, requiring constant attention, assessment of the degree of real threat, and the development of possible measures to counter and mitigate the effects of a collision. In many countries there are programs of discovery, cataloging, and classification of potentially dangerous space objects (Shustov and Rykhlova 2010). The constant monitoring of NEO with telescope robots in specialized astronomical observatories is being carried out.

Systematizing and synthesis of the gained knowledge allow to point out a number of problems, the study of which is important for a better understanding of the physics of individual processes and the development of computer technology, the use of which may be aimed at predicting the consequences of asteroid–comet interaction.

Computing technologies for this perspective are, essentially, the only tool to understand the processes of collision, dynamics, and cratering on a planetary scale. Many processes of hypervelocity impacts on the global scale such as melting, radiation, evaporation, and others are impossible to reproduce in laboratory on the Earth. The development of existing and construction of new physical and mathematical models to realistically simulate a collision of celestial bodies are a sufficiently relevant and complex issue of modern mathematical physics. The level of the development of computer technology for a full-scale simulation of the collision process is at a very early stage. It is due to the fact that the numerical calculations of two-dimensional and even three-dimensional physical and mathematical problems of the passage of a collapsing body through the atmosphere of planets and their impacts on the surface are very laborious and, consequently, can be reduced only to descriptions of certain stages and separate processes. The attempts to build approximate models look natural. One of such models carried to the user program was created in the University of Arizona (Collins et al. 2005).

One of the little-studied stages of the asteroid impact on the Earth's surface is the fall of a celestial body in the World's ocean and the formation of a tsunami wave. These waves are called "cosmogenic" tsunami. Taking into account the fact that the water surface is about two thirds of the Earth, the probability of an asteroid falling into the ocean is much higher than that on land. However, so far only 15–20 craters formed by the impact into the sea were found (Shuvalov and Trubestkaya 2002), which can be explained by the peculiarities of the underwater crater formation.

The description of a "cosmogenic" tsunami is quite difficult due to the mutual influence of a few steps (Korycansky and Lynett 2007), from which we can single out the main five:

1. The movement and characterization of the movement of an asteroid in the atmosphere. At this stage, in addition to assessing the likelihood of such an event, as well as determining the size, shape, speed, density of a meteorite, a series of complex processes such as rotation, thermochemistry, loss of velocity and mass, and others, which ultimately affect the final speed of a collision, should be taken into account.
2. The formation of a crater on the water surface. The process of interaction of an asteroid with water is strongly nonlinear. The formation of a crater is accompanied by a number of specific processes, such as evaporation, the release of water into the atmosphere, the collision with water of several fragments which leads to the formation of groups of foci, and others. At this stage it is necessary to define the relationship between the characteristics of a meteorite, bathymetry of the seabed, and the parameters perturbation source.
3. The motion of the asteroid in the deep of the ocean. A large enough asteroid on falling into the ocean is able to pass through its thickness freely and form a crater at the bottom. At this stage, it is necessary to determine the impact of the collapse of the gas bubble which is formed as an asteroid is passing through the water column, on the parameters of the tsunami source.

4. The destruction of the water of the crater, its impact on the formation of the initial wave, and the initial stage of spread near the source of the disturbance.
5. The propagation of a "cosmogenic" tsunami and its movement on the land. The description of this stage can be based on the nonlinear theory of shallow water which is well developed for the tsunami of seismic origin (Pelinovsky 1996; Levin and Nosov 2016). The properties of a "cosmogenic" tsunami are likely to be comparable to the seismic ones.

From the point of view of the mathematical modeling of a "cosmogenic" tsunami, the stage of the impact on the surface and formation of a cavity with its subsequent evolution, the formation of underwater bubbles, and their influence on the formation of the source is highly problematic. Experimental studies and modeling of the formation of a cavity, for the most part, are referred to solids or viscoelastic fluids, between which and the liquid only a slight analogy can be made, as there are significant differences.

2 Models to Describe "Cosmogenic" Tsunamis

Research in the area of a body falling into the water can be divided into two directions. The first of them is based on the theory of explosion, and the second is based on the mechanics of incompressible fluid. The use of incompressible fluid mechanics is made necessary because the pressure developed by the explosion is so great, that the strength and elastic–plastic properties of the medium in most cases, as well as frictional forces as compared to inertial forces can be ignored.

Moreover, if we ignore the compressibility of the medium, then we get a model of incompressible fluid in which, generally, it is possible to take into account the friction force by introducing viscosity. Using the methods of hydrodynamics it is possible to describe the phenomenon in general, later specifying it by adding the properties of the real environment.

When a meteorite falls into the water we can observe a number of specific effects, such as the movement of a meteorite in an aqueous medium, breaking waves, emission of water in the atmosphere, vaporization, and others, which will require the complication of the general model. Furthermore, unlike in the case of a hard surface, the impact of a body against a water surface a significant portion of it (between 25 and 100 % depending on the speed and the trajectory angle) remains unmelted. All these processes are a major obstacle for many methods of theoretical analysis.

Besides the waves generated directly by the impact against the surface, waves are also formed, extending outwardly from the crater due to its collapse, and filling with water. Some processes regarded in the model falling onto a hard surface can be ignored when a meteorite falls into the water. Such processes include dispersion and fragmentation of the meteorite, which generally will not affect the formation of a common wave pattern, unless the fragments are not large enough.

The brightest and most important process in the hydrosphere worth studying is a tsunami wave generated by the fall of a meteorite. The kinetic energy of a meteorite with the diameter of 100 m, falling at a speed of 20 km/s makes about 3×10^{17} J, which corresponds to the energy of a very strong tsunami of seismic origin. The energy of an asteroid with the diameter of 1 km is three orders of magnitude, which exceeds several times the energy of the greatest earthquake of twentieth century—the 1960 Chilean earthquake.

The evaluation of the kinetic energy (in megatons of trinitrotoluene equivalent) emitted as a result of the collision can be accomplished by the following formula (Marusek 2007):

$$E = 6.256 \times 10^{-8} d^3 v^2 \rho \quad (1)$$

where d is the diameter of the asteroid, v is the speed of the asteroid relative to the Earth, ρ is the density of the asteroid. Based on this formula (Marusek 2007) are given the estimation tables of typical values of the kinetic energy of the collision for short- and long-period comets of different diameters. A meteorite of 5–6 km in diameter falling into the center of the Atlantic Ocean will generate a tsunami with the height of several hundred meters, capable of destroying Europe and North America (Jack et al. 1999). According to other estimates, an asteroid with the diameter of 1.5 km, a collision which may occur once in half a million years, is able to destroy the existing civilization.

Asteroids larger than the depth of the ocean are able to generate a shock wave near the point with amplitude comparable to that depth. Of course, the fall of a large meteorite able to cause a tsunami wave with the height of one hundred meters, it seems unlikely, or even negligible, at least for the present civilization. Nevertheless, the prediction of this phenomenon still seems vital, as they took place before (Kharif and Pelinovsky 2005) and can cause significant damage if they occur again. Kharif and Pelinovsky (2005) give a historical sketch of "cosmogenic" tsunamis that occurred on our planet in the past (Fig. 3, taken from Kharif and Pelinovsky 2005).

The study of "cosmogenic" tsunami occurrences, in general, creates a great number of problems to solve which the construction of a single physical–mathematical model is hardly possible. The general model should take into account the processes of hydrodynamics, aerodynamics, mechanics of stress-deformity state, the dynamics of multi-component media, or in a more complex case, multiphase media. For this reason, when modeling a "cosmogenic" tsunami a number of subtasks are singled out for which specific models are constructed. The most common of these problems include the formation of source, i.e., directly the fall of a meteorite, crater formation on the ocean floor, the release of water into the air, wave propagation, and others.

To model the propagation of a "cosmogenic" tsunami the most important step is the formation and the initial stage of evolution of the water crater (the tsunami source) on the ocean surface. Prior to the main wave propagation, at the stage of the crater formation, there can be defined three related hydrodynamic flows. The first flow is an air flow resulting from the ejection of gas from the body cavity created

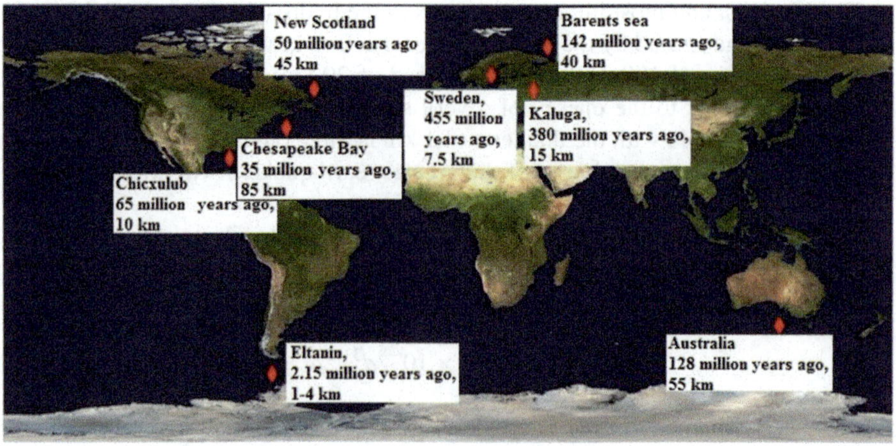

Fig. 3 Location of historical "cosmogenic" tsunamis

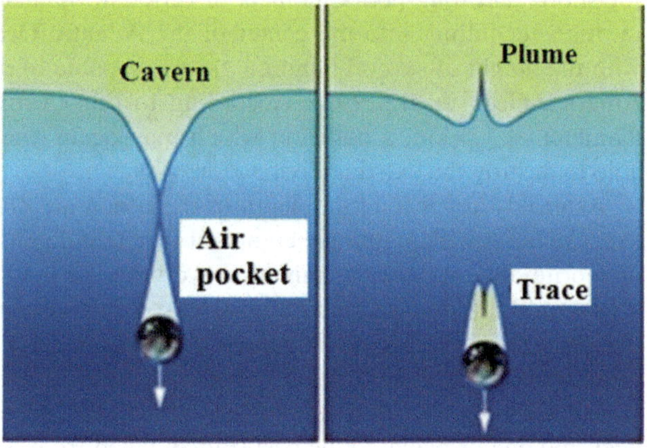

Fig. 4 The stages of a body falling into the water

by the fall. The rate of flow is essentially connected with the water flow (the second hydrodynamic flow), arising due to the collapse of walls of the crater cavity, which results in its moving upward.

The collapse of the walls of the water cavity results in forming a narrow neck, resembling a Laval nozzle (Fig. 4 from http://future-science.ru), the velocity of the air from which can reach enormous values. Experimental studies of gas flowing from the air cavity axisymmetric occurring upon the collision of the body with water are presented in Gekle et al. (2010). The rate of flow of air in the experiment, with the minimum diameter of the formed cavity of the "nozzle" has reached more than 300 m/s, which is comparable with the speed of sound in water.

The collapse of the air cavity can also be divided into two stages, which eventually will form a general picture on the surface of the water. In the first stage, after the final collapse of the "nozzle," a conical cavity is formed on the water surface (Fig. 4, left). Its further transformation due to the hydrostatic pressure and the air flow by that time liberated from the gas bubble in the form of a shock wave will result in the burst on the surface—"Sultan" (Fig. 4, right).

The second stage is the formation of a body trace—a pear-shaped gas bubble (Fig. 4, right) in which the body is moving. When lowering to the bottom, in the bubble there will be the increase in speeds, its diameter will quickly begin to rise up to a point, and the pressure will drop. The bubble breakthrough can occur either during the immersion of the body or on achieving the bottom. If this happens in the course of the immersion, the body forms a turbulent wake vortex with developed cavitation (the third hydrodynamic flow) (Fig. 4, right), which consists of a pair of large elongated vortices rotating in opposite directions. Subsequently, the body will continue to move to full immersion regardless of the bubble.

The mechanism of the evolution of the bubble in this case is not very clear but it is very likely to fit the description of the collapse of the bubble in the water, presented in Lavrentiev and Shabat (1973). According to this description, the bubble will flatten in its lower part, forming a characteristic recessed cap curved upwards, which subsequently will lead to a collapse (a collapse, apparently, may be the result of the process of separating the bladder from the body).

At the time of the collapse there occurs a water hammer, resulting in a stream having a cumulative character and directed vertically upwards. Generally speaking, this stream will also cause the jet of a second plume on the surface of the water, with its height not less than on the stage of the cavity transformation. There may be several "secondary" plumes on the surface due to the destruction of the main bubble into a number of smaller ones. This is true both for the small bodies of high density and of large bodies of low density.

The shape of the water crater prior to its transformation is likely to coincide with the shape of a similar crater on a hard surface, but the mechanism of its formation is quite different. The heights and shapes of the central and peripheral parts will differ. For the water crater, these heights will depend both on the size and shape of the asteroid, its speed and, which is much more important, the processes occurring in the water column.

All this, of course, will affect the height, shape, and speed of propagation of the tsunami wave from the source. Naturally, all these aspects must be considered in the fall of the body with the diameter much smaller than the depth of the ocean. If the size of the body is comparable with the depth of the ocean, it will break through the water column to the bottom, and the process of the collapse of the cavity will correspond to the process of transformation of the cone-shaped cavity on the surface as described above.

If a body falls in the deep ocean, a tsunami wave, according to calculations (Shuvalov and Trubetskaya 2007; Weiss et al. 2006; Shuvalov et al. 2012), will be formed as a result of the collapse of the intermediate water crater. With the diameter of an asteroid less than 0.5 km, the crater will not reach the bottom of the ocean. In case of a large asteroid reaching the bottom at a sufficient speed, the formation of the bottom structure may be accompanied by movement of its peripheral part and the structural bottom lift during the growth of the crater, which can also generate a

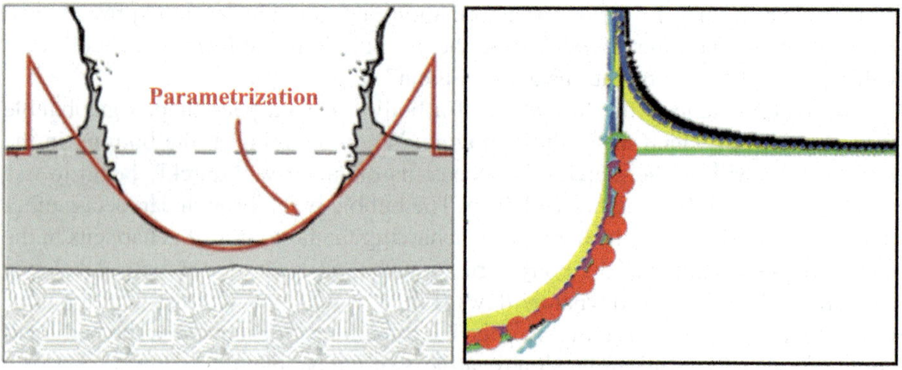

Fig. 5 Comparison of the results of modeling and a parameterized model (*left*) and numerical results with the experimental data (*red line*) (*right*)

tsunami. This process was observed in the formation of the crater Myolnir with the diameter of 40 km in the Barents Sea (Shuvalov and Trubestkaya 2002). This crater was formed 140 million years ago due to the fall of an asteroid with the diameter of 1–3 km into the sea of 300–500 m deep.

The evaluation of the bottoming asteroid and the possible formation of the bottom crater is given in Shuvalov et al. (2012). To determine the effect of the water column on the process crater formation on the bottom, we enter the parameters d/H (d is diameter of the asteroid and H is the depth of the sea). When $0.1 < d/H < 1$, the depth of the ocean significantly affects this process, the size, and the morphology of the crater. If the diameter of the asteroid is four times larger than the ocean depth, it will reach the bottom without any destruction or inhibition.

The first tsunami wave will propagate since the initial stage of the falling body—intermediate water crater (Fig. 5, left). It is this form of the source as the initial disturbance that is used in many papers on modeling the propagation of a "cosmogenic" tsunami (Levin and Nosov 2016; Kharif and Pelinovsky 2005). In this case, the stage of the fall and immersion of an asteroid is not considered. The shape of the cavity on the surface is given by the analytical formulas (Ward and Asphaug 2000).

According to Ward and Asphaug (2000), the analytical formula for the instantly formed asteroid cavity $u^{\mathrm{imp}}(\mathbf{r}_0)$ will lead to the vertical displacement of the water surface, which can be represented as:

$$u^{\mathrm{surf}}(\boldsymbol{r}, t) = \frac{\mathrm{Re}}{2\pi} \sum_{\mathrm{n}=-\infty}^{\infty} \int_0^{\infty} k\mathrm{d}k\mathrm{J_n}(kr)\mathrm{e}^{-i(\omega(k)t-n\theta)} \int_{\mathrm{r_0}} \mathrm{d}\boldsymbol{r}_0 \mathrm{u}^{\mathrm{imp}}(\boldsymbol{r}_0)\, J_{\mathrm{n}}(k\boldsymbol{r}_0)\, e^{-\mathrm{in}\theta_0} \quad (2)$$

where $r = |\boldsymbol{r}|$ is the radius vector of the point, $k = |\boldsymbol{k}|$ is wave number, $\omega(k)$ is frequency, h is the constant ocean depth, Re is the real part and $u^{\mathrm{imp}}(\boldsymbol{r}_0) \neq 0$, a $R = |\boldsymbol{r} - \boldsymbol{r}_0|$, $J_n(x)$ is the Bessel function. It is believed that the initial stage of a crater formation the meteorite creates a radially symmetrical cavity, which can be described by the function:

$$\begin{aligned} u^{\text{imp}}(\boldsymbol{r}) &= D_{\text{C}}\left(1 - r^2/R_{\text{C}}^2\right) r \le R_{\text{D}} \\ u^{\text{imp}}(\boldsymbol{r}) &= 0 r > R_{\text{D}} \end{aligned} \tag{3}$$

where D_c is the depth of the cavity, R_C and R_D are inner and outer radii of the cavity, respectively. In case of equal inner and outer radii of the cavity$R_C = R_{\text{D}}$, the surface is a cavity corresponding to the release of water and the atmosphere or its evaporation. In the case $R_{\text{D}} = \sqrt{2}R_C$, the water ejected from the cavity forms the outer splash—a ring structure characteristic of the fall of the object into the water, the volume of which corresponds exactly to the volume of water discharged from the cavity (Fig. 5).

Based on the evaluation of the potential energy of the initial disturbance, (Levin and Nosov 2016; Ward and Asphaug 2000) simple analytical formulas to calculate the radius and the depth of the cavity are defined:

$$D_C = \left(\frac{2\varepsilon\rho_I R_I^3 V_I^2}{\rho_w g R_C^2}\right)^{1/2}; \; R_C = R_I\left(2\varepsilon\frac{V_I^2}{gR_I}\right)^{\delta}\left(\frac{\rho_I}{\rho_w}\right)^{1/3}\left(\left(\frac{\rho_w}{\rho_I}\right)^{1/3-\delta}\left(\frac{1}{qR_I^{\alpha-1}}\right)^{2\delta}\right) \tag{4}$$

where ρ_w is density of water, g is the acceleration of gravity, ε is the proportion of the kinetic energy of the meteorite, turning into a tsunami energy, ρ_I, R_I, V_I are the density, radius, and the speed of the meteorite, and q and α are the factors associated with the properties of the meteorite and the aqueous layer.

The phase of the formation of the initial disturbance on the water surface in the studies related to the tsunami is the most problematic. The construction of an adequate model is a highly non-trivial task because of the complexity of the solution of the original equation, on the one hand, and the lack of experimental data, on the other. In the currently conducted experiments only internal geometric dimensions of the formed cavity are usually measured. The majority of these experiments deal with the barrier penetration and interaction of bodies at different speeds.

Experimental studies of the cavity formation on the surface of the water when a body falls at hypersonic speed are described in Pierazzo et al. (2008). The experiment evaluated the radius and depth of the cavity formed by the impact of a glass bead of 2 mm in diameter at a speed of 4.64 km/s with liquid. This rather complicated experiment used a high-speed centrifuge to create the gravitational field of several 100 g, in order to be able to reduce the linear dimensions of the experimental area accordingly, as compared with the real dimensions.

These studies were used to calibrate the numerical methods used to simulate the formation of the crater. The shape of the crater obtained numerically by using different numerical models, corresponds to the cavity shown in Fig. 5 (right), but the experimental data are presented only for the cavity shown in Fig. 5 (left). The use of these experiments is helpful on the stage of the calibration of numerical models. However, to explain the mechanism of formation of terrestrial craters the use of the data can be incorrect, since here strength craters of a small diameter are studied, and their morphology differs from large gravity craters (Shuvalov and Trubestkaya 2002).

It may be interesting to note that by using a numerical model that takes into account the properties of the atmosphere, the results of the calibration are substantially different from those in which the atmosphere is not taken into account, but not for the better. The numerical results obtained with the use of complex nonlinear models are consistent with the parameterization (3). The shape of the crater is described quite acceptably (Fig. 5). There is a mismatch, however, in the description of the outer ring structure (Fig. 5, left), consisting of short-wave component, which, are due to dispersion and dissipation, will not affect the wave pattern away from the source (Levin and Nosov 2016).

As follows from the above, the process of the fall of an asteroid into the ocean is followed by three stages, which may lead to the formation of a tsunami—the collapse of the intermediate water crater, the collapse of the underwater bubble generating a cumulative jet, and the formation of the bottom crater. As estimated, the formed tsunami wave can reach the height of several hundred meters or even several kilometers.

In the numerical simulation of forming an underwater structure "Eltanin" (Shuvalov and Trubetskaya 2007), the amplitude of a tsunami wave reaches more than a km in the place of the collision, 300 m at a distance of 70 km from the point of impact, and about 10 m off the coast of South America. To simulate tsunami propagation an ocean of constant depth is used. The analysis of the numerical calculations of the oblique fall of the asteroid (15° to the horizon) shows that tsunami waves are isotropic.

By now, not a single "cosmogenic" tsunami has been registered, which does not allow us to compare simulation results with the field data. The heights of the generated waves obtained numerically can be estimated by using empirical formulas deduced from the experimental data obtained in the course of underwater explosions (Shuvalov and Trubetskaya 2007):

$$h = 45\frac{H}{L}(Y)^{0.25} \tag{5}$$

where h is wave height, H is the depth of the ocean, L is the distance from the source, and Y is the collision energy of TNT. Evaluation of a tsunami wave height at different distances from the source in the simulation of an underwater structure "Eltanin" formation is in good agreement with the data obtained by the formula (5). However, this formula gives an adequate match only in the case of the generation of a tsunami wave as a result of the collapse of the intermediate water crater. If a tsunami is generated also by the advances of bottom that occur as a result of cratering, this formula gives a large error (up to several times).

In Marusek (2007) is given the table of "cosmogenic" tsunami wave heights, depending on the size of the asteroid, based on a formula derived by the Los Alamos National Laboratory to evaluate the tsunami height at a distance of 1000 km from the point of collision:

$$h = 1.0081 \times 10^{-5}\left(d^3 v^2 \rho\right)^{0.54} \tag{6}$$

where v is the speed of the asteroid relative to the Earth, ρ is the density of the asteroid.

The calibration of the numerical techniques designed to simulate the propagation of "cosmogenic" tsunamis can be carried out on some of the available experimental data. In Bukreev and Gusev (1996) were experimentally studied gravity waves at the free fall of the body into shallow water. The same experiments were carried out for the effects related to the finite-depth fluid to be more vividly manifested. In the above-mentioned experimental results, attention is drawn to the fact that during the vertical fall of a cubic body, in addition to the release of a vertical jet, a horizontal jet is also ejected out of the bulk of the liquid. The paper also evaluates the proportion of the energy of the initial disturbance, carried away at long distances. It is shown how difficult it is to describe gravitational waves theoretically, having only the data of the energy of a falling body, if it is large enough, and in the event of a celestial body it will be very large. Another conclusion drawn from these experiments is that in spite of the complexity of the processes near the body, a variety of gravity waves at large distances is small, i.e., in the propagation of waves of falling bodies with an unknown mass and velocity, limiting amplitude solitary waves can present the greatest danger. In this connection the issue of the possible number of solitary waves in the fall of an asteroid into the ocean becomes important. In the given experiment it was impossible to get more than two consecutive solitary waves, although in the experiment of offset portion of the basin bottom, a large number of solitary waves were observed.

The fall of an asteroid into the ocean will generate a wave that is different from the waves generated by an undersea earthquake. Tsunami waves of seismic origin are very long, their length is much greater than the depth of the ocean, they are propagated with very little loss of energy over long distances at a speed $c = \sqrt{gh}$. The carried out calculations show that the velocity of waves generated as a result of an asteroid fall is significantly less than $\sqrt{gh}$, while their length is twice less than the diameter of the intermediate aqueous crater (Gisler et al. 2011).

To generate a source of coherent waves (such waves can propagate over long distances without a significant loss of energy), the intermediate aqueous crater size should be 3–5 times greater than the ocean depth, i.e., for the ocean depth of 4 km the water crater should have the diameter of 20 km. On the basis of the empirical evaluation, the diameter of an asteroid should be 1 km (Paine 1999).

The numerical methods to simulate the propagation of a tsunami of seismic origin are based on the theory of shallow water (Pelinovsky 1996; Levin and Nosov 2016; Marusek 2007; Kharif and Pelinovsky 2005), in which pressure is strictly hydrostatic, dispersion properties of waves are not taken into account. In addition, in problems with large computational domains and complex structure of the bottom there will inevitably be areas where the shallow water approximation, strictly speaking, is not applicable because of the three-dimensional nature of the flow.

It is noted in Glimsdal et al. (2007) that the process of propagation of "cosmogenic" tsunamis when modeling the formation of the crater Myolnir in this approximation is described inadequately. In the fall of asteroids there are formed high and moderately long waves propagating under the influence of nonlinearity and dispersion. A wave generated by an asteroid fall can have amplitude of several km, which is much greater than the ocean depth, and its collapse can generate a wave train of the height comparable with that depth. Such waves are strongly nonlinear, and they rain down in the open ocean in areas of constant depth, when reaching a certain critical velocity of propagation they create packets of waves of shorter length. The effects of nonlinear propagation of such waves over long distances are difficult to compute and they can be very dangerous to the nearest coast. An interesting question arises, in this regard, as to what part of the wave energy and in what form is preserved after the collapse in the open ocean.

To describe this process it is necessary to use the full system of Navier–Stokes equations. However, the numerical integration of the system of the Navier–Stokes equations is very challenging. Currently, "cosmogenic" tsunamis are modeled by using the Euler equations, Boussinesq equations, using semi-empirical models, as well as shallow water equations taking into account the bottom friction. Different approaches to the modeling of a "cosmogenic" tsunami are considered in Badescu and Isvoranu (2011). It also analyzes the model of the formation of the source, the possibility of using the linear theory and the theory of shallow water for wave propagation. The comparative analysis of the results of "cosmogenic" tsunami simulation, previously obtained by other authors, often shows significant differences in some of the evaluations.

The strategy of numerical simulation of "cosmogenic" tsunamis, including all stages beginning with the formation of the initial disturbance on the surface up to the runup on land, is presented in Weiss et al. (2006). In these calculations, the depth of the ocean is assumed constant, and the runup is calculated for the model shelf with a certain angle. In the presented strategy, the formation of the initial wave, propagation, and runup were computed by using different programs.

The generated source of a tsunami wave is transmitted as a boundary condition in the program of calculation of its propagation, after which the runup was calculated by using a special program. The impact of a "cosmogenic" tsunami on the coast of the Black Sea is analyzed in Badescu and Isvoranu (2011). Here the analytical formula was used as a source, and the propagation was modeled by using two-dimensional hyperbolic equations of shallow water. The calculations use the real bathymetry of the Black Sea.

Simultaneous modeling of the stage of the initial disturbance and a tsunami waves propagation, on the example of the known historical underwater formations is considered in Shuvalov and Trubestkaya (2002), Shuvalov and Trubetskaya (2007), and Shuvalov et al. (2012). Here the Euler equations are used and the depth of the ocean is considered to be constant. In these papers a greater attention is paid to the crater formation at the bottom of the ocean, but not to tsunami waves. It is emphasized that the depth of the ocean has a significant impact on the processes that take place at the fall of an asteroid.

The simulation of "cosmogenic" tsunami propagation and their potential impact, including the assessment of the economic damage to the Pacific coast, was carried out in Yabushita (1994). Here the source was specified as an analytical formula. According to the carried out calculations the maximum height of the tsunami on the coast of Japan will turn out to be more than 60 m. The possible consequences for Europe and America after the fall of an asteroid with the diameter of 5 km in the middle of the Atlantic Ocean are discussed in Jack et al. (1999) and Yabushita (1994). According to the given forecast, the tsunami will flood two-thirds of the upper-part of the Eastern United States to the foothills of the Appalachian Mountains. The greatest impact in Europe will affect Portugal.

The complete description of the impact of an asteroid on the surface of the ocean is only possible when using a full-scale physical and mathematical model that takes into account, among other things, multiphase flows. In the study of "cosmogenic" tsunamis in the first phase it is possible to ignore the interphase interaction of the atmosphere and the ocean, restricting only to the role of density and pressure. To generate the source of "cosmogenic" tsunamis it is necessary to solve the complete system of Navier–Stokes equations considering the free surface. Below the implementation of the approach used in our research is described.

3 Simulation of a Meteorite in the Water

Let us consider the system of "air-water" as a set of two incompressible media separated by an interface. We are going to use the one-velocity approximation, in which the continuity equation and the equation of momentum conservation are the same for both water and air, and that are solved for the resulting medium the properties of which are linearly dependent on the volume fraction. This approach is quite common, and gives good results in solving problems with a free surface, including tsunami waves (Horrillo et al. 2013; Kozelkov et al. 2015b).

In this approximation the motion of the given system is described by the Navier–Stokes equations, including the equations of continuity, momentum conservation, and the equation for the phase volume fractions:

$$\begin{cases} \nabla \cdot \mathbf{u} = 0 \\ \frac{\partial}{\partial t}\sum\limits_k \alpha^{(k)}\rho^{(k)}\mathbf{u} = -\nabla \cdot \sum\limits_k \left(\alpha^{(k)}\rho^{(k)}\mathbf{u}\mathbf{u}\right) + \nabla \cdot \sum\limits_k \left(\alpha^{(k)}\mu^{(k)}\nabla \mathbf{u}\right) \\ -\nabla p + \sum\limits_k \alpha^{(k)}\rho^{(k)}\mathbf{g} \\ \frac{\partial}{\partial t}\alpha^{(k)}\rho^{(k)} + \nabla \cdot \left(\alpha^{(k)}\rho^{(k)}\mathbf{u}\right) = 0 \end{cases} \tag{7}$$

where $\mathbf{u}$ is a three-dimensional velocity vector, $\rho^{(k)}$ is density phase k, and $\alpha^{(k)}$ is its volume fraction ($\sum_k \alpha_k = 1$), p is pressure, $\mu^{(k)}$ is molecular viscosity of the phase k, $\mathbf{g}$ is the acceleration of gravity.

This system is solved without the use of the Reynolds averaging and the subsequent closure of equations for the moments of turbulence, i.e., turbulence is derived by the direct numerical simulation. This allows solving turbulent structures, the minimum scope of which is determined by the grid resolution. For grid models used in the calculations only enough large vortices are solved due to the fact that the purpose of the calculations is to analyze the parameters of the source prior to the collapse of the formed cavity and the formation of small vortices. In this formulation of the problem, such an approach can be considered fair. It should be noted here that, as only large vortex structures are derived, rather strict requirements for the scheme viscosity and dissipative numerical schemes to be used for the simulation of turbulent flows (Kozelkov et al. 2014, 2015c; Kozelkov and Kurulin 2015; Kozelkov et al. 2015d) are weakened.

For the numerical solution the resulting system of equations must be supplemented by initial and boundary conditions. On solid walls (e.g., the bottom of the ocean), the gradient of pressure and volume fractions as well as the speed are equal to zero:

$$\frac{\partial p}{\partial n} = 0, \quad \frac{\partial \alpha_k}{\partial n} = 0, \quad \mathbf{u} = 0 \tag{8}$$

On the "free" boundaries (the upper limit of the air layer) static pressure is fixed, the gradients of pressure and volume fractions are equal to zero:

$$\frac{\partial u}{\partial n} = 0, \quad \frac{\partial v}{\partial n} = 0, \quad \frac{\partial w}{\partial n} = 0, \quad \frac{\partial \alpha_k}{\partial n} = 0 \tag{9}$$

The body movement is modeled by a single phase by using the Immersed Boundary Method (Mittal and Iaccarino 2005). The model is implemented in the software package, LOGOS—a software product designed to solve three-dimensional problems of convective heat and mass transfer, aerodynamics, and hydrodynamics on parallel computers (Kozelkov et al. 2013; Betelin et al. 2014; Deryugin et al. 2015). Parallel implementation of the model is based on an algebraic multigrid method (Kozelkov et al. 2013), which allows using hundreds of cores and improving the convergence of the numerical method significantly.

The LOGOS software package is aimed at solving problems of computational fluid dynamics on arbitrary unstructured grids. The LOGOS software package has been successfully verified, and showed quite good results on a series of different hydrodynamic problems, including calculations of turbulent and unsteady flows (Kozelkov et al. 2014, 2015c; Kozelkov and Kurulin 2015; Kozelkov et al. 2015d). A separate study published in Kozelkov et al. (2015b, e) was devoted to the validation of the LOGOS software package to compute free surface problems shows good agreement with the experimental data and suggests that numerical schemes implemented to process the phase boundary are most suitable for calculations with the use of arbitrary unstructured grids.

Let us consider the fall of a solid spherical body with the diameter of 1 m into the water at different speeds of the fall and angles of attack. The density of the body is taken to be 3.3 g/cm^3, the parameters of the water and air are standard. The settlement area $160 \times 160 \times 50$ m^3 is an unstructured three-dimensional grid consisting of truncated polyhedra of arbitrary shape. The undisturbed water depth is 10 m. The height of the undisturbed air flow is 40 m. The calculations assume the initial velocity of the body entering the water varying in the range 10–200 m/s, the entrance angle varied from 10 to 90° with respect to the water surface.

For correct simulation of the body motion in the water was conducted the condensation of the computational grid in the cylindrical region to the typical cell size of 10 cm so that the diameter of the meteorite had at least 10 cells. This number of cells is optimal. The additionally conducted calculations with a larger number of cells had no effect on the result. This area is located with a slope corresponding to the angle of incidence of the body (the lower part of Fig. 6 for angle 20^0). The flight of the body in the air is not modeled, body movement starts from the surface of the water, and when it reaches the bottom, the body is automatically stopped by the application of a special boundary condition, which does not mean a complete stop of the body—on reaching the bottom the body can continue moving, i.e., it can rebound and fall not far from the original crash site.

As an illustration, Fig. 6 shows the pattern of change of the water surface in the modeling of a body falling at an angle of 20° (the body is flying from left to right). The pictures of the fall at other angles have a very similar configuration; the mapping for this angle is sufficient to understand the patterns of change in the parameters of the cavity. The detailed analysis of the vertical fall of the body into the water in this model is given in Kozelkov et al. (2015a).

The cavity formed when the body falls at an angle of 20° has an asymmetric structure with a splash stand out during the fall of the body—the front splash. The raising of the surface ("rear" splash) practically is not observed after the body at low speeds (about 10 m/s), while in the course of falling the water level offset reaches 2–4 m. The formed cavity has a tapered structure, and its width near the moment of collapse is about 8 m.

The height of the main emission increases with the increase in the speed of the fall. The height of the emission after the body and simultaneously the size of the cavity also increase. With increasing velocity the amplitudes of the front and rear splashes differ in 4–5 times. The front splash is crescent-shaped and, therefore, it will crumble both on the inner side of the cavity and on the exterior one during the fall of the body.

The increasing in the angle of incidence reduces the amplitude of the front splash and increase the amplitude of the rear one, along with it the size of the full cavity increases. The increase in the angle of incidence dramatically changes the shape of the front wave, which from the crescent-shaped changes into an upward vertical splash, and this pattern is enhanced by the increase in speed. The collapse this splash

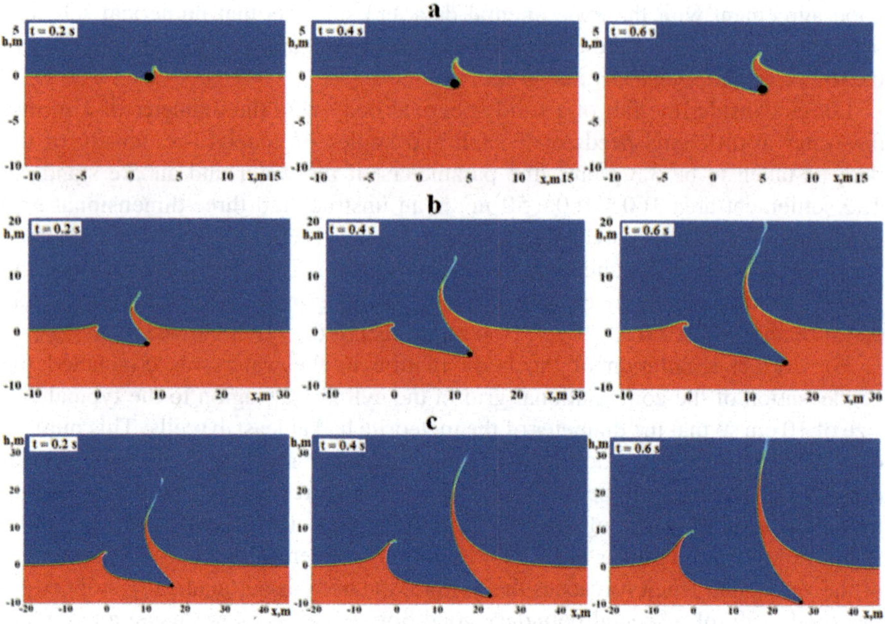

Fig. 6 The fall of the body at an angle of 20° at a speed of (**a**—10 m/s, **b**—60 m/s, **c**—200 m/s, the body—the *black circle*, the *red phase*—water, *blue*—air)

will occur according to the scheme of a body falling vertically down with the slant on the inside of the cavity, practically leaving out the outer side of its body during the fall. In the interior part of the cavity the rear splash will also crumble. Under these conditions of the fall the formation of the inner part of the cavity will continue even during the collapse of the front and rear splash and, eventually, it will lead to the formation of a large air cavity in the water column. As the angle of incidence increases, this pattern will continue. With a vertical drop the cavity will have a symmetrical structure (this case is studied in detail in Kozelkov et al. 2015a). The analysis shows that at the approach to the vertical direction the earlier collapse of the cavity, as compared with a smaller angle of incidence, should be expected. In this case the water displacement is of a much smaller amplitude (may vary at times) than the main emission of water at an angle of incidence.

It is possible to evaluate the pattern of the change in the cavity geometry using the following parameters as shown in Fig. 7: R_{in}– the internal radius of the cavity, R_{out}– the outer radius of the cavity, H_{in}– the cavity depth of the recess, and H_{out}– height of the cavity. It should be noted here that it is not possible to determine the values of these parameters accurately enough due to the blurring boundaries, so the approximate (estimated) data are given.

Figure 8 shows the normalized changes in the parameters for different speeds and angles of incidence. The change of the cavity parameters with the increase in the speed of the fall occurs more sharply, and the most significant change is observed at low angles of incidence.

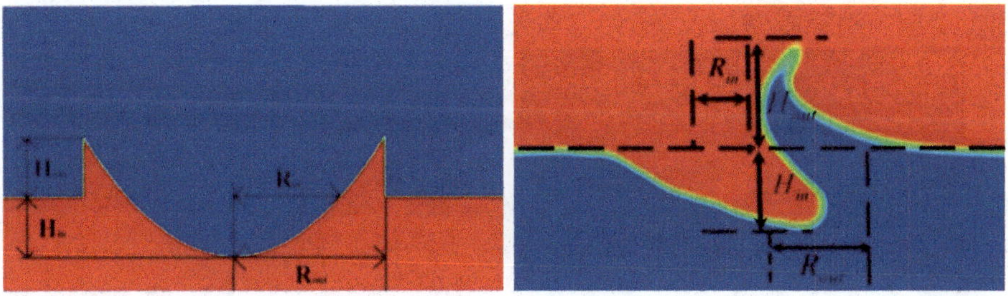

Fig. 7 The basic parameters of the cavity (*left*—symmetric cavity, *right*—oblique cavity formed by the fall of the body at an angle)

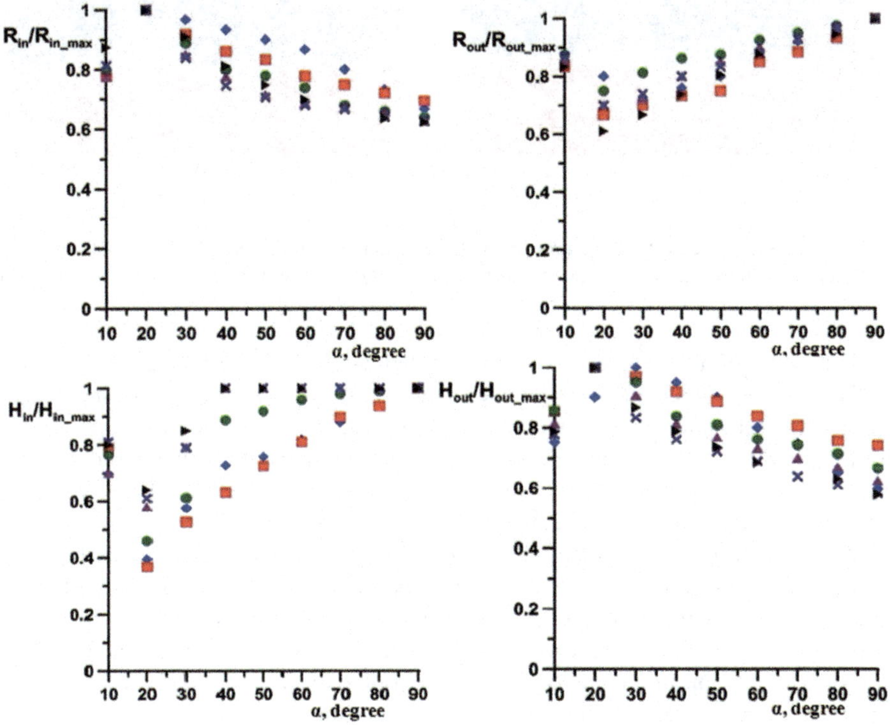

Fig 8 Normalized values of the cavity parameters (*rightpointing triangle*—200 m/s, *cross*—150 m/s, *filled triangle*—100 m/s, *filled circle*—60 m/s, *filled square*—30 m/s, *filled diamond* 10 m/s)

As can be seen, the change in the parameters relative to the vertical incidence is observed before and after 20°. The graph shows that the vicinity of this angle is a turning point in the parameter changes—the values R_{in} and H_{out} before this value of the angle show growth, then they begin to diminish, as for values R_{out} and H_{in}, they behave exactly the opposite—before this value they decrease, and after that they show growth.

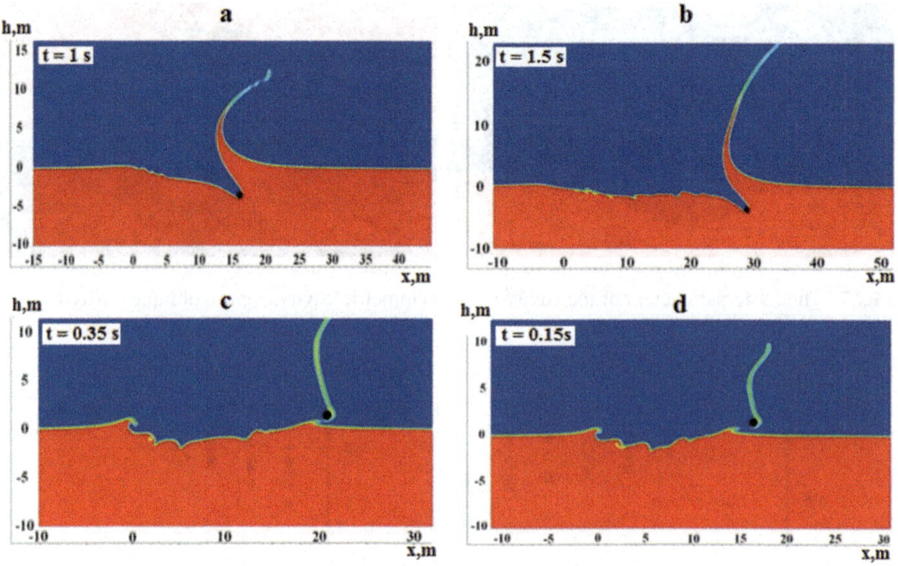

Fig. 9 The fall of the body at an angle of 5° at different velocities (**a**—30 m/s, **b**—60 m/s, **c**—100 m/s, **d**—150 m/s, the body—the *black circle*, the *red phase*—water, *blue*—air)

The values of the parameters of the cavity below 20° are out of the general picture of the changes. This is likely to be due to the fact that at small angles of incidence, there is no formation of a pronounced cavity, and the body for some time slides over the surface of the water (Fig. 9). Slightly plunging, the body then rapidly sinks. Such modes of incidence depend on many factors such as weight, shape, velocity, density, etc., and require a separate study.

With certain properties of the body composition and parameters of its flight, it may simply bounce off the surface of the water and can do this several times (as we know, when throwing a pebble into the water at certain angles a "frog" is formed, jumping on the surface and leaving characteristic circles). In Truscott et al. (2014) it is shown that the optimum angle of incidence for the body to rebound off the surface (or, rather, for the greatest number of bounces) is exactly 10–20°.

In this case, the body begins to bounce off the water surface at a speed of 50–60 m/s for the angle of incidence of 5°, wherein it is seen that no cavity is observed in this case (Fig. 9b). The cavity begins to break down due to the perturbations generated by sliding of the body over the surface of water keeping an impressive length. With the increase in speed, the rebound occurs earlier.

If the body begins to rebound, the release of water immediately becomes a stream of drops (Fig. 9c, d) which does not have a clearly defined structure and orientation. If the body is not going to bounce, the emission has a characteristic "arc" shape with the height of several times larger than the size of the body (Fig. 9a, b).

It should be noted here that for angles of incidence up to 10° it is possible to choose the velocity at which the body will rebound (with the parameters studied in this paper), but a further increase in the angle does not guarantee a rebound at the given density and shape of the body.

The change in the cavity parameters after 20° follows the quasi-linear law. This trend continues until the vertical drop and has the same character with a significant increase in speed. As can be seen from the graphs, the regularity of the parameter changes is approximately the same at different speeds. A more drastic change in the parameters at high speeds also indicates that for such modes of incidence it is necessary to consider the change of the thermodynamic properties of the liquid and the gas (in particular, the compressibility of the air should be considered, and at very high speeds the compressibility of water too).

The graphs show that the change in the parameters of the cavity at all angles and speeds tends to the same law of change after 200. Here the two parameters (R_{out}, H_{in}) show growth, and the other two (R_{in}, H_{out}) show decrease. In the graphs, however, the corresponding curves do not merge, so the dependence on the speed of the body entering the water remains, although relatively weak.

4 Conclusion

A brief review of the asteroid–meteorite hazard to the Earth is given. The mathematical model based on the Navier–Stokes equations, used in the study of the generation of disturbances in the water and on the surface, is described. It is shown that the change in the parameters of the cavity occurs most rapidly at angles of a body falling into the water of more than 20° and is subject to a quasi-linear law.

The intensity of change increases with the increase of speed, the trend of a line change is at that preserved. The fall of a body into the water at angles less than 20° takes a different scenario, and under certain conditions the body bounces off the surface of the water, and the area of the disturbance has very blurred boundaries.

Acknowledgement These results are obtained with the support of State Contract № 2014/133 and the RFBR grants (14-05-00092, 15-45-02061, 16-01-00267).

References

Adushkin, V.V., Nemchinov, I.V.: Catastrophic Events Caused by Cosmic Objects. Springer (2008)

Badescu, V., Isvoranu, D.: Dynamics and coastal effects of tsunamis generated by asteroids impacting the black sea. Pure Appl. Geophys. **168**, 1813–1834 (2011)

Benest, D., Froeschele, C. (eds.): Impacts on Earth. Lecture Notes in Physics, vol. 505. Springer, Heidelberg (1998)

Betelin, V.B., Shagaliev, R.M., Aksenov, S.V., Belyakov, I.M., Deryuguin, Y.N., Korchazhkin, D.A., Kozelkov, A.S., Nikitin, V.F., Sarazov, A.V., Zelenskiy, D.K.: Mathematical simulation of hydrogen–oxygen combustion in rocket engines using LOGOS code. Acta Astronaut. **96**, 53–64 (2014)

Bobrowsky, P.T., Rickman, H. (eds.): Comet/Asteroid Impacts and Human Society. An Interdisciplinary Approach, 546 pp. Springer, Berlin/Heidelberg (2007). doi:10.1007/978-3-540-32711-0. http://www.springer.com/us/book/9783540327097 [ISBN 978-3-540-32711-0]

Bukreev, V.I., Gusev, A.V.: Gravity waves at body fall into shallow water. Appl. Math. Tech. Phys. **37**, 90–98 (1996)

Chapman, C.R.: The hazard of near-Earth asteroid impacts on earth. Earth Planet. Sci. Lett. **222**, 1–15 (2004)

Collins, G.S., Melosh, H.J., Marcus, R.A.: Earth impact effects program: a web-based computer program for calculating the regional environmental consequences of a meteoroid impact on Earth. Meteorit. Planet. Sci. **40**, 817–840 (2005)

Deryugin, Y.N., Zhuchkov, R.N., Zelenskiy, D.K., Kozelkov, A.S., Sarazov, A.V., Kudimov, N.F., Lipnickiy, Y.M., Panasenko, A.V., Safronov, A.V.: Validation results for the LOGOS multifunction software package in solving problems of aerodynamics and gas dynamics for the lift-off and injection of launch vehicles. Math. Models Comput. Simulat. **7**, 144–153 (2015)

Earth Impact Database. http://www.passc.net/EarthImpactDatabase

Gekle, S., Peters, I.R., Gordillo, J.M., van der Meer, D., Lohse, D.: Supersonic air flow due to solid-liquid impact. Phys. Rev. Lett. **104**(2), 024501(1–4) (2010)

Gisler, G., Weaver, R., Gittings, M.: Calculations of asteroid impacts into deep and shallow water. Pure Appl. Geophys. **168**, 1187–1198 (2011)

Glimsdal, S., Pedersen, G.K., Langtangen, H.P., Shuvalov, V., Dypvik, H.: Tsunami generation and propagation from the Mjølnir asteroid impact. Meteorit. Planet. Sci. **42**, 1473–1493 (2007)

Hills, J.G., Goda, M.P.: Damage from the impacts of small asteroids. Planet. Space Sci. **46**, 219–229 (1998)

Horrillo, J., Wood, A., Kim, G.B., Parambath, A.: A simplified 3-D Navier-Stokes numerical model for landslide-tsunami: application to the Gulf of Mexico. J. Geophys. Res. Oceans **118**, 6934–6950 (2013)

Jack, J., Hills, J.G., Goda, M.P.: Damage from comet-asteroid impacts with earth. Physica D **133**, 189–198 (1999)

Kharif, C., Pelinovsky, E.: Asteroid impact tsunamis. C. R. Phys. **6**, 361–366 (2005)

Korycansky, D.G., Lynett, P.J.: Run-up from impact tsunami. Geophys. J. Int. **170**, 1076–1088 (2007)

Kozelkov, A.S., Kurulin, V.V.: Eddy resolving numerical scheme for simulation of turbulent incompressible flows. Comput. Math. Math. Phys. **55**, 1232–1241 (2015)

Kozelkov, A.S., Deryugin, Y.N., Lashkin, S.V., Silaev, D.P., Simonov, P.G., Tyatyushkina, E.S.: Implementation in LOGOS software of a computational scheme for a viscous incompressible fluid using the multigrid method based on an algorithm SIMPLE. VANT. Ser. Mat. Mod. Phys. Proc. **4**, 31–43 (2013)

Kozelkov, A.S., Kurulin, V.V., Tyatyushkina, E.S., Puchkova, O.L.: Modeling of incompressible viscous turbulent flows on unstructured grids using the DES model. Math. Models Comput. Simulat. **26**, 81–96 (2014)

Kozelkov, A.S., Kurkin, A.A., Pelinovskii, E.N., Kurulin, V.V.: Modeling the cosmogenic tsunami within the framework of the Navier–Stokes equations with sources of different types. Fluid Dyn. **50**, 306–313 (2015a)

Kozelkov, A.S., Kurkin, A.A., Pelinovskii, E.N., Kurulin, V.V., Tyatyushkina, E.S.: Modeling the disturbances in the lake Chebarkul caused by the fall of the meteorite in 2013. Fluid Dyn. **50**, 828–840 (2015b)

Kozelkov, A., Kurulin, V., Emelyanov, V., Tyatyushkina, E., Volkov, K.: Comparison of convective flux discretization schemes in detached-eddy simulation of turbulent flows on unstructured meshes. J. Sci. Comput. **89**, 1–16 (2015c)

Kozelkov, A.S., Krutyakova, O.L., Kurkin, A.A., Kurulin, V.V., Tyatyushkina, E.S.: Zonal RANS–LES approach based on an algebraic Reynolds stress model. Fluid Dyn. **50**, 621–628 (2015d)

Kozelkov, A.S., Kurkin, A.A., Sharipova, I.L., Kurulin, V.V., Pelinovsky, E.N., Tyatyushkina, E.S., Meleshkina, D.P., Lashkin, S.V., Tarasova, N.V.: Minimal basis tasks for validation of methods of calculation of flows with free surfaces. Trans. Nizhni Novgorod State Tech. Univ. **2**(109), 49–69 (2015e)

Lavrentiev, M.A., Shabat, B.V.: Hydrodynamics Problems and Mathematical Models. Fismatgiz, Moscow (1973) (in Russian)

Levin, B.V., Nosov, M.A.: Physics of Tsunami Waves, 327 pp. Springer, The Netherlands (2016). doi:10.1007/978-1-4020-8856-8. http://www.springer.com/us/book/9789048180073 [Softcover ISBN 978-90-481-8007-3]

Marusek, J.A.: Comet and Asteroid Threat Impact Analysis. Nuclear Physicist & Engineer, Impact, RR6, Box 442 (2007)

Mittal, R., Iaccarino, G.: Immersed boundary methods. Ann. Rev. Fluid Mech. **37**, 239 (2005)

Morrison, D.: Defending the earth against asteroids: the case for a global response. Sci. Glob. Secur. **13**, 87–103 (2005)

Near-Earth Object Program (NASA, JPL). http://neo.jpl.nasa.gov/

Paine, M.P.: Asteroid impacts: the extra hazard due to tsunami. Sci. Tsunami Haz. **17**, 155–166 (1999)

Pelinovsky, E.: Hydrodynamics of Tsunami Waves. Institute of Applied Physics, Nizhny Novgorod (1996) (in Russian)

Pierazzo, E., Artemieva, N., Asphaug, E., Baldwin, E.S., Cazamias, J., Coker, R., Collins, G.S., Crawford, D.A., Davison, T., Elbeshausen, D., Holsapple, K.A., Housen, K.R., Korycansky, D.J., Wunemann, K.: Validation of numerical codes for impact and explosion cratering: impacts on strengthless and metal targets. Meteorit. Planet. Sci. **43**, 1917–1938 (2008)

Shustov, B.M., Rykhlova, L.V.: Asteroid – Comet Danger: Yesterday, Today and Tomorrow. Fismatgiz, Moscow (2010) (in Russian)

Shuvalov, V.V., Trubestkaya, I.A.: Numerical modeling of marine target impacts. Solar Syst. Res. **36**, 417–430 (2002)

Shuvalov, V.V., Trubetskaya, I.A.: Numerical modeling of the formation of the Eltanin submarine impact structure. Solar Syst. Res. **41**, 56–64 (2007)

Shuvalov, V.V., Dypvik, H., Kalleson, E., Setsa, R., Riis, F.: Modeling the 2.7 km in diameter, shallow marine Ritland impact structure. Earth Moon Planets **108**, 175–188 (2012)

Truscott, T., Belden, J., Hurd, R.: Water-skipping stones and spheres. Phys. Today **67**(12), 70 (2014)

Ward, S.N., Asphaug, E.: Asteroid impact tsunami: a probabilistic hazard assessment. Icarus **145**, 64–78 (2000)

Weiss, R., Wunnemann, K., Bahlburg, H.: Numerical modelling of generation, propagation and run-up of tsunamis caused by oceanic impacts: model strategy and technical solutions. Geophys. J. Int. **167**, 77–88 (2006)

Yabushita, S.: On the possible hazard on the major cities caused by asteroid impact in the Pacific ocean. Earth Moon Planets **65**, 7–13 (1994)

The Donation Collections Routing Problem

Emmett J. Lodree, Derek Carter, and Emily Barbee

Abstract This paper introduces the donation collections problem (DCP), which is a network routing problem motivated by certain challenges that arise during the response phase immediately following large-scale disaster events. In particular, catastrophic events are often characterized by a dramatic surge of unsolicited donations and spontaneous volunteers that pose significant logistical problems for officials, and also inhibit the organized relief efforts of professional responders. The purpose of the DCP is to present a practical alternative for managing post-disaster logistics operations associated with material and volunteer convergence. The DCP is represented mathematically is an integer programming problem. We propose a host of common sense heuristic policies to generate routes for the DCP, and then evaluate the performance of these heuristic methods through computational experimentation. Our results indicate that longer routes are generally preferable to shorter ones based on the DCP objective function, and that network nodes characterized by rapid accumulation of donations should be served during the latter portion of a route's execution. Our findings also show that routing strategies that would be potentially appealing to inexperienced volunteers produce extraordinarily undesirable results. The collections routing literature almost entirely focuses on the development of optimal or near optimal solution algorithms. However, the humanitarian contexts that motivate the DCP warrant examination of simple heuristic policies that can be easily implemented in practice. This approach seems to represent a unique line of inquiry in the domain of collection routing.

Keywords Maximum collection routing • Time-dependent routing • Relief routing • Humanitarian logistics • Heuristic policies

---

E.J. Lodree (✉) • D. Carter • E. Barbee
Department of Information Systems, Statistics, and Management Science, Culverhouse College of Commerce, The University of Alabama, Tuscaloosa, AL 35487, USA
e-mail: ejlodree@cba.ua.edu; dacarter1@crimson.ua.edu; ecbarbee@crimson.ua.edu

I.S. Kotsireas et al. (eds.), *Dynamics of Disasters—Key Concepts, Models, Algorithms, and Insights*, Springer Proceedings in Mathematics & Statistics 185,
DOI 10.1007/978-3-319-43709-5_9

1 Introduction

This paper introduces the *donation collections problem* (DCP), which is defined as the following network routing problem. Consider an undirected graph $\mathcal{G} = \{\mathcal{V}, \mathcal{E}\}$, where $\mathcal{V}$ is the set of vertices and $\mathcal{E}$ is the set of edges, in which a single server visits a set of vertices $\mathcal{N} \subseteq \mathcal{V}$ while traversing some set of edges $\mathcal{A} \subseteq \mathcal{E}$. Similar to classical network optimization problems such as the traveling salesman problem (TSP) and the vehicle routing problem (VRP), each edge $(i,j) \in \mathcal{E}$, where $i,j \in \mathcal{V}$, is characterized by a travel time $c_{ij} = c_{ji}$, which can also be interpreted as travel distance or travel cost. However, the DCP is distinct from existing routing problems in that each vertex $i \in \mathcal{V}$ consists of a *donation accumulation rate* λ_i, and the server's goal is to maximize the quantity of donated items collected. There are no donations available for pickup at the beginning of the planning horizon (time $t = 0$), and the quantity of donated goods collected from vertex i at time t is $\lambda_i t$. Within this context, the server's objective is to maximize the total amount of donations collected by finding a route through the network that begins at the origin (a depot or warehouse with no accumulation rate), travels to each vertex in the set $\mathcal{N}$, and returns to the depot.

1.1 Motivation

The DCP is inspired by certain challenges that emerge during the response phase that follows large-scale disaster events. In particular, a major challenge that first responders and disaster survivors have to deal with is debilitating traffic congestion. This is caused by a common post-disaster phenomenon known as convergence. Within the context of disaster management, *convergence* refers to "the mass movement of people, messages, and supplies toward the disaster-struck area" (Fritz and Mathewson, 1957). Traffic congestion, as well as other issues that are direct consequences of convergence, inhibits the organized relief efforts of humanitarian organizations such as Red Cross, as well as professional responders such as police, fire fighters, and emergency medical technicians. In fact, Fritz and Mathewson (1957) cite several examples where officials reported traffic congestion as the most significant problem that they had to deal with. Traffic congestion issues that emerge following catastrophic events can primarily be attributed to volunteer and material convergence. Fritz and Mathewson (1957) refer to the influx of the helpers, curious, anxious, and exploiters to affected areas after the occurrence of a disaster as *personal convergence*. From this perspective, *volunteer convergence* can be considered a special case of personal convergence that pertains specifically to the helpers. Similarly, *material convergence* refers to the accumulation of emergency supplies and equipment in exorbitant quantities intended for disaster relief purposes (Holguín-Veras et al., 2012).

The DCP aims to help alleviate post-disaster traffic congestion by designing a process in which donors deliver goods to smaller satellite collection centers positioned at various safe locations outside of the affected areas. These satellite centers, which would be temporary constructs established specifically for the purpose of facilitating relief efforts, could be neighborhood schools, churches, or community centers. This network design is in contrast to the traditional setting where individual donors deliver their contributions to one or so large donation processing centers, which in turn propagates traffic congestion along main thoroughfares. In the proposed design, a small number of emergency management professionals or volunteers would be responsible for collecting donations from the satellite centers and transporting the collected goods to a larger primary relief center. This approach is likely to reduce the amount of traffic on main thoroughfares and enable official responders to reach target areas more quickly. Note that the proposed framework is based on the assumption that material and volunteer convergence will take place. The DCP is not intended to reduce or eliminate convergence, but rather present an alternative configuration to help manage the resulting traffic flow by reducing the number of vehicles traveling near affected areas.

1.2 Objective

The DCP applies to the following interpretation of the humanitarian relief scenario just described. Consider a donation collections effort that takes place at a nearby city, county, state, or providence where donors make their contributions locally as opposed to inundating the affected area. In this situation, donations retrieved from the satellite collection centers by emergency management professionals or volunteers would be transported to the disaster area upon completion of the collection route. The transport of material donations into the affected area may sometimes require alternate modes of transportation such as sea, air, or rail. From this perspective, the goal of the collections effort is to maximize the quantity of donations that reach the port, airport, or rail station prior to a scheduled departure so that sufficient quantities of donated supplies and equipment are available to meet demands that arise at the affected area.

1.3 Other DCP Applications

In addition to providing an alternative framework for managing post-disaster material and volunteer convergence, the DCP is also applicable within other contexts. Here, we describe two representative scenarios: (i) a commercial application that is somewhat general and (ii) a specific application that pertains to collections for food banks. In the commercial context, consider mail carriers and third party logistics providers who, in addition to making deliveries, also collect parcels that accumulate over time. Examples of pickup and delivery locations characterized by parcel accumulations include US Postal Service mailboxes, UPS stores, and FedEx stores. In each of these cases, parcels are collected and transported to a distribution center at the end of the daily route. Many of these parcels are then transferred to an aircraft for long-distance transport that may have a scheduled departure time. The DCP framework introduced in this paper would enable the quantity of items that reach the aircraft before its scheduled departure time to be maximized.

The DCP is also applicable to an annual food drive competition between two public universities located in the southeastern United States. The six-week long competition, which has been ongoing since 1994, begins each October and ends in late November on the Friday that precedes the Thanksgiving Holiday. Growing in size each year, nearly half of a million pounds of food are collected between the two institutions annually. The food drive competition has a significant impact in terms of ongoing humanitarian efforts to combat hunger. For example, food donations collected during the six-week competition by just one of the competing institutions can sustain 90,000 residents per month for a period of seven months. The relevance of the DCP to the above-mentioned food drive competition has to do with the process of collecting donations and delivering them to one of the participating food banks. During the competition, donated food items accumulate at specific locations on the university's campus as well as surrounding areas throughout the city. Although the competition lasts for six weeks, the majority of donations are collected during the last week with the largest influx occurring on the last day of competition. As such, the DCP concerns the design of a collection route that begins and ends during a single day in which the quantity of food donations collected and delivered to the food bank during the last day of the competition is maximized.

The remainder of the paper proceeds as follows. In the next section, we review relevant academic literature and discuss the unique contributions of this study. Next, a mathematical formulation of the DCP is presented followed by descriptions of several common sense heuristic solution approaches. The effectiveness of the heuristic policies are then evaluated in a computational study, and the implications of the results with respect to practice are discussed. Next, an extension of the basic DCP is examined. Finally, the paper concludes with a summary of our findings along with potential opportunities for future research.

2 Literature Review

The DCP is related to several routing problems that have been addressed in the academic literature, namely (i) routing in humanitarian relief, (ii) blood collection routing, (iii) specimen collection routing for clinical trials, and (iv) time-dependent maximum collection routing. The DCP also has similarities to mainstream routing problems such as (v) the orienteering problem and (vi) the inventory routing problem. In this section, we discuss representative studies in each of these areas and describe how they are related to the DCP.

2.1 Relief Routing

Vehicle routing is an integral part of the disaster relief process. As such, the subject has received considerable attention in the humanitarian logistics literature. Comprehensive surveys of relief routing research are presented in Caunhye et al. (2012), de la Torre et al. (2012), Ortuño et al. (2013), and Anaya-Arenas et al. (2014). The review by de la Torre et al. (2012) focuses exclusively on routing while Anaya-Arenas et al. (2014) consider distribution in a broader context that includes facility location, transportation (i.e., routing), and combined location / transportation models. The surveys by Caunhye et al. (2012) and Ortuño et al. (2013) have an even broader scope. They review operations research models for disaster relief logistics in general.

The routing component of disaster relief activities typically entails (i) last mile distribution, which is the delivery of emergency supply items from local distribution centers to disaster survivors in affected areas (e.g., Balcik et al., 2008), or (ii) casualty transportation, which refers to the transport of injured disaster survivors to medical facilities (Jotshi et al., 2009). The majority of relief routing studies address various aspects of the last mile distribution problem, while only the paper by Jotshi et al. (2009) addresses casualty transportation. However, Barbarosoğlu et al. (2002), Yi and Kumar (2007), and Yi and Özdamar (2007) consider last mile distribution and casualty transportation routing simultaneously.

More recent examples of relief routing that have been published subsequent to the above-mentioned survey articles include (i) a dynamic version of the integrated last mile distribution and casualty transportation problem with information updates (Najafi et al., 2014); (ii) routing with imperfect information updates obtained from social media outlets (Kirac et al., 2015); (iii) ambulance routing that accommodates the transport of large numbers of victims who at once require medical attention after a disaster event (Talarico et al., 2015); and (iv) the allocation of a repair crew to restore damaged arcs of a network following a disaster (Duque et al., 2016).

What distinguishes relief routing models from the vast array of traditional routing problems, which are primarily driven by commercial applications, is the objective function. While classical routing problems such as the TSP and VRP focus on minimizing the duration or cost of a route, customer focused performance measures are more pertinent in the emergency response context. The survey by de la Torre et al. (2012) enumerates relief routing objective functions that have been addressed in the academic literature. These include minimization of (i) unsatisfied demand, (ii) the maximum unsatisfied demand, (iii) latest arrival time of goods to beneficiaries, (iv) total response time, as well as (v) travel costs.

The DCP introduced in this paper contributes to the relief routing literature by expanding the list of objective functions that have been considered in a humanitarian logistics context (i.e., maximization of collected donations). In addition, existing relief routing studies address last mile distribution, casualty transportation, or both. Therefore by focusing on collections, the DCP also broadens the range of relief routing applications.

2.2 Blood Collection Routing

The DCP is closely related to the process of collecting donated blood. Yi (2003) is the first academic study to address the vehicle routing aspect of blood collections. Their research is based on the blood supply logistics operations of the American Red Cross, which entails the acquisition of blood from donors, the collection of donated blood from a predetermined set of donor sites, the transport of the collected blood to a facility for processing, and the distribution of processed blood to health care providers. The process of extracting blood from donors takes place in blood collection vehicles positioned at selected donor locations. However, the transport of donated blood involves more than simply driving each collection vehicle to the processing facility at the end of the day. Since blood is a perishable commodity, it must reach the processing facility within a certain amount of time after extraction. To facilitate this process, shuttles are routed among the collection vehicles to retrieve donated blood, and then transport the collected blood to the facility. The routing of the shuttles is referred to as *bring backs* and is the focus of the study by Yi (2003). Bring back routing is also considered in Doerner et al. (2008) who develop exact and approximate solution algorithms. In addition, Şahinyazan et al. (2015) extend the bring back routing framework described above to also include a schedule for collection vehicles (which they refer to as bloodmobiles). Their proposed design specifies the assignment of bloodmobiles to collection sites, the amount of time bloodmobiles remain at each site, the sequence in which bloodmobiles visit sites, and the routing of shuttles among bloodmobiles.

It should be noted that the blood collection process and its associated literature encompass a broad range of logistical decisions besides vehicle routing such as configuration of collection sites (Pratt and Grindon, 1982), allocation of staff (Brennan et al., 1992), and appointment scheduling of donors (Michaels et al., 1993). The interested reader may refer to the recent survey by Osorio et al. (2015) for a detailed discussion of blood collection research as well as other aspects of blood supply chains (i.e., production, storage/inventory, and distribution).

From a modeling perspective, the bring back operation represents a VRP with time windows and time-dependent rewards. Time windows are considered because the acquisition of blood from donors at collection vehicle sites is assumed to take place only during certain hours. Rewards are time-dependent because the amount of blood retrieved from a particular site depends on the time at which the shuttle visits that site during a tour. This time-dependent reward structure is what makes the bring back process of blood collections similar to the DCP. In both cases, the item being collected accumulates at each site while the route is being executed.

On the other hand, there are also several differences between the donation and blood collection routing problems. First, blood collection routing studies are concerned with minimizing distance traveled, or equivalently, minimizing the travel cost incurred that is directly proportional to route duration. In this sense, the blood collection routing literature is more closely related to commercial applications than to relief routing. Another difference is the relevance of time windows. In the context of blood collection, accumulations at collection sites are determined by scheduled blood donor appointments. Moreover, as mentioned previously, blood is a perishable product that must be transported to the processing facility within a limited time of each scheduled appointment. So there are meaningful time windows at the sites associated with bring backs in blood collection routing. By contrast, the process of providing relief to survivors immediately following large-scale disaster events is typically ongoing. As a result, the accumulation of donations due to material convergence occurs on a continuous basis. Furthermore, the composition of donated items primarily consists of nonperishable products such as canned food and clothing. From this perspective, distinct time windows at collection points are practically immaterial in the disaster relief context that motivates the DCP. Finally, each site is not visited during a single tour based on the blood collection routing models that have been proposed thus far. Shuttles are allowed to return to the depot after a single tour, and then be assigned to one or more additional tours during a single day. Furthermore, it is possible for sites to be included in multiple tours. The basic version of the DCP introduced in this paper ensures that each site is included exactly once during a single tour.

2.3 *Specimen Collection Routing*

Yücel et al. (2013) introduce a routing problem that is concerned with the collection of specimens for clinical trials, which they call the Collection for Processing Problem (CfPP). This problem is closely related to both donation and blood collection routing, and can be described as follows. Consider a company that runs clinical trials of specimen collected from physician offices, hospitals, and other sources. The task of retrieving specimen involves constructing multiple tours for a fleet of vehicles that visits each site at least once per day. Similar to the DCP and blood collection routing problems, the items collected accumulate over time and are brought to a depot for processing. However unlike blood collection routing models, the CfPP is characterized by an objective function that is similar to that of the DCP. In order to account for the high cost of idle time at the processing facility because of expensive equipment and skilled labor, the CfPP constructs tours in which the arrival rate of specimen to the facility is equal to order exceeds the facility's processing rate. As such, the primary objective of the CfPP is to maximize the amount of specimen processed by a given day's deadline, which depends on the amount retrieved during collection tours as well as the throughput of the facility. The CfPP also takes transportation costs into account as a second-level objective.

Of all the various routing problems that have been studied in the academic literature, the CfPP introduced in Yücel et al. (2013) seems to be the one that is closest to ours. Both the DCP and CfPP are concerned with routing a single vehicle through a network characterized by the accumulation of items at each node, where the objective is to maximize the quantity of items processed. However, there are differences. Similar to the blood collection routing problems described in Sect. 2.2, the CfPP allows multiple tours that must all be completed by a deadline. In addition, the CfPP is also characterized by a finite production rate at the depot that influences the objective function. The CfPP is actually a generalization of the DCP introduced in this paper. However, our focus is the evaluation of common sense heuristic policies that are relevant to the humanitarian context that motivate the DCP. On the other hand, CfPP studies are concerned with route optimization through mathematical programming techniques that are appropriate for the more technologically advanced commercial setting associated with the specimen collection routing.

2.4 Maximum Collection Problems with Time-Dependent Rewards

Erkut and Zhang (1996) introduce the maximum collection problem with time-dependent rewards (MCPTDR). In MCPTDR, each node i in a network is characterized by a decreasing reward function $r_i(t)$, as well as a service time v_i required to collect each reward. The objective is to construct a route consisting of a single tour that visits each node in the network such that the total reward collected is maximized. The MCPTDR problem is related to the DCP introduced in this paper in that the objective is to maximize collections based on a single tour of all nodes in the network. The differences are (i) in DCP, items accumulate over time whereas MCPTDR considers rewards that deteriorate over time; and (ii) MCPTDR includes a service time for collection at each node.

Tang et al. (2007) consider a variation of the MCPTDR problem that allows multiple tours and is referred to as the multiple tour maximum collection problem with time-dependent rewards (MTMCPTD). Another extension of MCPTDR is presented in Ekici and Retharekar (2013), in which routes are carried out by multiple agents.

2.5 Other Routing Problems

The DCP, blood collection routing, and CfPP are all related to more mainstream routing problems that have been studied extensively in the literature, namely inventory routing and orienteering problems. Node accumulations occur within the context of inventory routing problems (IRPs). However, these accumulations represent demands that are to be satisfied via distribution from a central depot as opposed to items that need to be collected and returned to a central depot as in the DCP, CfPP, and blood collections problem. Also, IRPs are generally concerned with the minimization of transportation and inventory holding costs, whereas the DCP seeks to maximize the quantity of items collected. On the other hand, orienteering problems (OPs) are related to the DCP in that the objective is to maximize collections. In particular, the fundamental orienteering problem, which is also known as the selective TSP, can be described as follows. Each node of a given network is characterized by a reward. The OP involves the construction of a route that covers a subset of nodes within a specified amount of time in an effort to maximize the total reward collected. OPs have the property that no tour exists that can cover all nodes of the network by the given deadline. The latter represents a noteworthy difference between OPs and the DCP. Although a deadline in terms of route duration can sometimes be relevant to the DCP, both of the applications that motivate the DCP generally seek to provide service to each node of the network.

Comprehensive surveys of the inventory routing and orienteering problems are presented in Andersson et al. (2010) and Vansteenwegen et al. (2011), respectively.

2.6 Additional Contributions of This Paper to the Literature

The DCP is related to several routing problems that have been studied in the academic literature. In all of these studies, the development of optimal and/or near optimal solution procedures is emphasized. In this paper, our focus is on assessing the performance of common sense heuristic policies that could be easily implemented in the humanitarian applications that motivate the DCP. In particular, the retrieval of donations within the context of the disaster relief and food bank scenarios described in Sect. 1 are primarily carried out by a volunteer workforce who lacks the technical expertise and supporting infrastructure to generate optimal routes using specialized commercial software. In these environments, route construction based on optimization techniques is an impractical option, and simple heuristic rules represent a more viable alternative. While the effectiveness of simple heuristic policies has been examined in routing problems such as the TSP (e.g., Rosenkrantz et al., 1977), time-dependent VRP (e.g., Malandraki and Daskin, 1992), and humanitarian oriented routing problems (e.g., Bartholdi III et al., 1983, "meals on wheels"), this paper seems to be the first to consider heuristic rules in the domain of collection routing.

3 Integer Programming Model

In this section, an integer programming (IP) formulation for a stylized version of the DCP is presented. First, the modeling assumptions are described followed by a detailed development of the IP model.

3.1 Modeling Assumptions

The IP model that will be developed in Sect. 3.2 is based on the following assumptions. We consider the relaxation of one or more of these assumptions to be an extension of the basic DCP.

A.1 The accumulation rates, λ_i, are constants and do not change over time.
A.2 Items accumulate continuously at each location i according to their respective accumulation rates, λ_i, for the duration of the server's route.
A.3 No deadline for the time it takes to complete any route through the network is considered. Therefore, the longest route through a network is always a possible solution to the DCP. However, an extension of the basic DCP that includes a route duration deadline is examined in Sect. 6 of this paper.
A.4 The service time at each node is negligible. Consequently, the server's arrival and departure times at each node are the same.
A.5 Travel times between vertices (i.e., along edges) are constant.
A.6 Vehicle capacities are not considered.

Assumptions A.1 through A.6 are intended to accentuate common characteristics across a range of possible real-world implementations of the disaster relief and food drive competition applications of the DCP described in Sect. 1. These principal features include the accumulation of donations at geographically dispersed locations, and the objective of collecting as much as possible. Other characteristics such as variable accumulation rates, imposing a deadline on route duration, or time windows at collection nodes, are not necessarily relevant in all practical situations. For instance, although the accumulation of donations is likely to vary with time, there is no empirical evidence to support this claim to the best of our knowledge. In order to estimate accumulation rates in practice, considerable manual effort would likely be required. This is because the process of handling material donations at collection sites typically takes place without the assistance information technologies such as RFID, which could be used to document the donation information needed to determine accumulation rates. Nevertheless, Assumption A.1 adopts a linear modeling approach to donation accumulations, which is consistent with the linear model used to represent reward deterioration in maximum collection problems (e.g., Erkut and Zhang, 1996).

On the other hand, assumptions A.2 and A.3 are simplifications of reality in some situations and representative of practice in others. For example, the process of collecting donations for the food drive competition must be completed before the food bank closes for the day. However, the nodes where food donations accumulate are actually large barrels that can be accessed at any time for donation or collection purposes. In this situation, the deadline pertaining to route duration is relevant whereas distinct time windows at collection points are not. Also, consider the disaster relief interpretation of the DCP. Since relief operations are typically ongoing during the response phase that immediately follows large-scale disaster events, time windows at collection nodes are not particularly relevant. Therefore, it is generally not necessary to consider time windows in the disaster relief context of the DCP.

Another motivation for the above simplifying assumptions is to facilitate the development of a stylized mathematical model that can be analyzed in depth. Case in point, if Assumption A.3 is relaxed, then a solution to the DCP would potentially have to specify not only a sequence of nodes, but also which nodes would and would not be serviced during the execution of a given route. Similarly, Assumption A.4 limits the scope of our study to the construction of a single route that covers all collection nodes in the network. Without Assumption A.4, the service time at each node would also have to be determined. It is also worth mentioning that traffic congestion, which actually motivates the DCP as described in Sect. 1.1, could be captured using nonlinear delays between vertices. In fact, nonlinear delay functions are particularly relevant in the disaster relief context (e.g., Haghani and Oh, 1996; Nagurney et al., 2015). However, we consider the standard approach in which travel times are assumed to be constant as pointed out in Assumption A.5. Finally, Assumption A.6 indicates that we do not consider vehicle capacity. This is a limitation of our study that can potentially be examined through future research using a similar approach to Sect. 6 of this paper in which the effect of a route completion deadline is addressed.

By focusing our attention on a stylized model, we hope to uncover fundamental properties of effective solutions to the DCP that translate into guiding management principles for practice. As mentioned in Sect. 2.6, the humanitarian applications that inspire the DCP are characterized by (i) very limited technological infrastructure relative to related routing problems in the commercial sector, and (ii) a predominantly volunteer workforce. Therefore in order to address the realities of the real-world environments that motivate the DCP, this paper emphasizes the performance of common sense heuristic solutions that do not require specialized optimization software for practical implementation. As such, the development of optimal algorithms for solving the DCP lies beyond the scope of this paper. Furthermore, by applying common sense heuristics to a stylized model, we also expect to identify the conditions in which the performance of simple heuristic rules is comparable to that of more sophisticated solution techniques.

3.2 Development of Integer Programming Model

The mathematical programming formulation of the DCP can be expressed in words as follows:

Maximize : *Total quantity of items collected from all nodes of a network* (1)

Subject to : *The arrival time at the* k^{th} *node is the sum of the arrival time at the* (2)

$(\mathrm{k}-1)^{\mathrm{st}}$ *node and the travel time between the* $(\mathrm{k}-1)^{\mathrm{st}}$ *and* k^{th} *nodes*

Each sequence position associated with any route contains exactly one node (3)

Each node is assigned to exactly one sequence position of a route (4)

Table 1 List of notations

Notation	Description
n	Number of nodes in the network, not including the depot
i, j	Labels of nodes in the network
k	Position in route sequence
λ_i	Constant accumulation rate of items at node i
t_k	Server arrival and departure time at node in route position k
x_{ik} (decision variables)	Binary variable set to 1 if node i is in route position k
c_{ij}	Travel time between nodes i and j
$\mathcal{J}_k$	Set of nodes in route positions $1, \ldots, k$
$\mathcal{J}'_m$	Complement of $\mathcal{J}_k$
j_m	Node in position $k+1$ using heuristic $m \in \{1, \ldots, 8\}$ if current position is k

In order to rigorously define the DCP, we develop a mathematical representation of the optimization problem shown in (1) through (4). To formulate the objective function given by (1), let λ_k denote the accumulation rate of the node that occupies position k of a given route, t_k the server's arrival at this node, and n is both the number of nodes in the network (excluding the depot) as well as the number of sequence positions associated with any route.[1] Then the objective function (1) can be written as

$$\lambda_1 t_1 + \lambda_2 t_2 + \cdots + \lambda_k t_k + \cdots + \lambda_n t_n. \tag{5}$$

To derive the arrival time constraints described in (2), consider that all DCP routes begin and end at the depot. Notationally, let node $i = 0$ represent the depot. Then the arrival time at the node in sequence position 1 of a given route is c_{0j}, where $j \in \{1, 2, \ldots, n\}$ and c_{0j} is the travel time between nodes 0 and j. For example, the arrival time to the node in route position 1 is $t_1 = c_{02}$ if node 2 is in position 1; the arrival time is $t_1 = c_{03}$ if node 3 is in position 1, etc. Now let x_{ik} be a binary variable defined as follows:

$$x_{ik} = \begin{cases} 1, & \text{ifnodeiisassignedtoroutepositionk;} \\ 0, & \text{otherwise.} \end{cases} \tag{6}$$

Then a mathematical representation of Constraint (2) when $k = 1$ is

$$t_1 = \sum_{i=1}^{n} c_{0i} x_{i1}. \tag{7}$$

[1]A complete list of the mathematical notations used in this paper is shown in Table 1.

Note that the c_{ij} values are the travel times associated with a given network, and that the x_{ij}'s are decision variables whose values represent a solution to the optimization problem outlined in (1) through (4). Also note that Eq. (7) is valid only if exactly one of the network's n nodes is assigned to route position 1, which is a special case of the condition posited by Constraint (3).

We now proceed to develop Constraint (2) by generalizing Eq. (7). First, consider another special case, t_2, the server's arrival time to the second node visited after the depot. If node i is in route position 1 and node j in position 2, then $x_{i1} = 1, x_{j2} = 1$, and $t_2 = t_1 + x_{i1}x_{j2}c_{ij}$. Since both i and j can take on any value from 1 to n, with the exception that $i \neq j$, it follows that

$$\begin{aligned} t_2 &= t_1 + x_{11}(c_{12}x_{22} + c_{13}x_{32} + \cdots + c_{1n}x_{n2}) + x_{21}(c_{21}x_{12} + c_{23}x_{32} + \cdots + c_{2n}x_{n2}) + \cdots \\ &\quad + x_{i1}(c_{i1}x_{12} + c_{i2}x_{22} + \cdots + c_{in}x_{n2}) + \cdots + x_{n1}(c_{n1}x_{12} + c_{n2}x_{22} + \cdots + c_{n,n-1}x_{n-1,2}) \\ &= x_{11}\sum_{j\neq 1} c_{1j}x_{j2} + x_{21}\sum_{j\neq 2} c_{2j}x_{j2} + \cdots + x_{i1}\sum_{j\neq i} c_{ij}x_{j2} + \cdots + x_{n1}\sum_{j\neq n} c_{nj}x_{j2}. \end{aligned}$$

We express the general case of Constraint (2) for $k = 2, \ldots, n$ in a more compact form:

$$t_k = t_{k-1} + \sum_i \sum_{j\neq i} c_{ij}x_{i,k-1}x_{jk}. \tag{8}$$

Constraints (3) and (4) are the usual constraints that appear in classical routing or sequencing problems when the decision variables are defined as in Eq. (6). The mathematical form of these two constraints, respectively, are

$$\sum_{i=1}^{n} x_{ik} = 1, \quad k = 1, \ldots, n \tag{9}$$

$$\sum_{k=1}^{n} x_{ik} = 1, \quad i = 1, \ldots, n. \tag{10}$$

Equation (8) is nonlinear because of the second order nonlinear terms $x_{i,k-1}x_{jk}$, which is the multiplication of decision variable pairs. In order to develop an equivalent linear model, define z_{ijk} as follows:

$$z_{ijk} = \begin{cases} 1, & \text{if node } i \text{ is in position } k-1 \text{ and node } j \text{ in position } k; \\ 0, & \text{otherwise.} \end{cases} \tag{11}$$

Alternatively, z_{ijk} can be thought of as binary value equal to 1 if arc (i,j) is assigned to route position k, where $i = 0$ for $k = 1$ and i,j each belong to the set $\{1, \ldots, n\}$ for $k = 2, \ldots, n$. Vander Wiel and Sahinidis (1996) refer to

z_{ijk} as *transition variables*. Using these variables, the arrival time at each node $k = 2, \ldots, n$ given by Eq. (8) can be replaced by the following equivalent set of equations that do not involve any nonlinear terms:

$$t_k = t_{k-1} + \sum_i \sum_{j \neq i} c_{ij} z_{ijk}, \qquad k = 2, \ldots, n. \tag{12}$$

$$\sum_{i=1}^{n} z_{ijk} = x_{jk}, \qquad j, k = 1, \ldots, n \tag{13}$$

$$\sum_{k=1}^{n} z_{ijk} = x_{i,k-1}, \qquad i, k = 1, \ldots, n \tag{14}$$

$$z_{ijk} \geq 0, \quad i, j, k = 1, \ldots, n. \tag{15}$$

The equivalence of Eq. (8) and the above set of equations (12) through (15) is based on the fact that $z_{ijk} = x_{i,k-1} x_{jk}$ is the unique solution to equations (13), (14), (9) (10), and (6), as shown in Vander Wiel and Sahinidis (1995).

So the full IP formulation of the DCP is as follows:

$$\begin{aligned}
\text{Maximize}: \quad & \sum_{k=1}^{n} \lambda_k t_k && && (5)\\
\text{Subject to}: \quad & t_k = \sum_{j=1}^{n} c_{0j} x_{j1}, && k = 1 && (7)\\
& t_k = t_{k-1} + \sum_i \sum_{j \neq i} c_{ij} z_{ijk}, && k = 2, \ldots, n && (12)\\
& \sum_{i=1}^{n} z_{ijk} = x_{jk}, && j, k = 1, \ldots, n && (13)\\
& \sum_{k=1}^{n} z_{ijk} = x_{i,k-1}, && i, k = 1, \ldots, n. && (14)\\
& \sum_{i=1}^{n} x_{ik} = 1, && k = 1, \ldots, n && (9)\\
& \sum_{k=1}^{n} x_{ik} = 1, && i = 1, \ldots, n. && (10)\\
& x_{ik} \in \{0, 1\}, && i, k = 1, \ldots, n && (6)\\
& z_{ijk} \geq 0, && i, k = 1, \ldots, n. && (15)
\end{aligned}$$

4 Methodology

This section describes several approaches to solving the DCP and seeks to determine which methods are most effective under various experimental conditions. First, a number of common sense heuristic policies are proposed that would be easy to implement in practice. Next, a diverse set of problem instances is constructed through an experimental design. Finally, the results of the computational study are analyzed along with a discussion of insights and lessons learned.

4.1 Solution Approaches

We describe eight options for generating solutions to the DCP, all of which can be classified as heuristic methods. The proposed approaches are heuristics based on (i) travel times between nodes, (ii) accumulation rates at nodes, (iii) a combination of travel times and accumulation rates, or (iv) neither travel times nor accumulation rates. The solution methods are summarized in Table 2.

With the exception of Method 8, each of the solution approaches delineated in Table 2 is intended to be easily implementable options for the DCP applications mentioned in Sect. 1. The random assignment heuristic (method 1) is the base case that is assumed to represent current practice. Methods 2 and 3 are essentially myopic policies based on travel times, while methods 4 and 5 are myopic with respect to accumulation rates. These methods are myopic in the sense that if the server is currently in the kth position of a route, then the node in position $k + 1$ is selected without any consideration for subsequent positions $k + 2, k + 3, \ldots, n$. Furthermore, the straightforward form of each method 2 through 5 is conducive to practical implementation. For instance, if k is the current route position, then rule 2

Table 2 Proposed methods for solving the DCP

Method	Name	Parameters	Description
1	Random assignment	N/A	Randomly assigns nodes to route positions
2	Shortest time	c_{ij}	Node in position $k + 1$ is closest to node in position k
3	Longest time	c_{ij}	Node in position $k + 1$ is furthest from node in position k
4	Largest λ	λ_i	Arrange nodes such that $\lambda_{k+1} \leq \lambda_k$
5	Smallest λ	λ_i	Arrange nodes such that $\lambda_{k+1} \geq \lambda_k$
6	Largest ratio	c_{ij}, λ_i	Node in position $k + 1$ has largest $\frac{\lambda_{k+1}}{c_{k,k+1}}$ relative to k
7	Smallest ratio	c_{ij}, λ_i	Node in position $k + 1$ has smallest $\frac{\lambda_{k+1}}{c_{k,k+1}}$ relative to k
8	Optimal	N/A	Solution generated by commercial software

simply selects the closest available node to occupy route position $k + 1$. Similarly, rule 3 prioritizes the most distant node; rule 4 prioritizes the node with the largest accumulation rate; and rule 5 favors the smallest accumulation rate.

Heuristics 6 and 7 are slightly more sophisticated myopic rules that incorporate both travel times and accumulation rates into routing decisions. In particular, the two ratio rules, 6 and 7, in some sense attempt to merge methods 2 and 4, and methods 3 and 5, respectively. It is evident that the Largest Ratio and Largest λ policies (methods 4 and 6, respectively) are closely related since both prioritize larger accumulation rates. Similarly, the Largest Ratio rule (method 6) is also based on the same logic as the Shortest Time rule (method 2). To see this, consider that larger ratios are consistent with smaller c_{ij} values. In other words, the shorter travel times prioritized in method 2 achieve the larger ratios prioritized in method 6. Consequently, method 6 essentially combines methods 2 and 4. An analogous discourse would reveal that a similar relationship exists among methods, 3, 5, and 7; specifically, that method 7 represents a composite of methods 3 and 5. From a practical standpoint, routes based on rules 6 and 7 can be easily constructed once the ratios have been determined by simple division. Of course, real-world implementation of rules 2 through 5 is slightly more straightforward relative to rules 6 and 7 since no calculations are required.

Unlike the experiential focus of the seven above-mentioned heuristic route construction rules, the primary goal of method 8 is to find the theoretically optimal route. In this study, the commercial software package Excel Solver™ is used for this purpose. However, the implementation of Solver™ used for the computational experiments in this paper does not directly solve the IP model presented in Sect. 3.2. Instead, we generate route sequences using the Excel Solver™ "alldifferent" constraint and "Evolutionary" solving method. Consequently, the resulting solutions produced by method 8 are not necessarily optimal, which means that method 8 can also be referred to as a heuristic approach. This strategy will allow us to apply method 8 to large problem instances.

It should also be noted that methods 2 and 4 are perhaps the most intuitively appealing choices. In practice, individuals are inclined to choose the closest location to their current position, which is equivalent to rule 2. This reasoning is most likely to occur within the humanitarian contexts described in Sect. 1 where the collector is often a volunteer, not an adept professional. It also seems reasonable to associate accumulation rates with the relative importance of collection points; i.e., locations with larger accumulation rates are more important than those with smaller ones. From this perspective, a logical course of action would be to assign higher priority locations (those with larger accumulation rates in this case) to earlier route positions, which is the essence of method 4. Given the instinctual qualities of methods 2 and 4 (and perhaps, method 6), the performance of these approaches in terms of the computational experiments are of particular interest.

5 Computational Study

5.1 Experimental Design

Several numerical examples are developed in order to test the effectiveness of the solution methods described in Sect. 4.1. These problem sets are intended to reflect a variety of scenarios based on the parameters that essentially define a DCP problem instance. DCP problem parameters include the number of network nodes, n; travel times between nodes, c_{ij}; and accumulation rates, λ_i. These three parameters are used as a basis to construct an experimental design with the following five factors: (i) n; (ii) the mean of the n accumulation rates, denoted μ_{λ_i}; (iii) standard deviation of accumulation rates, σ_{λ_i}; (iv) the mean travel time between nodes, $\mu_{c_{ij}}$; and (v) standard deviation of travel times, $\sigma_{c_{ij}}$. We begin with the 2^5 factorial design shown in Table 3.

Note that for illustrative purposes, the units of λ can be thought of as pounds per minute, and the units of the c_{ij} as minutes. Also, recall that n is the number of collection points, not including the depot. So the total number of nodes in the network, which consists of both the depot and collection points, is $n + 1$.

A few remarks are in order to explain the preliminary design shown in Table 3. First, $\sigma_{\lambda_i} = 0$ and $\sigma_{c_{ij}} = 0$ represent a couple of special cases that we want to explore. In particular, $\sigma_{\lambda_i} = 0$ is the case in which each node has the same accumulation rate, and $\sigma_{c_{ij}} = 0$ leads to equal travel times between all nodes. Second, it is not necessary to consider the two above-mentioned special cases simultaneously. The reason is that $\sigma_{\lambda_i} = \sigma_{c_{ij}} = 0$ is the trivial case in which all routes have the same objective function value. There are eight cases where this occurs within the context of the 2^5 factorial design in Table 3. After removing these eight cases from the preliminary design, the remaining 24 form the experimental conditions examined in this study as shown in Table 8 of the Appendix. Ten problem instances are randomly generated for each of the 24 cases resulting in a total of 240 test problems, each of which is solved by the eight approaches described in Sect. 4.1. Further details regarding the generation of test problems are given in the Appendix.

Table 3 Preliminary 2^5 factorial design

Factor	n	μ_{λ_i}	σ_{λ_i}	$\mu_{c_{ij}}$	$\sigma_{c_{ij}}$
Low	10	1	0	5	0
High	50	5	$\mu_{\lambda_i}/2$	25	$\mu_{c_{ij}}/2$

5.2 Results

5.2.1 Special Cases

Of the 24 problem scenarios associated with the experimental design shown in Table 8 of the Appendix, 16 can be considered special cases. These include the experiments where either $\sigma_{\lambda_i} = 0$ or $\sigma_{c_{ij}} = 0$. When $\sigma_{\lambda_i} = 0$, all accumulation rates are the same throughout the network (i.e., $\lambda_i = \lambda$ for each vertex $i = 1, \ldots, n$). Similarly, $\sigma_{c_{ij}} = 0$ corresponds to equal travel times along each edge of the network, including those that are directly connected to the depot. The latter can be expressed symbolically as $c_{ij} = c$ for $i = 0, 1, \ldots, n$ and $j = 1, \ldots, n$.

For the two aforementioned special cases considered in this study, the ratio rules are equivalent to other heuristic policies. To see this, recall that if node i occupies route position k, then the two ratio rules select the node j that will occupy position $k + 1$ according to $\max_{j \in \mathcal{J}'}\{\lambda_j / c_{ij}\}$ and $\min_{j \in \mathcal{J}'}\{\lambda_j / c_{ij}\}$, respectively, where $\mathcal{J}'$ represents the set of nodes not yet visited during a route. When $c_{ij} = c$ for all (i, j), the ratio becomes λ_j / c. Thus $\max_{j \in \mathcal{J}'}\{\lambda_j / c_{ij}\} = (1/c) \max_{j \in \mathcal{J}'} \lambda_j$, which is equivalent to the Largest λ rule described in Table 2. Analogously, the Smallest Ratio rule is the same as the Smallest λ rule whenever $c_{ij} = c$ for all (i, j). Using similar arguments, it can be shown that the Largest Ratio and Shortest Time methods are equivalent if $\lambda_i = \lambda$ for all nodes i, and that the Smallest Ratio and Longest Time rules are also equivalent. For these reasons, the ratio rules are not included in the results that pertain to the special cases, namely Tables 4 and 5. Also, it is obvious that the min c_{ij} and max c_{ij} rules need not be included since all of the c_{ij} values are equal.

Table 4 shows the average objective function value for each heuristic method and experimental condition that corresponds to the special case $\sigma_{c_{ij}} = 0$ (the best performing heuristic shown in bold). In particular, the values displayed in Table 4 represent averages over the 10 randomly generated problem instances for each of the eight scenarios shown. The results reveal that the Smallest λ rule, denoted min λ in Table 4, outperforms each of the other heuristic policies in all cases. In addition, the

Table 4 Results for special case where $\sigma_{c_{ij}} = 0$

Exp	Random	max λ	min λ	Excel™	% Deviation
3	265	183	**335**	335	0.00 %
5	1325	915	**1676**	1676	0.00 %
9	1754	1463	**2001**	2001	0.00 %
11	8770	7314	**10007**	10007	0.00 %
15	5698	3560	**7816**	7813	−0.03 %
17	28492	17801	**39079**	39069	−0.03 %
21	38939	31226	**46551**	46551	0.00 %
23	194696	156130	**232757**	232751	0.00 %
Avg	34992	27234	**42528**	42525	0.00 %

% Deviation column in Table 4 shows that the Smallest λ policy produces results that are comparable to the pseudo optimal solutions given by Excel Solver™. In fact, the min λ rule actually outperforms Excel Solver™ for the 50 node problem scenarios, which are indicated by negative values in the % Deviation column.

The superior performance of the min λ rule for the $\sigma_{c_{ij}} = 0$ special case can be explained as follows. First, recall that $\sigma_{c_{ij}} = 0$ implies that $c_{ij} = c$ for all relevant arcs (i, j) as previously mentioned. Consequently, the duration of any route through the network is the same, which means that the objective function is completely determined by the accumulation rates, λ_i. Now consider that the min λ rule prioritizes locations with small λ values, which in turn postpones the service of locations with larger λ values until the latter part of the route. This approach effectively utilizes the more productive locations (i.e., locations with larger values of λ) for longer periods of time, and utilizes the less productive locations (i.e., those with smaller values of λ) for shorter periods of time. From this perspective, it is evident that the min λ rule should work well for the $\sigma_{c_{ij}} = 0$ special case.

The other special case addressed in this study is $\sigma_{\lambda_i} = 0$, which means that each node has the same accumulation rate. Similar to Table 4, the values in Table 5 represent averages over the 10 randomly generated problem instances for each of the eight experiments listed. But this time, the max c_{ij} rule outperforms all the others, on average. The max c_{ij} rule refers to the Longest Time policy described in Table 2. The results are comparable to the solutions generated by Excel Solver™ as evidenced by the low percentage deviations shown in the next to last column of Table 5. Furthermore, the negative values in the % Deviation column indicate that the max c_{ij} heuristic policy actually outperforms Excel Solver™ on the larger networks with 50 nodes (experimental conditions 13, 14, 19, and 20).

The max c_{ij} rule, or equivalently, Longest Time rule, generates lengthy routes relative to the other heuristic policies as shown in Table 9 in the Appendix. However, an important insight is that although the longest route is often best in terms of the maximizing the quantity of donations collected, it would be erroneous to conclude that the longest route is *always* the best route. A few counterexamples are shown in the last column of Table 5. Specifically, "# Not Best" refers to how many replications

Table 5 Results for special case where $\sigma_{\lambda_i} = 0$

Exp	Random	min c_{ij}	max c_{ij}	Excel™	% Deviation	# Not best
1	333	190	**480**	485	1.11 %	2
2	1882	1317	**2464**	2488	0.96 %	2
7	1670	967	**2395**	2428	1.37 %	2
8	9409	6583	**12322**	12445	0.98 %	2
13	8061	3248	**12414**	12186	−1.87 %	0
14	44898	25547	**62399**	61486	−1.48 %	0
19	40307	16242	**62070**	61226	−1.38 %	0
20	224492	127735	**311996**	308146	−1.25 %	0
Avg	41382	22729	**58318**	57611	−0.002 %	

Table 6 General results

Exp	Random	max λ	min λ	min c_{ij}	max c_{ij}	min $\frac{\lambda_j}{c_{ij}}$	max $\frac{\lambda_j}{c_{ij}}$	Excel™	% Deviation	# Not best
4	311	217	362	185	**443**	321	220	504	12.17 %	7
6	1750	1218	2088	1275	**2285**	1762	1449	2629	13.08 %	8
10	2079	1755	2200	1213	**2980**	2554	1359	3159	5.68 %	5
12	11725	9870	12682	8344	**15364**	13185	9065	16363	6.10 %	6
16	6884	4444	9926	2942	**11014**	8862	3458	14035	21.52 %	10
18	38668	24900	**55463**	23160	55356	46835	25382	71836	22.79 %	10
22	47359	38645	58553	19918	**75385**	67272	22129	84501	10.79 %	10
24	265987	216637	327445	156826	**378914**	344768	171161	428613	11.60 %	9
Avg	46846	37211	58590	26733	**67718**	60695	29278	77705	12.85 %	8.13

of the 10 problem instances for each of the experiments shown in Table 5 in which the longest route does not correspond to the best DCP objective function. This phenomenon is more pronounced in the general case discussed in Sect. 5.2.2.

5.2.2 General Case

Unlike the special cases examined in the previous section, the eight experimental scenarios analyzed here are each characterized by heterogeneous accumulation rates and travel times. For these scenarios, the two ratio rules (methods 6 and 7 in Table 2) do not reduce to simpler rules as they did for the special cases. Consequently, the ratio rules, denoted max $\frac{\lambda_j}{c_{ij}}$ and min $\frac{\lambda_j}{c_{ij}}$ respectively, are included in the analysis which is summarized in Table 6.

Results for the general case reveal that the Longest Time rule, max c_{ij}, outperforms each of the other common sense heuristic policies, which is in agreement with the $\sigma_{\lambda_i} = 0$ special case results shown in Table 5. However, the general and special cases differ in several respects. First, the Excel Solver™ solution outperforms the max c_{ij} policy for each of the scenarios shown in Table 6, on average, which is not the case when $\sigma_{\lambda_i} = 0$. Furthermore, the % Deviation column of Table 6 shows that the superior performance of Excel Solver™ relative to the Longest Time rule is more pronounced compared to the $\sigma_{\lambda_i} = 0$ special case, especially in experiments 4, 6, 16, and 18. A common feature of the latter scenarios, which can be determined by observing Table 8 in the Appendix, is that σ_{λ_i} is a high factor without μ_{λ_i} being a high factor. So it is apparent that the performance of the Longest Time policy deteriorates with increasing variation in accumulation rates. However, it is also worth mentioning that the quality solutions generated by Solver™ come at the expense of computation time. The average time it took for Solver™ to generate solutions to 10-node problem instances was approximately 55 seconds. The average time was about 135 seconds (2.25 minutes) for 50-node instances, with a maximum recorded time of 239 seconds (just under 4 minutes). Although the above-mentioned computation times are modest, they are significantly worse than the seven common sense methods shown in Table 2. Computation times were practically instantaneous when simultaneously generating routes for all seven heuristic policies across all 10 problem instances associated with each of the 24 experimental conditions.

The general results also differ from the $\sigma_{\lambda_i} = 0$ special case with regard to the number of problem instances in which the largest quantity of collected donations was not achieved by the longest route. The "#Not Best" column of Table 6 shows that for the majority of problem instances (8.13 in 10 on average), the longest route and best route in terms of the DCP objective function did not coincide. However, for each of the problem instances referred to in the last column of Table 6, the max c_{ij} policy produced longest route and the pseudo optimal solution generated by Excel Solver™ produced the next to longest route. So overall, the best route was always characterized by either the longest or next to longest duration route. These findings suggest that although longer routes are preferable, the longest route is not necessarily the most beneficial.

Given that the DCP generally favors longer routes, it seems logical to expect poor performance from the heuristics that induce shorter routes. Table 6 confirms that this is actually the case. The Shortest Time rule (denoted min c_{ij} in Table 6) ranks last among all of the other solution approaches, including Random Assignment. In fact, this rule produces results that deviate from the best solution by more than 60 %, on average. This finding is of particular interest because as mentioned at the end of Sect. 4.1, the Shortest Time rule is intuitively appealing and is likely to be used by untrained volunteers in practice. The same can be said concerning the Largest λ policy. Intuitively, it seems as though prioritizing nodes in the network with large accumulation rates would yield favorable results. However, Table 6 shows that the max λ_j rule also performs quite poorly relative to the other heuristics, and that the min λ_j approach is significantly better. These results suggest that heuristic rules known to perform well in many other contexts produce extraordinarily undesirable results within the conditions defined in this paper.

Lastly, we were surprised by the fact that the performance of the ratio rules were not more competitive. Unlike the overall best common sense heuristic policy, max c_{ij}, the ratio rules use both travel time and accumulation rate data to construct routes. The min $\frac{\lambda_j}{c_{ij}}$ ratio policy ranks second overall to the max c_{ij} rule in terms of the heuristic methods, but it still surprising that this rule is, on occasion, outperformed by yet another single parameter heuristic, namely the min λ_j rule. Moreover, the Smallest λ policy is better than the Minimum Ratio policy in half of the general case experiments, namely experiments 10, 12, 22, and 24. A final anomaly worth mentioning is one that illudes any apparent logical explanation: the min λ rule outperformed the overall best heuristic policy (max c_{ij}), on average, for experimental scenario 18.

6 The DCP with Route Duration Deadline

This section examines an extension of the basic DCP in which there is a time limit on the duration of the route. A deadline at the depot impacts the DCP in the following way. First, if the deadline is such that all routes can be completed within the allotted time, then the deadline is irrelevant and all of the results described in Sect. 5 remain in tact. So a deadline is only meaningful insomuch as all nodes cannot be reached during a single tour. Similar to the analysis carried out in Sect. 5, our goal is to evaluate the performance of the common sense heuristic policies presented in Table 2 in the presence of a route deadline. The development of more sophisticated heuristic approaches characterized by features such as multiple tours, multiple servers, or strategic selection of which nodes to include in the route are beyond the scope of this study.

Our analysis of the DCP with a route deadline is guided by the following research question:

Question 1. *How does route deadline affect the choice of which heuristic rule should be used?*

Based on our analysis of the basic DCP presented in Sect. 5, we have determined that the overall best option among common sense heuristic policies is the max c_{ij} rule. In this section, we investigate whether or not imposing a deadline on route duration alters this result.

In order to address the above research question, we conduct another computational experiment in which the route duration deadline is varied within the context of the experimental design described in Sect. 5.1 and Table 8. More specifically, an additional 2,160 problem instances is considered by incorporating a deadline into each of the 240 problems associated with Sect. 5.1 and Table 8. For each problem instance, the deadline is varied as a scalar multiple of the average travel time and number of nodes. In particular, the deadline for experimental condition y, where $y = 1, \ldots, 24$, is computed as

$$d_y = \alpha n \mu_{c_{ij}}, \tag{16}$$

where α varies from 0.25 to 2.25 in increments of 0.25. So each of the 240 problem instances is solved for nine different values of α resulting in $9 \times 240 = 2,160$ additional problems.

6.1 *Special Case* $\sigma_{c_{ij}} = 0$

Indeed, route duration does affect the best choice in terms of heuristic rule selection. The results are very definitive for the special case in which all travel times are equal throughout the network (i.e., when $\sigma_{c_{ij}} = 0$). Figure 1 shows the results for

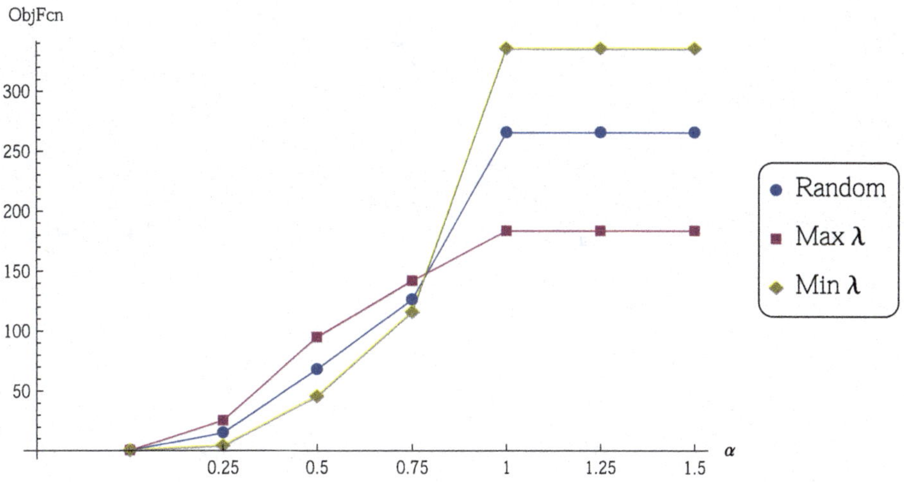

Fig. 1 The effect of route deadline on choice of heuristic rule for Experiment 3

Experiment 3, which is one of the eight experiments that represents the $\sigma_{c_{ij}} = 0$ special case.[2] The objective function values shown in Fig. 1 represent the average of 10 replications for each value of α. In fact, the objective function values and route durations in each figure depicted in this paper are also average values over 10 replications.

Figure 1 shows that the max λ policy gives the best results for tight deadline scenarios (i.e., smaller α values). The best rule then switches to the min λ rule when $\alpha = 1$, which is consistent with the result of Sect. 5 when no deadline is considered. The insight here is intuitive. If the deadline is tight and only a small number of nodes can be covered in a single tour, then the tour should include collection points with the highest accumulation rates. Otherwise, nodes with larger accumulation rates should be assigned to later route positions if the deadline is such that the majority of the nodes can be covered during a tour. This will enable items to accumulate in larger quantities as discussed in Sect. 5. Interestingly, the turning point at which the change in the best heuristic occurs is precisely when the deadline is such that each node in the network can be served. Figure 2 shows that this occurs when $\alpha = 1$.

The magnitude of the difference between the best and worst performing heuristic policies is also worthy of discussion. As shown in Fig. 1, the three heuristic rules yield similar results for tighter deadlines ($\alpha \leq 0.75$). On the other hand, the disparity between the objective function values is more significant when $\alpha \geq 1$. This finding suggests that the penalty for not choosing the right heuristic rule is more substantial when the route deadline is less restrictive. Results for the other seven $\sigma_{c_{ij}} = 0$ special cases are identical.

[2]The eight experiments that correspond to the special case $\sigma_{c_{ij}} = 0$ are 3, 5, 9, 11, 15, 17, 21, and 23.

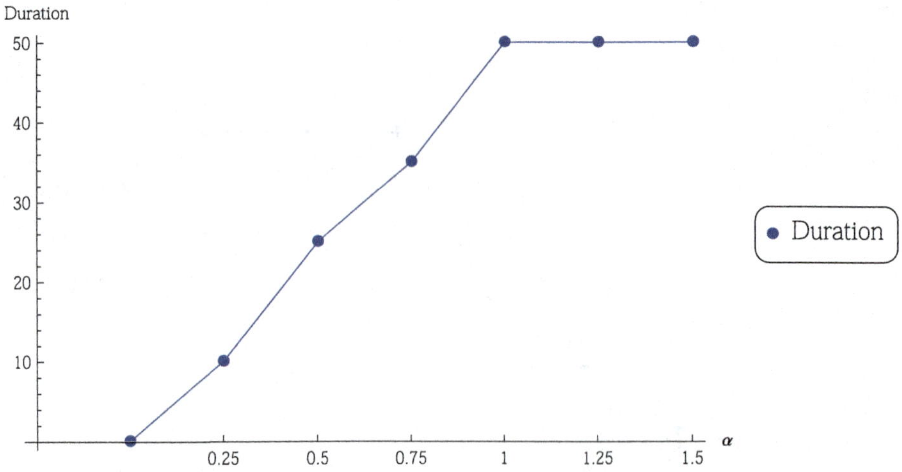

Fig. 2 The effect of route deadline on route duration for Experiment 3

6.2 Special Case $\sigma_{\lambda_i} = 0$

There is also some consistency among the results for the special case in which the accumulation rate at each node is the same (i.e., $\sigma_{\lambda_i} = 0$).[3] Figure 3 shows how route deadline affects the common sense heuristic policies for Experiment 20. In this case, there are two turning points: $\alpha = 1.25$ and $\alpha = 1.75$. There is slight variation in where the turning points occur among the eight $\sigma_{\lambda_i} = 0$ special case experiments (in the interest of space, graphs for the other seven experiments are not shown in this paper). However, the first turning point almost always occurs between $\alpha = 1.0$ and $\alpha = 1.25$ (the only exception is Experiment 14 where it occurs at $\alpha = 1.5$), and the second occurs between $\alpha = 1.5$ and $\alpha = 1.75$ (again, the only exception is Experiment 14 where the second turning point occurs at $\alpha = 2.0$). These eight cases are consistent in that the sequence of the best heuristic policies is $\min c_{ij}$ up until the first turning point; then random assignment between the first and second turning points; and finally $\max c_{ij}$ after the second turning point.

These findings are also somewhat intuitive. When the deadline is tight, use the $\min c_{ij}$ rule to include as many nodes in the tour as possible. Under a more flexible deadline, the $\max c_{ij}$ extends the route duration which allows items to accumulate for longer periods of time. In between these two extreme cases, both $\min c_{ij}$ and $\max c_{ij}$ are outperformed by the random assignment approach.

It is also important to note that the turning points do not coincide with the minimum deadline such that all nodes are covered, which is contrary to what

[3]The eight experiments that correspond to the special case of $\sigma_{\lambda_i} = 0$ are 1, 2, 7, 8, 13, 14, 19, and 20.

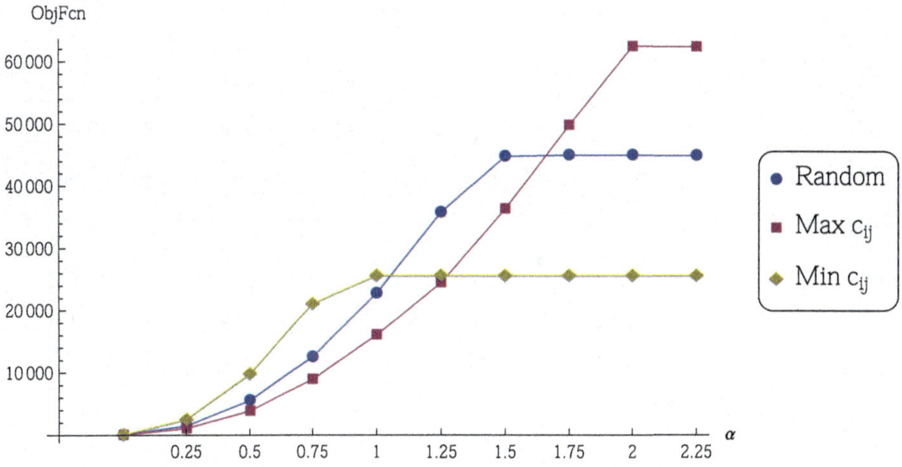

Fig. 3 The effect of route deadline on choice of heuristic rule for Experiment 20

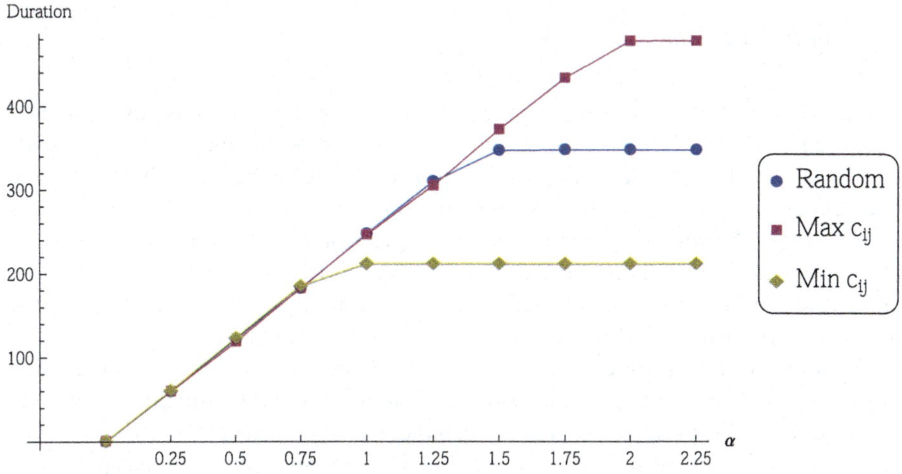

Fig. 4 The effect of route deadline on route duration for Experiment 20

happened for the $\sigma_{c_{ij}} = 0$ special case. This result is inferred from Fig. 4, which also shows that the longest route does not always lead to the best objective function value.

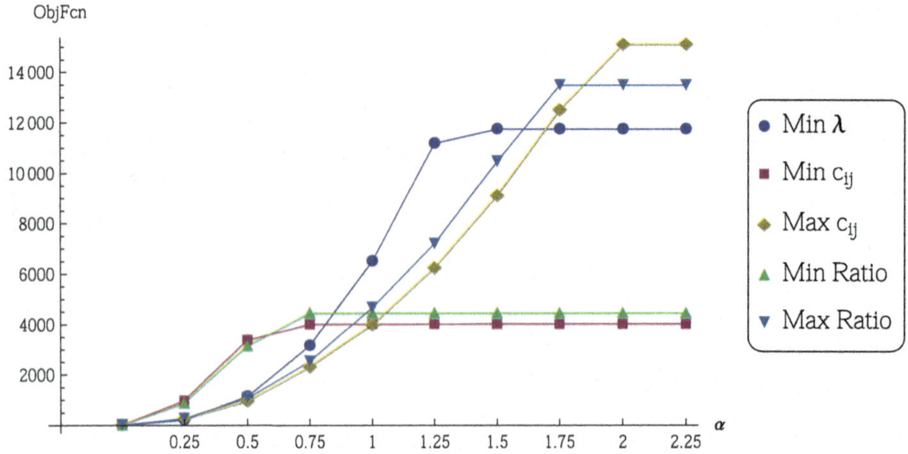

Fig. 5 The effect of route deadline on choice of heuristic rule for Experiment 22

6.3 *General Case*

There is less consistency in the results among the eight general case experimental conditions.[4] Dissimilarities exist in terms of the number of turning points, when they occur, and even which heuristic policies are preferable. Results for the 10 node experiments are also quite different compared to the results from the 50 nodes experiments. The number of turning points ranges from three to four, and they occur at various values of α between 0.5 and 2.0. Again, the graphs that support these claims are not included in this paper in the interest of preserving space. Instead, the results for Experiment 22 are shown in Fig. 5 as a representative case.

A summary of the progression of how the best heuristic rule changes as a function of α for each of the eight general case experiments is presented in Table 7. We see that (i) min λ is the preferred heuristic when the deadline is moderate, and (ii) max c_{ij} outperforms the other rules when the deadline is irrelevant in all cases except for one (Experiment 18), where the performance of max c_{ij} is a close second to min λ_i. In addition, experiments 4 and 6 are consistent in terms of the sequence in which the best rule changes with α. Experiments 22 and 24 are also consistent in this sense, with Experiment 10 having a similar progression. Otherwise, guiding principles do not seem to exist for the general experiments. The results are fairly inconclusive.

[4]The eight general case experiments are 4, 6, 10, 12, 16, 18, 22, and 24.

Table 7 Progression of the best heuristic rule as route deadline increases

	Exp.	4	6	10	12	16	18	22	24
Turn. Pt.		max λ	max λ	min c_{ij}	min ratio	min c_{ij}	min ratio	min c_{ij}	min c_{ij}
	1	min c_{ij}	min c_{ij}	max λ	min c_{ij}	min ratio	min c_{ij}	min ratio	min ratio
	2	min λ	min λ	min λ	min λ	min λ	min ratio	min λ	min λ
	3	max c_{ij}	max c_{ij}	max c_{ij}	max c_{ij}	max c_{ij}	min λ	max c_{ij}	max c_{ij}
	4			max c_{ij}				max c_{ij}	max c_{ij}

7 Summary

The donations collection problem (DCP) is a routing optimization problem characterized by the accumulation of items at each node of a network in which a server aims to maximize the quantity of items collected during a single route. It is inspired by a potential process for managing convergence following large-scale disaster events whereby donors make contributions at safe locations such as neighborhood churches, community centers, or schools. A responder simultaneously retrieves donations from these collection points and returns them to a large disaster relief warehouse where items are processed for eventual distribution to disaster survivors. The DCP is also related to collection routing within the context of collecting donations for food banks.

We describe the DCP mathematically as a linear integer programming model and present several common sense heuristic policies for solving it. Through computational experimentation, the performance of the proposed heuristic methods is assessed. Our findings indicate that longer routes are preferable to shorter ones, but the longest route doesn't always give best solution. The results also indicate that collection points with large accumulation rates should be served during the latter part of the route as opposed to the beginning. At first, these conclusions appear to be somewhat counterintuitive. In many practical applications, shorter routes are typically preferable and often the most efficient. Moreover, it is natural to associate accumulation rates to relative importance, and consequently prioritize locations with larger accumulation rates by assigning them to earlier route positions. On the other hand, the longer routes suggested by our results allow more time for items to accumulate at each location before collections take place, which is consistent with the objective of maximizing the quantity of donations collected. Similarly, postponing collections at nodes with larger accumulation rates produces more abundant stockpiles compared to servicing nodes with lower rates at the end of the route. These perspectives suggest that our results are indeed plausible.

The basic DCP introduced in this paper is based on several simplifying assumptions. Consequently, practice could be better served by studying variations of the DCP in which one or more of these assumptions is relaxed. We take a step in this direction by examining an extension of the basic DCP in which there is a deadline associated with route completion. Although no general guiding management principles emerged from our analysis, there was some consistency in the results pertaining to special cases. Finally, optimal policies for the DCP and its extensions would also be of interest.

Table 8 Experimental design parameters for computational study

Experiment	High factors	n	μ_{λ_i}	σ_{λ_i}	$\mu_{c_{ij}}$	$\sigma_{c_{ij}}$	a_{λ_i}	b_{λ_i}	$a_{c_{ij}}$	$b_{c_{ij}}$
1	$\sigma_{c_{ij}}$	10	1	0	5	2.5	1	1	2.26	7.74
2	$\mu_{c_{ij}}, \sigma_{c_{ij}}$	10	1	0	25	12.5	1	1	18.88	31.12
3	σ_{λ_i}	10	1	0.5	5	0	−0.2247	2.2247	55	
4	$\sigma_{\lambda_i}, \sigma_{c_{ij}}$	10	1	0.5	5	2.5	−0.2247	2.2247	2.26	7.74
5	$\sigma_{\lambda_i}, \mu_{c_{ij}}$	10	1	0.5	25	0	−0.2247	2.2247	25	25
6	$\sigma_{\lambda_i}, \mu_{c_{ij}}, \sigma_{c_{ij}}$	10	1	0.5	25	12.5	−0.2247	2.2247	18.88	31.12
7	$\mu_{\lambda_i}, \sigma_{c_{ij}}$	10	5	0	5	2.5	5	5	2.26	7.74
8	$\mu_{\lambda_i}, \mu_{c_{ij}}, \sigma_{c_{ij}}$	10	5	0	25	12.5	5	5	18.88	31.12
9	$\mu_{\lambda_i}, \sigma_{\lambda_i}$	10	5	2.5	5	0	2.26	7.74	5	5
10	$\mu_{\lambda_i}, \sigma_{\lambda_i}, \sigma_{c_{ij}}$	10	5	2.5	5	2.5	2.26	7.74	2.26	7.74
11	$\mu_{\lambda_i}, \sigma_{\lambda_i}, \mu_{c_{ij}}$	10	5	2.5	25	0	2.26	7.74	25	25
12	All except n	10	5	2.5	25	12.5	2.26	7.74	18.88	31.12
13	$n, \sigma_{c_{ij}}$	50	1	0	5	2.5	1	1	2.26	7.74
14	$n, \mu_{c_{ij}}, \sigma_{c_{ij}}$	50	1	0	25	12.5	1	1	18.88	31.12
15	n, σ_{λ_i}	50	1	0.5	5	0	−0.2247	2.2247	5	5
16	$n, \sigma_{\lambda_i}, \sigma_{c_{ij}}$	50	1	0.5	5	2.5	−0.2247	2.2247	2.26	7.74
17	$n, \sigma_{\lambda_i}, \mu_{c_{ij}}$	50	1	0.5	25	0	−0.2247	2.2247	25	25
18	$n, \sigma_{\lambda_i}, \mu_{c_{ij}}, \sigma_{c_{ij}}$	50	1	0.5	25	12.5	−0.2247	2.2247	18.88	31.12
19	$n, \mu_{\lambda_i}, \sigma_{c_{ij}}$	50	5	0	5	2.5	5	5	2.26	7.74
20	$n, \mu_{\lambda_i}, \mu_{c_{ij}}, \sigma_{c_{ij}}$	50	5	0	25	12.5	5	5	18.88	31.12
21	$n, \mu_{\lambda_i}, \sigma_{\lambda_i}$	50	5	2.5	5	0	2.26	7.74	5	5
22	$n, \mu_{\lambda_i}, \sigma_{\lambda_i}, \sigma_{c_{ij}}$	50	5	2.5	5	2.5	2.26	7.74	2.26	7.74
23	$n, \mu_{\lambda_i}, \sigma_{\lambda_i}, \mu_{c_{ij}}$	50	5	2.5	25	0	2.26	7.74	25	25
24	All	50	5	2.5	25	12.5	2.26	7.74	18.88	31.12

Appendix: Problem Instances for Computational Experiment

In Table 8, a_{λ_i} and b_{λ_i} represent the lower and upper bounds of a uniform distribution associated with randomly generated λ values. Similarly, $a_{c_{ij}}$ and $b_{c_{ij}}$ are lower and upper bounds for randomly generated c_{ij} values. In particular, recall that the computational experiment conducted in Sect. 5 consists of 240 problem instances—10 replications for each of the 24 scenarios shown in Table 8. Each of the 10 problem instances for each scenario is based on a randomly generated accumulation rate λ from a uniform distribution with parameters $(a_{\lambda_i}, b_{\lambda_i})$, and randomly generated travel times $c_{ij}, i, j = 1, \ldots, n$ from a uniform distribution with parameters $(a_{c_{ij}}, b_{c_{ij}})$. The uniform distribution parameters $(a_{\lambda_i}, b_{\lambda_i})$ and $(a_{c_{ij}}, b_{c_{ij}})$

Table 9 Route durations for special case of $\sigma_{\lambda_i} = 0$

Exp	Random	max λ	min λ	min c_{ij}	max c_{ij}	Excel™
1	62	58	58	41	**84**	85
2	345	333	333	261	**435**	439
7	62	58	58	41	**84**	85
8	345	333	333	261	**435**	439
13	310	305	305	141	**472**	468
14	1737	1716	1716	1058	**2388**	2371
19	310	305	305	141	**472**	470
20	1737	1716	1716	1058	**2388**	2371
Avg	613	603	603	375	**845**	840.96

in Table 8 are obtained by solving the following system of equation for the given values of $(\mu_{\lambda_i}, \sigma_{\lambda_i})$ and $(\mu_{c_{ij}}, \sigma_{c_{ij}})$, respectively:

$$\frac{a+b}{2} = \mu$$

$$\frac{(b-a)^2}{12} = \sigma^2.$$

Sample Route Duration Results

See Table 9.

References

Anaya-Arenas, A.M., Renaud, J., Ruiz, A.: Relief distribution networks: a systematic review. Ann. Oper. Res. **223**(1), 53–79 (2014)

Andersson, H., Hoff, A., Christiansen, M., Hasle, G., Løkketangen, A.: Industrial aspects and literature survey: Combined inventory management and routing. Comput. Oper. Res. **37**(9), 1515–1536 (2010)

Balcik, B., Beamon, B.M., Smilowitz, K.: Last mile distribution in humanitarian relief. J. Intell. Transp. Syst. **12**(2), 51–63 (2008)

Barbarosoğlu, G., Özdamar, L., Cevik, A.: An interactive approach for hierarchical analysis of helicopter logistics in disaster relief operations. Eur. J. Oper. Res. **140**(1), 118–133 (2002)

Bartholdi III, J.J., Platzman, L.K., Collins, R.L., Warden III, W.H.: A minimal technology routing system for meals on wheels. Interfaces **13**(3), 1–8 (1983)

Brennan, J.E., Golden, B.L., Rappoport, H.K.: Go with the flow: Improving red cross bloodmobiles using simulation analysis. Interfaces **22**(5), 1–13 (1992)

Caunhye, A.M., Nie, X., Pokharel, S.: Optimization models in emergency logistics: A literature review. Socio Econ. Plan. Sci. **46**(1), 4–13 (2012) de la Torre, L.E., Dolinskaya, I.S., Smilowitz, K.R.: Disaster relief routing: Integrating research and practice. Socio Econ. Plan. Sci. **46**(1), 88–97 (2012)

Doerner, K.F., Gronalt, M., Hartl, R.F., Kiechle, G., Reimann, M.: Exact and heuristic algorithms for the vehicle routing problem with multiple interdependent time windows. Comput. Oper. Res. **35**(9), 3034–3048 (2008)

Duque, P.A.M., Dolinskaya, I.S., Sörensen, K.: Network repair crew scheduling and routing for emergency relief distribution problem. Eur. J. Oper. Res. **248**(1), 272–285 (2016)

Ekici, A., Retharekar, A.: Multiple agents maximum collection problem with time dependent rewards. Comput. Ind. Eng. **64**(4), 1009–1018 (2013)

Erkut, E., Zhang, J.: The maximum collection problem with time-dependent rewards. Nav. Res. Logist. **43**(5), 749–763 (1996)

Fritz, C.E., Mathewson, J.H.: Convergence Behavior in Disasters: A Problem in Social Control: a Special Report Prepared for the Committee on Disaster Studies. National Academy of Sciences National Research Council, Washington, D.C. (1957)

Haghani, A., Oh, S.C.: Formulation and solution of a multi-commodity, multi-modal network flow model for disaster relief operations. Transp. Res. A Policy Pract. **30**(3), 231–250 (1996)

Holguín-Veras, J., Jaller, M., Van Wassenhove, L.N., Pérez, N., Wachtendorf, T.: Material convergence: Important and understudied disaster phenomenon. Nat. Hazards Rev. **15**(1), 1–12 (2012)

Jotshi, A., Gong, Q., Batta, R.: Dispatching and routing of emergency vehicles in disaster mitigation using data fusion. Socio Econ. Plan. Sci. **43**(1), 1–24 (2009)

Kirac, E., Milburn, A.B., Wardell III, C.: The traveling salesman problem with imperfect information with application in disaster relief tour planning. IIE Trans., **47**, 783–799 (2015)

Malandraki, C., Daskin, M.S.: Time dependent vehicle routing problems: Formulations, properties and heuristic algorithms. Transp. Sci. **26**(3), 185–200 (1992)

Michaels, J.D., Brennan, J.E., Golden, B.L., Fu, M.C.: A simulation study of donor scheduling systems for the American Red Cross. Comput. Oper. Res. **20**(2), 199–213 (1993)

Nagurney, A., Masoumi, A.H., Yu, M.: An integrated disaster relief supply chain network model with time targets and demand uncertainty. In: Regional Science Matters, pp. 287–318. Springer, New York (2015)

Najafi, M., Eshghi, K., de Leeuw, S.: A dynamic dispatching and routing model to plan/re-plan logistics activities in response to an earthquake. OR Spectr. **36**(2), 323–356 (2014)

Ortuño, M., Cristóbal, P., Ferrer, J., Martín-Campo, F., Muñoz, S., Tirado, G., Vitoriano, B.: Decision aid models and systems for humanitarian logistics. a survey. In: Decision Aid Models for Disaster Management and Emergencies, pp. 17–44. Springer, New York (2013)

Osorio, A.F., Brailsford, S.C., Smith, H.K.: A structured review of quantitative models in the blood supply chain: a taxonomic framework for decision-making. Int. J. Prod. Res. **53**(24), 7191–7212 (2015)

Pratt, M.L., Grindon, A.J.: Computer simulation analysis of blood donor queueing problems. Transfusion **22**(3), 234–237 (1982)

Rosenkrantz, D.J., Stearns, R.E., Lewis, II, P.M.: An analysis of several heuristics for the traveling salesman problem. SIAM J. Comput. **6**(3), 563–581 (1977)

Şahinyazan, F.G., Kara, B.Y., Taner, M.R.: Selective vehicle routing for a mobile blood donation system. Eur. J. Oper. Res. **245**(1), 22–34 (2015)

Talarico, L., Meisel, F., Sörensen, K.: Ambulance routing for disaster response with patient groups. Comput. Oper. Res. **56**, 120–133 (2015)

Tang, H., Miller-Hooks, E., Tomastik, R.: Scheduling technicians for planned maintenance of geographically distributed equipment. Transp. Res. E Logist. Transp. Rev. **43**(5), 591–609 (2007)

Vander Wiel, R.J., Sahinidis, N.V.: Heuristic bounds and test problem generation for the time-dependent traveling salesman problem. Transp. Sci. **29**(2), 167–183 (1995)

Vander Wiel, R.J., Sahinidis, N.V.: An exact solution approach for the time-dependent traveling-salesman problem. Nav. Res. Logist. **43**(6), 797–820 (1996)

Vansteenwegen, P., Souffriau, W., Van Oudheusden, D.: The orienteering problem: A survey. Eur. J. Oper. Res. **209**(1), 1–10 (2011)
Yi, J.: Vehicle routing with time windows and time-dependent rewards: A problem from the American Red Cross. Manuf. Serv. Oper. Manag. **5**(1), 74–77 (2003)
Yi, W., Kumar, A.: Ant colony optimization for disaster relief operations. Transp. Res. E Logist. Transp. Rev. **43**(6), 660–672 (2007)
Yi, W., Özdamar, L.: A dynamic logistics coordination model for evacuation and support in disaster response activities. Eur. J. Oper. Res. **179**(3), 1177–1193 (2007)
Yücel, E., Salman, F.S., Gel, E.S., Örmeci, E., Gel, A.: "Optimizing specimen collection for processing in clinical testing laboratories. Eur. J. Oper. Res. **227**(3), 503–514 (2013)

Network Criticality and Network Complexity Indicators for the Assessment of Critical Infrastructures During Disasters

Evangelos Mitsakis, Josep Maria Salanova, Iraklis Stamos, and Emmanouil Chaniotakis

Abstract Network criticality indicators, such as the Unified Network Performance Measure, provide powerful tools for interested entities who aim to assess those parts of the network, the closure of which would mostly affect its overall performance. The progress in complex networks analysis on the premises of graph theory allowed for advances on the identification of network characteristics and alternative sets of indicators for the evaluation of network performance. The aim of the present paper is to lay out the contribution of network analytics (in the form of complexity and criticality indicators) in disaster management, with the road network of the Peloponnese region, Greece, acting as a case study. Findings show that adopting interdisciplinary advances can provide useful insights to entities, responsible for the mitigation, preparedness, response, and reconstruction phases of disaster management and support them in the complex decision-making process.

Keywords Criticality • Centrality • Disaster management • Transportation networks • Network efficiency • Wild fires

1 Introduction

According to Baker (1991), disasters are events that occur scarcely, have a random nature and require immediate actions, exacerbating the situation faced by decision makers, who are bounded by their limited experience, cognitive limitations, and perception (Marakas 2003). In this direction, disaster management (DM) is the process of mitigating, preparing, responding, and reconstructing for and from a disaster (Fagel 2011, Lindell et al. 2007). This process, often referred to as the disaster life cycle, or comprehensive emergency management, is defined as a

E. Mitsakis • J.M. Salanova • I. Stamos (✉) • E. Chaniotakis
Centre for Research and Technology Hellas – Hellenic Institute of Transport, 6th km Charilaou-Thermi Road, 57001 Thermi, Thessaloniki, Greece
e-mail: stamos@certh.gr

I.S. Kotsireas et al. (eds.), *Dynamics of Disasters—Key Concepts, Models, Algorithms, and Insights*, Springer Proceedings in Mathematics & Statistics 185,
DOI 10.1007/978-3-319-43709-5_10

circular process to denote its ongoing and never ending character (Fagel 2011). In all phases of disasters, decision makers are found within a complex and highly uncertain situation, governed by multiple objectives that can be described from multiple perspectives (Fagel 2011). Such situations are commonly characterized as hard decisions (Clemen and Reilly 2013).

The first two phases (mitigation and preparedness) have a preparatory character for the succeeding ones (response and reconstruction) and aim at the alleviation and/or elimination of disasters' effects (Lindell et al. 2001). The response phase of DM is one of the most complicated and demanding phases, as therein lies a significantly high number of actions that need to be taken, often simultaneously. This high complexity of decision making is reflected in Shan et al. (2012), who model the emergency response to disasters using Petri models and conclude with a stochastic Petri network model for disaster response that has 29 places (states where actions should be taken) and 20 transitions between places (from one state to another) in its simplified version. Based on the temporal sequence governing DM, the reconstruction phase starts immediately after the disaster response and aims at restoring conditions (social, economic, etc.) to normal. Although it does not have the emergent character of response, it is also a complicated phase that requires prioritization of needs and efficient planning and acting (Schwab 1998).

During the response and reconstruction, transportation networks play a crucial role. In the response phase, the use of the transportation network for actions such as evacuation of the affected population, transportation of goods, and hazard mitigation is crucial for alleviating the effects of a disaster, while in reconstruction, one of the first actions that are commonly taken is the restoration of the critical transportation routes that would allow for the adequate provision of aid to those affected. The complicated character of DM and the importance of the transportation network in case of a disaster shape the imperative need to integrate indicators to be used for the evaluation of the transportation network by decision makers.

Such indicators lie in the field of network analysis (node-centric centrality indicators (Freeman 1978)) and their contribution to more adequate decision making in the DM process for the following cases is herein explored:

- Identification of important network segments for safeguarding
- Prioritization of reconstruction actions.

This paper sets the ground for the identification of the potential improvement of DM based on the use of network indicators in the areas specified above. The approach followed focuses firstly on the integration of knowledge from other scientific fields in DM and secondly on the interpretation and assessment of the contribution that those indicators can have to better decision making for DM.

The remainder of the paper is structured as follows. In Sect. 2, a literature review on the uses of network analytics indicators (centrality and criticality) is conducted. In Sect. 3, the case study area is briefly presented, followed by the calculation of network indicators (Sect. 4). Finally, conclusions are drawn and future research steps are presented in Sect. 5.

2 Literature Review

2.1 *Network Centrality Indicators*

The development of complex networks analysis on the premises of graph theory has lately allowed for advances on the identification of network characteristics and the definition of alternative sets of indicators for the evaluation of performance (Dorogovtsev and Mendes 2002; Rodrigue et al. 2013). A popular set of indicators is the set of the centrality indicators (Freeman 1978).

Centrality indicators have been previously used in transportation, with extensive work to take place on the analysis of maritime (Ducruet et al. 2010; Hu and Zhu 2009), air (Amaral et al. 2000; Bagler 2008; Guimerá et al. 2005), and railway transportation networks (Goncalves et al. 2009; Ouyang et al. 2014). Regarding surface transportation networks, complex network analysis and the use of graph indicators are limited, mainly due to the spatial constraints that road networks impose (Barabási and Bonabeau 2003; Erath et al. 2009; Xie and Levinson 2007). Barabási and Bonabeau (2003) used the example of the US highway network to illustrate the randomness of the road network, where the distribution degree follows a binomial distribution. Xie and Levinson (2007) argued that the analysis of complex networks is node-centric, while road networks are link-centric and suggested three indicators for structural analysis of networks (heterogeneity, connection pattern, and continuity). Erath et al. (2009) investigated the development of the Swiss road and railway network in 50 years on centralities, saturation, efficiency, and density and found regular network patterns for the network.

This type of analysis of spatial transportation networks, however, does not take into account the eminent character of DM and the need to use indicative indicators that could rapidly increase situational awareness of the decision maker. In DM, centrality indicators have received little attention, mainly in two directions: (a) the evaluation of processes and coordination of DM by examining the inter-organizational relations in a network setting (Lassa 2004; Moore et al. 2003; Uddin and Hossain 2011) (b) the allocation of emergency response units (ERU) (Seokcheon 2012). Given the main body of literature on network centrality analysis, the actual advantages and uses of centrality indicators are found to be investigated to a limited scope that is believed to reflect little of their potential.

In this paper, we focus on degree centrality, betweenness centrality, and closeness centrality, as they are believed to the most relevant indicators. Let $G(N, L)$ be a graph representation of a network, with N a set of nodes and L a set of links connecting those nodes. For $G(N, L)$ following centrality indicators can be defined (Freeman 1978):

Node Centrality (or node degree) (k) is defined as the number of neighboring nodes to which a node is connected.

Closeness Centrality of the random node i is defined as the inverse of the summation of length of shortest path between a random node i and random node j $(di, j \, \forall i, j \in N)$ (Eq. 1)

$$C_c(i) = \left[\sum_{i=1}^{N} d\,(i,j) \right]^{-1} \tag{1}$$

Betweenness Centrality is defined as the summation of the ratios of the number of the shortest paths that pass through the node examined v $(\sigma_{i,j}.(v))$ for the random pair of nodes (i,j) to the total number of shortest paths between the two random nodes $(\sigma_{i,j}.)$ for all random nodes of the network for all nodes in $G(N,L)$ (Eq. 2).

$$C_b = \sum_{i \neq v \neq j} \frac{\sigma_{i,j}(v)}{\sigma_{ij}} \tag{2}$$

2.2 Link Criticality

Link criticality is a network performance indicator developed in transportation-related research to assess the performance of the network components. It defines the importance of the links that comprise a network (Knoop et al. 2012). Criticality differs from vulnerability—based on the definition of Knoop et al. (2012)—as vulnerability illustrates the weak links of a network while criticality the important ones. The approach of Nagurney and Qiang's (2008) is adopted here and extensively presented in Sect. 3.3. It estimates the criticality of each link based on the difference of the travel times at network level before and after the closure of each link.

Link criticality is considered crucial in DM given the processes that take place (Mitsakis et al. 2014b). In case of a disaster, the closure of links characterized as critical affects the actual performance of the network, which can deteriorate the DM performance and even be life threatening.

The link categorization that link criticality introduces is based on the effect that each link closure has on the total generalized cost (i.e., travel time) of the network. The steps followed for the estimation of the importance of the link that comprise the transportation network examined are (Nagurney and Qiang 2008):

Step 1. A user equilibrium assignment is implemented for the given transportation demand and supply.

In addition, we also presented a cooperative, system-optimized model, in which the total costs of both the disaster relief organization and those of the freight service providers are minimized. We discussed the price of anarchy, which relates the total costs under our competitive model solution and its cooperative, system-optimized counterpart.

This paper adds to the growing literature on transportation in disaster relief and humanitarian operations but with the distinctive focus on freight service provision, which has only minimally been addressed rigorously, thus far. The contributions in this paper are also relevant to the corporate domain, especially in the case of healthcare crises in which, for example, vaccines or medicines must be delivered and pharmaceutical companies, or even governments, wish to determine the most cost-effective means for delivery.

Acknowledgements The author acknowledges the constructive comments and suggestions of the anonymous reviewer on an earlier version of this paper

The author thanks Professor Panos M. Pardalos of the University of Florida and Professor Ilias Kootsireas of Wilfrid Laurier University for the wonderful collaboration on the co-organization of the 2nd International Conference on Dynamics of Disasters, which took place in Kalamata, Greece, June 29–July 2, 2015.

References

Ap, T.: Ebola crisis: WHO slammed by Harvard-convened over slow response. CNN, 23 Nov 2015

Apex: Introducing apex emergency response freight services? The calm during the storm, Lafayette, LA (2015)

Balcik, B., Ak, D.: Supplier selection for framework agreements in humanitarian relief. Prod. Oper. Manag. **23**(6), 1028–1041 (2014)

Balcik, B., Beamon, B.M., Smilowitz, K.: Last mile distribution in humanitarian relief. J. Intell. Transp. Syst. **12**(2), 51–63 (2008)

Barbarosoglu, G., Arda, Y.: A two-stage stochastic programming framework for transportation planning in disaster response. J. Oper. Res. Soc. **55**(1), 43–53 (2004)

Barbarosoglu, G.L., Ozdamar, A., Cevik, A.: An interactive approach for hierarchical analysis of helicopter logistics in disaster relief operations. Eur. J. Oper. Res. **140**(1), 118–133 (2002)

Bertsekas, D.P., Gafni, E.M.: Projection methods for variational inequalities with application to the traffic assignment problem. In: Nondifferential and Variational Techniques in Optimization. Mathematical Programming Study, vol. 27, pp. 139–159. Springer, Berlin (1982)

Centers for Disease Control and Prevention: 2014 Ebola outbreak in west Africa - Case counts. http://www.cdc.gov/vhf/ebola/outbreaks/2014-west-africa/case-counts.html (2016)

Dafermos, S.: Traffic equilibrium and variational inequalities. Transp. Sci. **14**, 42–54 (1980)

Dafermos, S.C., Sparrow, F.T.: The traffic assignment problem for a general network. J. Res. Natl. Bur. Stand. **73B**, 91–118 (1969)

Falasca, M., Zobel, C.W.: A two-stage procurement model for humanitarian relief supply chains. J. Humanitarian Logist. Supply Chain Manag. **1**(2), 151–169 (2011)

Fischer, W.A. II, Hynes, N.A., Perl, T.M.: Protecting healthcare workers from Ebola: personal protective equipment is critical but not enough. Ann. Intern. Med. **161**(10), 753–754 (2014)

Gabay, D., Moulin, H.: On the uniqueness and stability of Nash equilibria in noncooperative games. In: Bensoussan, A., Kleindorfer, P., Tapiero, C.S. (eds.) Applied Stochastic Control of Econometrics and Management Science, pp. 271–294. North-Holland, Amsterdam (1980)

Hoxtell, W., Norz, M., Teicke, M.: Business engagement in humanitarian response and disaster risk management. Global Public Policy Institute, Berlin, Germany, May 2015

Huang, M., Smilowitz, K., Balcik, B.: Models for relief routing: equity, efficiency and efficacy. Transp. Res. E **48**, 2–18 (2012)

International Federation of Red Cross and Red Crescent Societies: Scope of service. http://www.ifrc.org/en/what-we-do/logistics/procurement/supply-services/ (2016)

Kinderlehrer, D., Stampacchia, G.: An Introduction to Variational Inequalities and Their Applications. Academic, New York (1980)

Knobler, S., Mahmoud, A., Lemon, S., Pray, L. (eds.): The Impact of Globalization on Infectious Disease Emergence and Control: Exploring the Consequences and Opportunities. The National Academies Press, Washington, DC (2006)

Kumar, S.: Managing risks in a relief supply chain in the wake of an adverse event. Oper. Res. Insight **24**(2), 131–157 (2011)

Lodree, E.J., Carter, D., Barbee, E.: The donation collections routing problem. In: Kotsireas, I.S., Nagurney, A., Pardalos, P.M. (eds.) Dynamics of Disasters. Springer International Publishing, Cham (2016)

Mete, H.O., Zabinsky, Z.B.: Stochastic optimization of medical supply location and distribution in disaster management. Int. J. Prod. Econ. **126**, 76–84 (2010)

Miller-Hooks, E., Sorrel, G.: The maximal dynamic expected flows problem for emergency evacuation planning. Transp. Res. Rec. **2089**, 26–34 (2008)

Na, H.S., Banerjee, A.: A disaster evacuation network model for transporting multiple priority evacuees. IIE Trans. **47**(11), 1287–1299 (2015)

Nagurney, A.: Network Economics: A Variational Inequality Approach, second and revised edition. Kluwer Academic, Dordrecht (1999)

Nagurney, A.: Supply Chain Network Economics: Dynamics of Prices, Flows, and Profits. Edward Elgar Publishing, Cheltenham (2006)

Nagurney, A., Masoumi, A.H.: Supply chain network design of a sustainable blood banking system. In: Boone, T., Jayaraman, V., Ganeshan, R. (eds.) Sustainable Supply Chains: Models, Methods and Public Policy Implications, pp. 49–72. Springer, London (2012)

Nagurney, A., Nagurney, L.S.: A mean-variance disaster relief supply chain network model for risk reduction with stochastic link costs, time targets, and demand uncertainty. In: Kotsireas, I.S., Nagurney, A., Pardalos, P.M. (eds.) Dynamics of Disasters—Key Concepts, Models, Algorithms, and Insights. Springer International Publishing, 231–255 (2016)

Nagurney, A., Qiang, Q.: Fragile Networks: Identifying Vulnerabilities and Synergies in an Uncertain World. Wiley, Hoboken, NJ (2009)

Nagurney, A., Qiang, Q.: Fragile networks: identifying vulnerabilities and synergies in an uncertain age. Int. Trans. Oper. Res. **19**, 123–160 (2012)

Nagurney, A., Dong, J., Zhang, D.: A supply chain network equilibrium model. Transp. Res. E **38**, 281–303 (2002)

Nagurney, A., Yu, M., Qiang, Q.: Multiproduct humanitarian healthcare supply chains: a network modeling and computational framework. In: Proceedings of the 23rd Annual POMS Conference, Chicago, IL (2012)

Nagurney, A., Yu, M., Floden, J., Nagurney, L.S.: Supply chain network competition in time-sensitive markets. Transp. Res. E **70**, 112–127 (2014)

Nagurney, A., Masoumi, A.H., Yu, M.: An integrated disaster relief supply chain network model with time targets and demand uncertainty. In: Nijkamp, P., Rose, A., Kourtit, K. (eds.) Regional Science Matters: Studies Dedicated to Walter Isard, pp. 287–318. Springer International Publishing, Cham (2015a)

Nagurney, A., Saberi, S., Shivani, S., Floden, J.: Supply chain network competition in price and quality with multiple manufacturers and freight service providers. Transp. Res. E **77**, 248–267 (2015b)

Patriksson, M.: The Traffic Assignment Problem. VSP, Utrecht (1994)

Pedraza Martinez, A.J., Stapleton, O., Van Wassenhove, L.N.: Field vehicle fleet management in humanitarian operations: a case-based approach. J. Oper. Manag. **29**(5), 404–421 (2011)

Qiang, Q., Nagurney, A.: A bi-criteria indicator to assess supply chain network performance for critical needs under capacity and demand disruptions. Transp. Res. A **46**(5), 801–812 (2012)
Regnier, E.: Public evacuation decisions and hurricane track uncertainty. Manag. Sci. **54**(2), 16–28 (2008)
Rottkemper, B., Fischer, K., Blecken, A.: A transshipment model for distribution and inventory relocation under uncertainty in humanitarian operations. Socio Econ. Plan. Sci. **46**, 98–109 (2012)
Roughgarden, T.: Selfish Routing and the Price of Anarchy. MIT Press, Cambridge, MA (2005)
Saadatseresht, M., Mansourian, A., Taleal, M.: Evacuation planning using multiobjective evolutionary optimization approach. Eur. J. Oper. Res. **198**, 305–314 (2009)
Sheffi, Y., Mahmassani, H., Powell, W.B.: A transportation network evacuation model. Transp. Res. A **16**(3), 209–218 (1982)
Sherali, H.D., Carter, T.B., Hobeika, A.G.: A transportation network evacuation model. Transp. Res. A **16**(3), 209–218 (1991)
Sheu, J.B.: An emergency logistics distribution approach for quick response to urgent relief demand in disasters. Transp. Res. E **43**(6), 687–709 (2007)
The World Bank: Cost to export (US$ per container). http://data.worldbank.org/indicator/IC.EXP.COST.CD (2016)
Tzeng, G.-H., Cheng, H.-J., Huang, T.: Multi-objective optimal planning for designing relief delivery systems. Transp. Res. E **43**(6), 673–686 (2007)
United Nations High Commissioner for Refugees: Doing business with UNHCR. UNHCR Global Service Centre, Budapest (2015)
United States Department of Commerce: Access costs everywhere. http://acetool.commerce.gov/shipping (2016)
UNOPS: Procurement manual, revision 5, May 1. Sustainable Practice Procurement Group (2014)
Van Wassenhove, L.N.: Blackett memorial lecture. Humanitarian aid logistics: supply chain management in high gear. J. Oper. Res. Soc. **57**(5), 475–489 (2006)
Vitoriano, B., Ortuño, M., Tirado, G., Montero, M.: A multi-criteria optimization model for humanitarian aid distribution. J. Glob. Optim. **51**, 189–208 (2011)
Vogiatzis, C., Pardalos. P.M.: Evacuation modeling and betweenness centrality. In: Kotsireas, I.S., Nagurney, A., Pardalos, P.M. (eds.) Dynamics of Disasters. Springer International Publishing, Cham (2016)
Vogiatzis, C., Walteros, J.L., Pardalos, P.M.: Evacuation through clustering techniques. In: Goldengorin, B., Kalyagin, V.A., Pardalos, P.M. (eds.) Models, Algorithms, and Technologies for Network Analysis, pp. 185–198. Springer, New York (2013)
Wilson, D.: CE: inside an Ebola ET: a nurses' report. Am. J. Nurs. **115**(12), 28–38 (2015)
Woods, R.: DHL, Qatar overcome logistics challenges in Nepal. Air Cargo World, 1 June 2015
World Health Organization: Health worker Ebola infections in Guinea, Liberia and Sierra Leone: a preliminary report, Geneva, 21 May 2015
Yi, W., Kumar, A.: Ant colony optimization for disaster relief operations. Transp. Res. E **43**(6), 660–672 (2007)

A Mean-Variance Disaster Relief Supply Chain Network Model for Risk Reduction with Stochastic Link Costs, Time Targets, and Demand Uncertainty

Anna Nagurney and Ladimer S. Nagurney

Abstract In this paper, we develop a mean-variance disaster relief supply chain network model with stochastic link costs and time targets for delivery of the relief supplies at the demand points, under demand uncertainty. The humanitarian organization seeks to minimize its expected total operational costs and the total risk in operations with an individual weight assigned to its valuation of the risk, as well as the minimization of expected costs of shortages and surpluses and tardiness penalties associated with the target time goals at the demand points. The risk is captured through the variance of the total operational costs, which is relevant to the reporting of the proper use of funds to stakeholders, including donors. The time goal targets associated with the demand points enable prioritization as to the timely delivery of relief supplies. The framework handles both the pre-positioning of relief supplies, whether local or nonlocal, as well as the procurement (local or nonlocal), transport, and distribution of supplies post-disaster. The time element is captured through link time completion functions as the relief supplies progress along paths in the supply chain network. Each path consists of a series of directed links, from the origin node, which represents the humanitarian organization, to the destination nodes, which are the demand points for the relief supplies. We propose an algorithm, which yields closed form expressions for the variables at each iteration, and

Presented at the 2nd International Conference on Dynamics of Disasters, Kalamata, Greece, June 29–July 2, 2015; revised December 2015. To appear in *Dynamics of Disasters*, I.S. Kotsireas, A. Nagurney, and P.M. Pardalos, Eds., Springer International Publishing Switzerland.

A. Nagurney (✉)
Department of Operations and Information Management, Isenberg School of Management, University of Massachusetts, Amherst, MA 01003, USA
e-mail: nagurney@isenberg.umass.edu

L.S. Nagurney
Department of Electrical and Computer Engineering, University of Hartford, West Hartford, CT 06117, USA

I.S. Kotsireas et al. (eds.), *Dynamics of Disasters—Key Concepts, Models, Algorithms, and Insights*, Springer Proceedings in Mathematics & Statistics 185,
DOI 10.1007/978-3-319-43709-5_12

demonstrate the efficacy of the framework through a series of illustrative numerical examples, in which trade-offs between local versus nonlocal procurement, post- and pre-disaster, are investigated. The numerical examples include a case study on hurricanes hitting Mexico.

Keywords Supply chains • Disaster relief • Humanitarian logistics • Network optimization • Risk reduction • Undertainty • Time constraints • Variational inequalities

1 Introduction

Natural disasters, such as earthquakes, hurricanes, tsunamis, floods, tornadoes, fires, and droughts, invoke all phases of the disaster management cycle from preparedness and mitigation to response and recovery. Notable recent examples of disasters include: Hurricane Katrina in 2005 and Superstorm Sandy in 2012, the two costliest disasters to strike the USA, the earthquake in Haiti in 2010, the triple disaster in Fukushima, Japan in 2011, and the devastating earthquake in Nepal in 2015. As noted in Nagurney and Qiang (2009), the number of disasters is growing as well as the number of people affected by disasters. Hence, the development of appropriate analytical tools that can assist humanitarian organizations and nongovernmental organizations as well as governments in the various disaster management phases has become a challenge to both researchers and practitioners.

Recently, there has been growing interest in constructing integrated frameworks that can assist in multiple phases of disaster management. Network-based models and tools, which allow for graphical depiction of disaster relief supply chains and the flexibility of adding nodes and links, coupled with effective computational procedures, in particular, offer promise. Such models necessarily have to be optimization-based and must incorporate stochastic elements since in disaster situations there is uncertainty associated with the demand for relief supplies and also uncertainty associated with various link costs along with variances.

In addition, as noted in Nagurney et al. (2015), time plays a critical role in disaster relief supply chains and, therefore, time must be a fundamental element in disaster relief models. The US Federal Emergency Management Agency (FEMA) has identified key benchmarks to response and recovery, which emphasize time and are: to meet the survivors' initial demands within 72 h, to restore basic community functionality within 60 days, and to return to as normal of a situation within 5 years (Fugate 2012). Walton et al. (2011) further reinforce the importance of speed in emergency response guidelines for disaster relief operations (see USAID 2005; UNHCR 2007). Timely and efficient delivery of relief supplies to the affected population not only decreases the fatality rate but may also prevent chaos. In the case of cyclone Haiyan, the strongest typhoon ever recorded in terms of wind speed, which devastated areas of Southeast Asia, especially, the Philippines, where 11 million people were affected, slow relief delivery efforts forced people to seek any possible means to survive. Several relief trucks were attacked and had food stolen, and some areas were reported to be on the brink of anarchy (Chicago Tribune 2013; CBS News 2013). In Nepal, post the April 2015 7.8 magnitude earthquake, there was near chaos at the Katmandu airport with relief airplanes not able to land, with numerous Nepalese citizens seeking to leave while Nepalese expatriates attempted to return to help their families (Luke and McVicker 2015). The BBC News (2015) reported that the slow distribution of aid led to clashes between protesters and riot police.

Furthermore, humanitarian relief organizations, for the most part, receive their primary funding and support from donors. Hence, they are responsible to these and other stakeholders in terms of accountability of the use of their financial funds (see Toyasaki and Wakolbinger 2014). It has been estimated that logistics accounts for about 80 % of the total costs in disaster relief (Van Wassenhove 2006). Thus, humanitarian organizations must utilize their resources in the most effective and efficient way while delivering relief supplies in a timely manner. As noted by Tzeng et al. (2007), once a disaster strikes, effective disaster relief efforts can mitigate the damage, reduce the number of fatalities, and bring relief to the survivors. For additional background, see the recent edited volume on disaster management and emergencies by Vitoriano et al. (2013), which includes a survey on decision aid models for humanitarian logistics by Ortuño et al. (2013).

In this paper, we develop a mean-variance disaster relief supply chain network model with stochastic link costs and time targets for delivery of the relief supplies at the demand points, under demand uncertainty. The model is inspired by the supply chain network integration model for risk reduction in the case of mergers and acquisitions developed by Liu and Nagurney (2011), coupled with the integrated disaster relief framework of Nagurney et al. (2015). Liu and Nagurney (2011) used a mean-variance (MV) approach for the measurement of risk associated with link supply chain network costs, but in a corporate, not a humanitarian, setting. That work also assessed synergies associated with mergers and acquisitions.

The MV approach to risk reduction dates to the work of the Nobel laureate Markowitz (1952, 1959) and is still relevant in finance (Schneeweis et al. 2010), in supply chains (Chen and Federgruen 2000; Kim et al. 2007), as well as in disaster relief and humanitarian operations, where the focus, to-date, has been on inventory management (Ozbay and Ozguven 2007; Das 2014). However, the model constructed here is the first to integrate preparedness and response in a supply chain network framework with a mean-variance approach for risk reduction under demand and cost uncertainty and time targets plus penalties for shortages and surpluses. Bozorgi-Amiri et al. (2013) developed a model with uncertainty on the demand side and also in procurement and transportation using expected costs and variability with associated weights but did not consider the critical time elements as well as the possibility of local versus nonlocal procurement post- or pre-disaster.

In addition, Boyles and Waller (2010) developed an MV model for the minimum cost network flow problem with stochastic link costs and emphasized that an MV approach is especially relevant in logistics and distribution problems with critical implications for supply chains. They noted that a solution that only minimizes expected cost and not variances may not be as reliable and robust as one that does.

In our model, the humanitarian organization seeks to minimize its expected total operational costs and the total risk in operations with an individual weight assigned to its valuation of the risk, as well as the minimization of expected costs of shortages and surpluses and tardiness penalties associated with the target time goals at the demand points. The risk is captured through the variance of the total operational costs, which is of relevance also to the reporting of the proper use of funds to stakeholders, including donors. The time goal targets associated with the demand points enable prioritization of demand points as to the timely delivery of relief supplies. This framework handles both the pre-positioning of relief supplies, whether local or nonlocal, as well as the procurement (local or nonlocal), transport, and distribution of supplies post-disaster. There is growing empirical evidence showing that the use of local resources in humanitarian supply chains can have positive impacts (see Matopoulos et al. 2014). Earlier work on procurement with stochastic components did not distinguish between local or nonlocal procurement (see Falasca and Zobel 2011).

The time element in our model is captured through link time completion functions as the relief supplies progress along paths in the supply chain network. Each path consists of a series of directed links, from the origin node, which represents the humanitarian organization, to the destination nodes, which are the demand points for the relief supplies.

The literature on humanitarian operations and disaster relief has been growing. Below we highlight publications that are relevant to aspects of supply chain network activities, such as procurement, transportation, storage, and distribution. Hale and Moberg (2005) proposed a set covering location model to identify secure sites for the storage of emergency supplies. Balcik and Beamon (2005) studied facility location in humanitarian relief. Beamon and Kotleba (2006) developed a stochastic inventory control model determining optimal order quantities and reorder points for a long-term emergency relief response. Barbarosoglu and Arda (2004) and Falasca and Zobel (2011) proposed two-stage stochastic models for the procurement and transportation of disaster relief items. Also, Mete and Zabinsky (2010) introduced a two-stage stochastic model for the storage and distribution of medical supplies to be used in case of emergencies. Huang et al. (2012) presented performance measures for the efficiency, efficacy, and equity of relief distribution.

Nagurney and Qiang (2012) proposed network robustness and performance measures in addition to synergy assessment of supply chain network integration in the case of humanitarian partnerships (see also Nagurney and Qiang 2009; Nagurney et al. 2012a). The synergy measure can be used to determine the potential benefits of horizontal cooperation and coordination between humanitarian organizations. Qiang and Nagurney (2012) introduced a bi-criteria indicator for performance evaluation of supply chains of critical needs products under capacity and demand disruptions. Rottkemper et al. (2012) presented a bi-criteria mixed-integer programming model for the inventory relocation of relief items. Ortuño et al. (2011) and Vitoriano et al.

(2011) developed goal programming frameworks for the distribution of relief goods while considering targets for attributes such as the cost and travel time.

The paper is organized as follows. In Sect. 2, we construct the mean-variance supply chain network model for disaster relief and provide its variational inequality formulation, with nice features for computations. In Sect. 3, we present the Euler method, which yields closed form expressions for the variables at each iteration, and then apply it to solve two sets of numerical examples. The first set consists of a small example with five variants whereas the second set consists of a larger example focusing on Mexico, and a variant. We have identified Mexico as an appropriate setting for the larger set of examples due to its natural disaster risk profile in terms of hurricanes, storms, floods, earthquakes, and droughts. Specifically, we focus on multiple hurricanes hitting Mexico, as happened in 2013, with two hurricanes, Manuel and Ingrid, making landfall within 24 h of each other and affecting Acapulco and the Mexico City area, respectively. In Sect. 4, we summarize the results and present our conclusions.

2 The Mean-Variance Disaster Relief Supply Chain Network Model for Risk Reduction

In this section, we construct the mean-variance disaster relief supply chain network model in which the humanitarian organization seeks to minimize its expected total operational costs and the total risk in operations with an individual weight assigned to its valuation of the risk, as well as the expected costs of shortages and surpluses and tardiness penalties associated with the target time goals at the demand points. The risk is captured through the variance of the total operational costs. The time goal targets associated with the demand points enable prioritization of demand points as to the timely delivery of relief supplies. The framework handles both the pre-positioning of relief supplies and the procurement, transport, and distribution of supplies post-disaster, whether local or nonlocal. The time element is captured through link time completion functions as the relief supplies progress via paths in the supply chain network. The paths consist of a series of directed links, from the origin node to the destination nodes, which are the demand points for the relief supplies.

2.1 Model Foundations and Notation

The network topology of the mean-variance disaster relief supply chain network is given in Fig. 1 and is denoted by $G = [N, L]$, where N denotes the set of nodes and L the set of links. The organization is associated with node 1, which also serves as the (abstract) origin node. The demand points, which receive the disaster relief

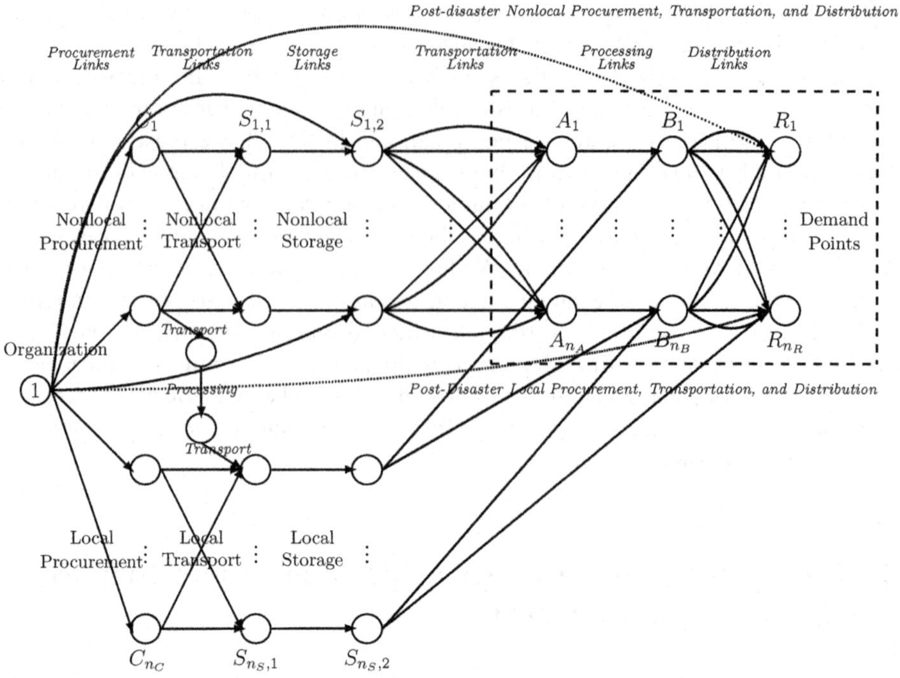

Fig. 1 Network topology of the mean-variance disaster relief supply chain

supplies, are denoted by nodes $R_1, \ldots, R_{n_R}$. We emphasize that the supply chain network topology may be modified/adapted for specific instances and situations. It, nevertheless, reflects the essential elements of a disaster relief supply chain and the associated activities of procurement, transportation, storage, processing, and, finally, the ultimate distribution to the demand points, as reflected by the links in Fig. 1. Also, the progression of time in Fig. 1 is reflected in the link directions from left to right.

Specifically, links joining node 1 to nodes $C_1, \ldots, C_{n_C}$ are procurement links. Procurement, depending upon the scenario, may be done locally or not, as depicted in Fig. 1. Transportation links connect the procurement nodes to storage nodes denoted by $S_1, \ldots, S_{n_S,1}$. Storage is reflected by the links joining the latter nodes to nodes: $S_{1,2}, \ldots, S_{n_S,2}$. Also, the links connecting node 1 to nodes $S_{1,2}, \ldots, S_{n_S,2}$ represent nonlocal procurement post-disaster and, hence, obviate the need for storage on links: $S_{1,1}$ to $S_{1,2}$, through $S_{n_S,1}$ to $S_{n_S,2}$. Joining the storage nodes are transportation links with individual links corresponding to a specific mode of transportation. In humanitarian operations it is important to distinguish among modes of transportation since relief supplies might be airlifted, arrive via ground transportation or even maritime transport, depending on the geography and the status of the critical infrastructure. The nodes: $A_1, \ldots, A_{n_A}$ are the arrival portals with the links emanating from such nodes reflecting processing links. In the case of imports across national boundaries there might be customs inspections, import duties and fees, and other processing prior to the ultimate consolidation for final distribution of supplies (see, e.g., Lorch 2015; Harris 2015). The processing facilities are denoted by nodes: $B_1, \ldots, B_{n_B}$. The links joining the nodes $B_1, \ldots, B_{n_B}$ in Fig. 1 with the demand point nodes $R_1, \ldots, R_{n_R}$ are the distribution links, which include the last mile distribution operations. The supply chain network topology revealed in Fig. 1 is a substantive generalization of the one in Nagurney et al. (2015) to include the options of local procurement, transportation, and distribution post-disaster as reflected by the links joining node 1 to the demand point nodes as well as the partitioning of pre-disaster choices according to whether they are local or not.

We assume that there exists at least one path in the disaster relief supply chain network connecting the origin (node 1) with each demand point: $R_1, \ldots, R_{n_R}$.

The links in the supply chain network are denoted by a, b, c, etc. The paths are denoted by p, q, etc., with the set of paths joining origin node 1 with demand point k denoted by $\mathcal{P}_k$, and the set of paths joining the node 1 with all demand points denoted by $\mathcal{P}$ with this set having $n_{\mathcal{P}}$ elements.

The notation for the model is summarized in Table 1.

The notation is similar to that in Nagurney et al. (2015) but with appropriate additions to capture link total cost uncertainty.

2.2 Formulation of the Mean-Variance Disaster Relief Supply Chain Network Model with Risk Reduction

Before constructing the objective function, we recall some preliminaries.

In the model, the demand is uncertain due to the unpredictability of the actual demand at the demand points. The literature contains examples of supply chain network models with uncertain demand and associated shortage and surplus penalties (see, e.g., Dong et al. 2004; Nagurney et al. 2011, 2015; Nagurney and Masoumi 2012). For example, the probability distribution of demand might be derived using census data and/or information gathered during the disaster preparedness phase. Since d_k denotes the actual (uncertain) demand at destination point k, we have

$$P_k(D_k) = P_k(d_k \leq D_k) = \int_0^{D_k} \mathcal{F}_k(u)du, \qquad k = 1, \ldots, n_R, \tag{1}$$

where P_k and $\mathcal{F}_k$ denote the probability distribution function, and the probability density function of demand at point k, respectively.

Recall from Table 1 that v_k is the "projected demand" for the disaster relief item at demand point $k; k = 1, \ldots, n_R$. The amounts of shortage and surplus at destination node k are calculated, respectively, according to:

$$\Delta_k^- \equiv \max\{0, d_k - v_k\}, \qquad k = 1, \ldots, n_R, \tag{2a}$$

$$\Delta_k^+ \equiv \max\{0, v_k - d_k\}, \qquad k = 1, \ldots, n_R. \tag{2b}$$

Table 1 Notation for the mean-variance disaster relief model

Notation	Definition
x_p	The nonnegative flow of the relief item on path p. We group the flows on all paths into the vector $x \in R_+^{n_\mathcal{P}}$.
f_a	The flow of the relief item on link a; $a \in L$.
v_k	The projected demand for the disaster relief item at point k; $k = 1, \ldots, R_{n_R}$.
d_k	The actual (uncertain) demand at point k; $k = 1, \ldots, R_{n_R}$.
Δ_k^-	The amount of shortage of the relief item at demand point k; $k = 1, \ldots, R_{n_R}$.
Δ_k^+	The amount of surplus of the relief item at demand point k; $k = 1, \ldots, R_{n_R}$.
λ_k^-	The unit penalty corresponding to a shortage of the relief item at demand point k; $k = 1, \ldots, R_{n_R}$.
λ_k^+	The unit penalty corresponding to a surplus of the relief item at demand point k; $k = 1, \ldots, R_{n_R}$.
$\tau_a(f_a)$	The completion time of the activity on link a; $a \in L$, with $\tau_a(f_a) = \hat{t}_a f_a + t_a$, where $\hat{t}_a$ and t_a are ≥ 0, $\forall a \in L$.
T_k	Target for the completion time of the activities on paths corresponding to demand point k determined by the organization's decision-maker where $k = 1, \ldots, n_R$.
T_{kp}	The target time for demand point k with respect to path $p \in \mathcal{P}_k$. $T_{kp} = T_k - t_p$, where $t_p = \sum_{a \in L} t_a \delta_{ap}$, where $\delta_{ap} = 1$, if link a is contained in path p, and is equal to 0, otherwise.
z_p	The amount of deviation with respect to target time T_{kp} associated with late delivery of the relief item to k on path p, $\forall p \in \mathcal{P}$. We group the z_ps into the vector $z \in R_+^{n_\mathcal{P}}$.
$\gamma_k(z)$	The tardiness penalty function corresponding to demand point k; $k = 1, \ldots, n_R$.
ω_a	An exogenous random variable affecting the total operational cost on link a; $a \in L$.
$\hat{c}_a(f_a, \omega_a)$	The total operational cost on link a; $a \in L$.

The expected values of shortage and surplus at each demand point are, hence:

$$E(\Delta_k^-) = \int_{v_k}^{\infty} (u - v_k)\mathcal{F}_k(u)du, \qquad k = 1, \ldots, n_R, \tag{3a}$$

$$E(\Delta_k^+) = \int_{0}^{v_k} (v_k - u)\mathcal{F}_k(u)du, \qquad k = 1, \ldots, n_R. \tag{3b}$$

The expected penalty incurred by the humanitarian organization due to the shortage and surplus of the relief item at each demand point is equal to:

$$E(\lambda_k^- \Delta_k^- + \lambda_k^+ \Delta_k^+) = \lambda_k^- E(\Delta_k^-) + \lambda_k^+ E(\Delta_k^+), \qquad k = 1, \ldots, n_R. \tag{4}$$

We have the following two sets of conservation of flow equations. The projected demand at destination node k, v_k, is equal to the sum of flows on all paths in the set $\mathcal{P}_k$, that is,

$$v_k \equiv \sum_{p \in \mathcal{P}_k} x_p, \qquad k = 1, \ldots, n_R. \tag{5}$$

The flow on link a, f_a, is equal to the sum of flows on paths that contain that link:

$$f_a = \sum_{p \in \mathcal{P}} x_p \, \delta_{ap}, \qquad \forall a \in L, \tag{6}$$

where δ_{ap} is equal to 1 if link a is contained in path p and is 0, otherwise.

The objective function faced by the organization's decision-maker, which he seeks to minimize, is the following:

$$E\left[\sum_{a \in L} \hat{c}_a(f_a, \omega_a)\right] + \alpha \mathit{Var}\left[\sum_{a \in L} \hat{c}_a(f_a, \omega_a)\right] + \sum_{k=1}^{n_R} (\lambda_k^- E(\Delta_k^-) + \lambda_k^+ E(\Delta_k^+)) + \sum_{k=1}^{n_R} \gamma_k(z)$$

$$= \sum_{a \in L} E\left[\hat{c}_a(f_a, \omega_a)\right] + \alpha \mathit{Var}\left[\sum_{a \in L} \hat{c}_a(f_a, \omega_a)\right] + \sum_{k=1}^{n_R} (\lambda_k^- E(\Delta_k^-) + \lambda_k^+ E(\Delta_k^+)) + \sum_{k=1}^{n_R} \gamma_k(z), \tag{7}$$

where E denotes the expected value, *Var* denotes the variance, and α represents the risk aversion factor (weight) for the organization that the organization's decision-maker places on the risk as represented by the variance of the total operational costs. The objective function (7) includes the expected total operational costs on all the links, the weighted variance of those costs, the expected costs due to shortages or surpluses at the demand points, and the sum of tardiness penalties at the demand points in the disaster relief supply chain network.

Here we consider total operational link cost functions of the form:

$$\hat{c}_a = \hat{c}_a(f_a, \omega_a) = \omega_a \hat{g}_a f_a + g_a f_a, \quad \forall a \in L, \tag{8}$$

where $\hat{g}_a$ and g_a are positive-valued for all links $a \in L$. We permit ω_a to follow any probability distribution and the ωs of different supply chain links can be correlated with one another. As noted in Liu and Nagurney (2011), the term $\hat{g}_a f_a$ in (8) represents the part of the total link operational cost that is subject to variation of ω_a with $g_a f_a$ denoting that part of the total cost that is independent of ω_a. The random variables ω_a, $a \in L$ can capture various elements of uncertainty, due, for example, to disruptions because of the disaster, and price uncertainty for storage, procurements, transport, processing, and distribution services.

The goal of the decision-maker is, thus, to minimize the following problem, with the objective function in (7), in lieu of (8), taking the form in (9) below:

$$\text{Minimize} \sum_{a \in L} E(\omega_a)\hat{g}_a f_a + \sum_{a \in L} g_a f_a + \alpha Var(\sum_{a \in L} \omega_a \hat{g}_a f_a)$$
$$+ \sum_{k=1}^{n_R} (\lambda_k^- E(\Delta_k^-) + \lambda_k^+ E(\Delta_k^+)) + \sum_{k=1}^{n_R} \gamma_k(z) \tag{9}$$

subject to constraint (6) and the following constraints:

$$x_p \geq 0, \quad \forall p \in \mathcal{P}, \tag{10}$$

$$z_p \geq 0, \quad \forall p \in \mathcal{P}, \tag{11}$$

$$\sum_{q \in \mathcal{P}} \sum_{a \in L} \hat{t}_a x_q \delta_{aq} \delta_{ap} - z_p \leq T_{kp}, \quad \forall p \in \mathcal{P}_k; k = 1, \ldots, n_R, \tag{12}$$

with the T_ks defined in Table 1. Constraint (10) guarantees that the relief item path flows are nonnegative. Constraint (10) guarantees that the path deviations with respect to target times on the respective paths are nonnegative, and (12) captures the goal target information for the paths.

In view of constraint (6) we can reexpress the objective function in (9) in path flows (rather than in link flows and path flows) to obtain the following optimization problem:

$$\text{Minimize} \sum_{a \in L} \left[E(\omega_a)\hat{g}_a \sum_{q \in \mathcal{P}} x_q \delta_{aq} + g_a \sum_{q \in \mathcal{P}} x_q \delta_{aq} \right] + \alpha Var(\sum_{a \in L} \omega_a \hat{g}_a \sum_{q \in \mathcal{P}} x_q \delta_{aq})$$
$$+ \sum_{k=1}^{n_R} (\lambda_k^- E(\Delta_k^-) + \lambda_k^+ E(\Delta_k^+)) + \sum_{k=1}^{n_R} \gamma_k(z) \tag{13}$$

subject to constraints: (10)–(12).

Let K denote the feasible set:

$$K \equiv \{(x, z, \mu) | x \in R_+^{n_{\mathcal{P}}}, z \in R_+^{n_{\mathcal{P}}}, \text{ and } \mu \in R_+^{n_{\mathcal{P}}}\}, \tag{14}$$

where recall that x is the vector of path flows of the relief item, z is the vector of time deviations on paths, and μ is the vector of Lagrange multipliers corresponding to the constraints in (12) with an individual element corresponding to path p denoted by μ_p.

Before presenting the variational inequality formulation of the optimization problem immediately above, we review the respective partial derivatives of the expected values of shortage and surplus of the disaster relief item at each demand point with respect to the path flows, derived in Dong et al. (2004), Nagurney et al. (2011, 2012b). In particular, they are given by:

$$\frac{\partial E(\Delta_k^-)}{\partial x_p} = P_k\left(\sum_{q\in\mathcal{P}_k} x_q\right) - 1, \quad \forall p \in \mathcal{P}_k; \quad k = 1,\ldots,n_R, \tag{15a}$$

and,

$$\frac{\partial E(\Delta_k^+)}{\partial x_p} = P_k\left(\sum_{q\in\mathcal{P}_k} x_q\right), \quad \forall p \in \mathcal{P}_k; \quad k = 1,\ldots,n_R. \tag{15b}$$

We now present the variational inequality formulation of the mean-variance disaster relief supply chain network problem for risk reduction. We assume that the underlying functions in the model are convex and continuously differentiable. The proof is immediate following the proof of Theorem 1 in Nagurney et al. (2015).

Theorem 1. *The optimization problem (13), subject to its constraints (10)–(12), is equivalent to the variational inequality problem: determine the vector of optimal path flows, the vector of optimal path time deviations, and the vector of optimal Lagrange multipliers* $(x^*, z^*, \mu^*) \in K$, *such that:*

$$\sum_{k=1}^{n_R}\sum_{p\in\mathcal{P}_k}\Bigg[\sum_{a\in L}(E(\omega_a)\hat{g}_a + g_a)\delta_{ap} + \alpha\frac{\partial Var(\sum_{a\in L}\omega_a\hat{g}_a\sum_{q\in\mathcal{P}} x_q^*\delta_{aq})}{\partial x_p}$$

$$+\lambda_k^+ P_k(\sum_{q\in\mathcal{P}_k} x_q^*) - \lambda_k^-(1 - P_k(\sum_{q\in\mathcal{P}_k} x_q^*)) + \sum_{q\in\mathcal{P}}\sum_{a\in L}\mu_q^* g_a\delta_{aq}\delta_{ap}\Bigg] \times [x_p - x_p^*]$$

$$+\sum_{k=1}^{n_R}\sum_{p\in\mathcal{P}_k}\left[\frac{\partial\gamma_k(z^*)}{\partial z_p} - \mu_p^*\right] \times [z_p - z_p^*]$$

$$+\sum_{k=1}^{n_R}\sum_{p\in\mathcal{P}_k}\left[T_{kp}+z_p^* - \sum_{q\in\mathcal{P}}\sum_{a\in L} g_a x_q^*\delta_{aq}\delta_{ap}\right] \times [\mu_p - \mu_p^*] \geq 0, \qquad \forall (x,z,\mu) \in K. \tag{16}$$

Variational inequality (16) can be put into standard form (Nagurney 1999) as follows: determine $X^* \in \mathcal{K}$ such that:

$$\left\langle F(X^*), X - X^*\right\rangle \geq 0, \qquad \forall X \in \mathcal{K}, \tag{17}$$

where $\langle\cdot,\cdot\rangle$ denotes the inner product in n-dimensional Euclidean space. If the feasible set is defined as $\mathcal{K} \equiv K$, and the column vectors $X \equiv (x, z, \mu)$ and $F(X) \equiv (F_1(X), F_2(X), F_3(X))$, where:

$$F_1(X) = \left[\sum_{a\in L}(E(\omega_a)\hat{g}_a + g_a)\delta_{ap} + \alpha\frac{\partial Var(\sum_{a\in L}\omega_a\hat{g}_a\sum_{q\in\mathcal{P}}x_q\delta_{aq})}{\partial x_p}\right.$$

$$\left.+\lambda_k^+ P_k(\sum_{q\in\mathcal{P}_k}x_q) - \lambda_k^-(1-P_k(\sum_{q\in\mathcal{P}_k}x_q)) + \sum_{q\in\mathcal{P}}\sum_{a\in L}\mu_q g_a\delta_{aq}\delta_{ap}, \quad p\in\mathcal{P}_k; \quad k=1,\ldots,n_R\right],$$

$$F_2(X) = \left[\frac{\partial\gamma_k(z)}{\partial z_p} - \mu_p, \quad p\in\mathcal{P}_k; \quad k=1,\ldots,n_R\right],$$

and

$$F_3(X) = \left[T_{kp} + z_p - \sum_{q\in\mathcal{P}}\sum_{a\in L}g_a x_q\delta_{aq}\delta_{ap}, \quad p\in\mathcal{P}_k; \quad k=1,\ldots,n_R, \right], \tag{18}$$

then variational inequality (16) can be reexpressed as standard form (17).

We utilize variational inequality (16) for our computations to obtain the optimal path flows and the optimal path time deviations. Then we use (6) to calculate the optimal link flows of disaster relief items in the supply chain network.

3 The Algorithm and Numerical Examples

In this section, we present the Euler method, which is induced by the general iterative scheme of Dupuis and Nagurney (1993) and then apply it to compute solutions to several numerical examples to illustrate the modeling framework. The realization of the Euler method for the solution of mean-variance disaster relief supply chain network problem governed by variational inequality (16) results in subproblems that can be solved explicitly and in closed form. Specifically, recall that at an iteration τ of the Euler method (see also Nagurney and Zhang 1996) one computes:

$$X^{\tau+1} = P_{\mathcal{K}}(X^\tau - a_\tau F(X^\tau)), \tag{19}$$

where $P_{\mathcal{K}}$ is the projection on the feasible set $\mathcal{K}$ and F is the function that enters the variational inequality problem: determine $X^* \in \mathcal{K}$ such that

$$\langle F(X^*), X - X^*\rangle \geq 0, \quad \forall X \in \mathcal{K}, \tag{20}$$

where $\langle\cdot,\cdot\rangle$ is the inner product in n-dimensional Euclidean space, $X \in R^n$, and $F(X)$ is an n-dimensional function from $\mathcal{K}$ to R^n, with $F(X)$ being continuous.

As shown in Dupuis and Nagurney (1993); see also Nagurney and Zhang (1996), for convergence of the general iterative scheme, which induces the Euler method, among other methods, the sequence $\{a_\tau\}$ must satisfy: $\sum_{\tau=0}^{\infty} a_\tau = \infty$, $a_\tau > 0$,

$a_\tau \to 0$, as $\tau \to \infty$. Specific conditions for convergence of this scheme can be found for a variety of network-based problems, similar to those constructed here, in Nagurney and Zhang (1996) and the references therein.

3.1 Explicit Formulae for the Euler Method Applied to the Disaster Relief Supply Chain Network Variational Inequality (16)

The elegance of this procedure for the computation of solutions to the disaster relief supply chain network problem modeled in Sect. 2 can be seen in the following explicit formulae. Specifically, (19) for the supply chain network problem governed by variational inequality problem (16) yields the following closed form expressions for the product path flows, the time deviations, and the Lagrange multipliers, respectively:

$$x_p^{\tau+1} = \max\{0, x_p^\tau + a_\tau(\lambda_k^-(1 - P_k(\sum_{q\in\mathcal{P}_k} x_q^\tau)) - \lambda_k^+ P_k(\sum_{q\in\mathcal{P}_k} x_q^\tau) - \sum_{a\in L}(E(\omega_a)\hat{g}_a + g_a)\delta_{ap}$$
$$-\alpha \frac{\partial Var(\sum_{a\in L}\omega_a\hat{g}_a\sum_{q\in\mathcal{P}} x_q^\tau \delta_{aq})}{\partial x_p} - \sum_{q\in\mathcal{P}}\sum_{a\in L}\mu_q^\tau g_a\delta_{aq}\delta_{ap})\}, \quad \forall p \in \mathcal{P}_k; \quad k = 1, \ldots, n_R, \tag{21}$$

$$z_p^{\tau+1} = \max\{0, z_p^\tau + a_\tau(\mu_p^\tau - \frac{\partial \gamma_k(z^\tau)}{\partial z_p})\}, \qquad \forall p \in \mathcal{P}_k; \quad k = 1, \ldots, n_R, \text{ and} \tag{22}$$

$$\mu_p^{\tau+1} = \max\{0, \mu_p^\tau + a_\tau(\sum_{q\in\mathcal{P}}\sum_{a\in L} g_a x_q^\tau \delta_{aq}\delta_{ap} - T_{kp} - z_p^\tau\}, \qquad \forall p \in \mathcal{P}_k; \quad k = 1, \ldots, n_R. \tag{23}$$

In view of (21), we can define a generalized marginal total cost on path p; $p \in \mathcal{P}$, denoted by $G\hat{C}'_p$, where

$$G\hat{C}'_p \equiv \sum_{a\in L}(E(\omega_a)\hat{g}_a + g_a)\delta_{ap} + \alpha \frac{\partial Var(\sum_{a\in L}\omega_a\hat{g}_a\sum_{q\in\mathcal{P}} x_q\delta_{aq})}{\partial x_p}. \tag{24}$$

In our numerical examples, we provide explicit formulae for the link generalized marginal total costs, from which the general marginal total cost on each path, as in (24), can be constricted by summing up the former on links that comprise each given path.

3.2 Numerical Examples

In order to fix ideas and concepts, we first present a smaller example for clarity purposes, along with variants, and then construct a larger example, also with a variant. We implemented the Euler method, as described above, in FORTRAN, using a Linux system at the University of Massachusetts, Amherst. The convergence criterion was $\epsilon = 10^{-6}$; that is, the Euler method was considered to have converged if, at a given iteration, the absolute value of the difference of each variable (see (21)–(23)) differed from its respective value at the preceding iteration by no more than ϵ. The sequence $\{a_\tau\}$ was: $0.1(1, \frac{1}{2}, \frac{1}{2}, \frac{1}{3}, \frac{1}{3}, \frac{1}{3} \ldots)$. We initialized the algorithm by setting each variable equal to 0.00.

3.3 Example 1 and Variants

The disaster relief supply chain network topology for Example 1 and its variants is given in Fig. 2. This might correspond to an island location that is subject to major storms. The humanitarian relief organization is depicted by node 1 and there is a single demand point for the relief supplies denoted by R_1, which is located on the island. The organization is considering two options, that is, strategies, reflected by the two paths connecting node 1 with node R_1 with path p_1 consisting of the links: 1, 2, 3, and 4, and path p_2 consisting of the links: 5, 6, 7, and 8. Path p_1 consists of nonlocal post-disaster procurement, transport, processing, and ultimate distribution, whereas path p_2 consists of the activities: local procurement, local transport and local storage, pre-disaster, followed by local transport and distribution. The local transport and distribution are done by ground transport. However, the transport on link 2 is done by air.

The covariance matrix associated with the link total cost functions $\hat{c}_a(f_a, \omega_a)$, $a \in L$, is the 8×8 matrix $\sigma^2 I$. In the variants of Example 1 we explore different values for σ^2 and also different values for α, the risk aversion factor (see (13)). The organization's risk aversion factor $\alpha = 1$ in Example 1 and its Variants 1, 2, and 3.

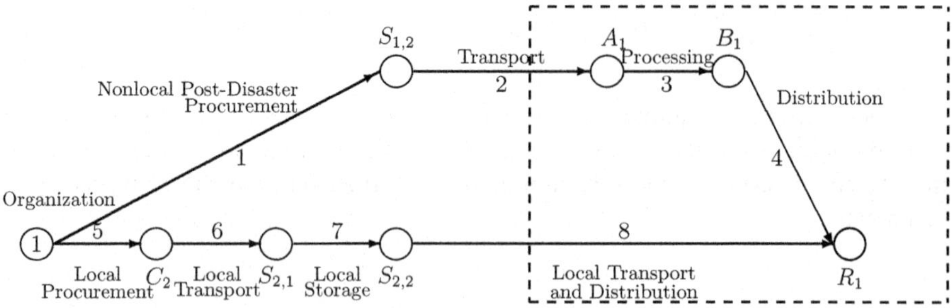

Fig. 2 Disaster relief supply chain network topology for Example 1 and its variants

The demand for the relief item at the demand point R_1 (in thousands of units) is assumed to follow a uniform probability distribution on the interval [10, 20]. The path flows and the link flows are also in thousands of units. Therefore,

$$P_{R_1}(\sum_{p\in\mathcal{P}_1} x_p) = \frac{\sum_{p\in\mathcal{P}_1} x_p - 10}{20-10} = \frac{x_{p_1} + x_{p_2} - 10}{10}.$$

We now describe how we construct the marginalized total link costs for the numerical examples from which the marginalized total path costs as in (24) are then constructed.

For our numerical examples, we have that:

$$\sum_{a\in L} \sigma^2 \hat{g}_a^2 f_a^2 = Var(\sum_{a\in L} \omega_a \hat{g}_a f_a) = Var(\sum_{a\in L} \omega \hat{g}_a \sum_{q\in\mathcal{P}} x_q \delta_{aq}), \tag{25}$$

so that:

$$\frac{\partial Var(\sum_{a\in L} \omega_a \hat{g}_a \sum_{q\in\mathcal{P}} x_q \delta_{aq})}{\partial x_p} = 2\sigma^2 \sum_{a\in L} \hat{g}_a^2 f_a \delta_{ap}. \tag{26}$$

In view of (26) and (24) we may define the generalized marginal total cost on a link a, $\hat{gc}'_a$, as:

$$\hat{gc}'_a \equiv E(\omega_a)\hat{g}_a + g_a + \alpha 2\sigma^2 \hat{g}_a^2 f_a, \tag{27}$$

so that

$$\hat{GC}'_p = \sum_{a\in L} \hat{gc}'_a \delta_{ap}, \quad \forall p \in \mathcal{P}. \tag{28}$$

Table 2 contains the link total operational cost functions, the expected value of the random variable associated with the total operational cost on each link, and the marginal generalized total link cost, as well as the link time completion functions, and the optimal link flows for Example 1 with $\sigma^2 = 0.1$ and for Variant 1 with $\sigma^2 = 1$. The time target at demand point R_1, $T_1 = 48$ (in hours). The link time completion functions for links: 5, 6, and 7 are 0.00 since these are completed prior to the disaster and the supplies on the path with these links are, hence, immediately available for local transport and distribution. Also, we set $\lambda_1^- = 1000$ and $\lambda_1^+ = 100$. The organization is significantly more concerned with a shortage of the relief item than with a surplus. The tardiness penalty function $\gamma_{R_1}(z) = 3(\sum_{p\in\mathcal{P}_{R_1}} z_p^2)$.

The optimal flow on path p_1, $x^*_{p_1}$, in Example 1 with $\sigma^2 = 0.1$ is 4.70. and that for path p_2, $x^*_{p_2}$, is 14.18, with the projected demand $v_{R_1} = x^*_{p_1} + x^*_{p_2} = 18.88$. In Variant 1 of Example 1 with $\sigma^2 = 1$, the new optimal path flow on path p_1, $x^*_{p_1} = 4.90$, and on path p_2, $x^*_{p_2} = 12.84$, with $v_{R_1} = x^*_{p_1} + x^*_{p_2} = 17.74$. The values of $z^*_{p_1}$ and $z^*_{p_2}$

Table 2 Link total cost, expected value of random link cost, marginal generalized link total cost, and time completion functions for Example 1 and Variant 1 and optimal link flows

Link a	$\hat{c}_a(f_a, \omega_a)$	$E(\omega_a)$	Marginal generalized total link cost $g\hat{c}'_a$	$\tau_a(f_a)$	f_a^*;$\alpha = 1$; $\sigma^2 = 0.1$	f_a^*; $\alpha = 1$; $\sigma^2 = 1$
1	$\omega_1 3f_1 + f_1$	1	$\alpha 18\sigma^2 f_1 + 4$	$f_1 + 1$	4.70	4.90
2	$\omega_2 2f_2 + f_2$	1	$\alpha 8\sigma^2 f_2 + 3$	$f_2 + 2$	4.70	4.90
3	$\omega_3 .5f_3 + f_3$	1	$\alpha .5\sigma^2 f_3^2 + 1.5$	$f_3 + .5$	4.70	4.90
4	$\omega_4 .4f_4 + f_4$	1	$\alpha .32\sigma^2 f_4 + 1.4$	$f_4 + 1$	4.70	4.90
5	$\omega_5 2f_5 + f_5$	1	$\alpha 8\sigma^2 f_5 + 3$	0.00	14.18	12.84
6	$\omega_6 .1f_6 + f_6$	1	$\alpha .02\sigma^2 f_6 + 1.1$	0.00	14.18	12.84
7	$\omega_7 f_7 + f_7$	1	$\alpha 2\sigma^2 f_7 + 2$	0.00	14.18	12.84
8	$\omega_8 .5f_8 + f_8$	1	$\alpha .5\sigma^2 f_8 + 1.5$	$.2f_8 + 2$	14.18	12.84

are both 0.00 for both these examples and the Lagrange multipliers $\mu_{p_1}^*$ and $\mu_{p_2}^*$ are also both 0.00 since the time target for delivery at R_1, post-disaster, is met by both paths for R_1.

One can see from the optimal solution to Example 1 and Variant 1 that, as the variance–covariance term σ^2 increases from 0.1 to 1, the amount of optimal flow on path p_2, which corresponds to local procurement, transport, and storage, decreases whereas the amount procured nonlocally post-disaster, increases. Given increased uncertainty as to the operational costs locally since the disaster may impact the storage location(s), for example, and local transport routes as well, it is better to pre-position less of the relief item locally. Also, interestingly, when $\sigma^2 = 1$, less of the relief item is provided (17.74) than when $\sigma^2 = 0.1$ (18.88). The humanitarian relief organization must report to its stakeholders, including donors, and, hence, it must adhere to the minimization of its objective function and with greater variability, there are greater associated costs.

Variants 2 and 3 of Example 1 are constructed as follows and the data are reported in Table 3. For Variant 2, we retain the data for Example 1 with $\sigma^2 = 0.1$ but now assume that air transport, due to the expected storm damage of the island airport, is no longer possible. Maritime transport is, nevertheless, available, so link 2 in Fig. 2 now corresponds to maritime transport rather than air transport. All the data, hence, for Variant 2 are as for Example 1 except that the total operational cost data and the time completion data for link 2 change as reported in Table 3.

Variant 3 is constructed from Variant 2 but with $\sigma^2 = 1$ (as in Variant 1 of Example 1). The optimal solutions for Variants 2 and 3 are reported in Table 3. In Variant 2, only the pre-positioning of relief items locally with local procurement as a strategy is optimal since $x_{p_1}^* = 0.00$ and $x_{p_2}^* = 18.84$. The maritime transport is simply too costly. The time target is met with the pre-positioning strategy and, hence, the time deviations on the paths, $z_{p_1}^*$ and $z_{p_2}^*$, are equal to 0.00 as are the path Lagrange multipliers: $\mu_{p_1}^*$ and $\mu_{p_2}^*$. In Variant 3, on the other hand, as the covariance σ^2 term increases from 0.1 to 1, there is diversification of risk, with both strategies now being applied, that is, maritime transport, post-disaster, and the pre-positioning

Table 3 Link total cost, expected value of random link cost, marginal generalized link total cost, and time completion functions for Example 1 Variants 2 and 3 and optimal link flows

Link a	$\hat{c}_a(f_a, \omega_a)$	$E(\omega_a)$	Marginal generalized total link cost $g\hat{c}'_a$	$\tau_a(f_a)$	$f_a^*; \alpha = 1; \sigma^2 = 0.1$	$f_a^*; \alpha = 1; \sigma^2 = 1$
1	$\omega_1 3f_1 + f_1$	1	$\alpha 18\sigma^2 f_1 + 4$	$f_1 + 1$	0.00	0.51
2	$\omega_2 12f_2 + 10f_2$	1	$\alpha 288\sigma^2 f_2 + 3$	$3f_2 + 10$	0.00	0.51
3	$\omega_3 .5f_3 + f_3$	1	$\alpha .5\sigma^2 f_3 + 1.5$	$f_3 + .5$	0.00	0.51
4	$\omega_4 .4f_4 + f_4$	1	$\alpha .32\sigma^2 f_4 + 1.4$	$f_4 + 1$	0.00	0.51
5	$\omega_5 2f_5 + f_5$	1	$\alpha 8\sigma^2 f_5 + 3$	0.00	18.84	16.90
6	$\omega_6 .1f_6 + f_6$	1	$\alpha .02\sigma^2 f_6 + 1.1$	0.00	18.84	16.90
7	$\omega_7 f_7 + f_7$	1	$\alpha 2\sigma^2 f_7 + 2$	0.00	18.84	16.90
8	$\omega_8 .5f_8 + f_8$	1	$\alpha .5\sigma^2 f_8 + 1.5$	$.2f_8 + 2$	18.84	16.90

Table 4 Link total cost, expected value of random link cost, marginal generalized link total cost, and time completion functions for Example 1 Variants 4 and 5 and optimal link flows

Link a	$\hat{c}_a(f_a, \omega_a)$	$E(\omega_a)$	Marginal generalized total link cost $g\hat{c}'_a$	$\tau_a(f_a)$	$f_a^*; \alpha = 10; \sigma^2 = 1$	$f_a^*; \alpha = 100; \sigma^2 = 1$
1	$\omega_1 3f_1 + f_1$	1	$\alpha 18\sigma^2 f_1 + 4$	$f_1 + 1$	3.17	0.68
2	$\omega_2 2f_2 + f_2$	1	$\alpha 8\sigma^2 f_2 + 3$	$f_2 + 2$	3.17	0.68
3	$\omega_3 .5f_3 + f_3$	1	$\alpha .5\sigma^2 f_3 + 1.5$	$f_3 + .5$	3.17	0.68
4	$\omega_4 .4f_4 + f_4$	1	$\alpha .32\sigma^2 f_4 + 1.4$	$f_4 + 1$	3.17	0.68
5	$\omega_5 2f_5 + f_5$	1	$\alpha 8\sigma^2 f_5 + 3$	0.00	8.10	1.74
6	$\omega_6 .1f_6 + f_6$	1	$\alpha .02\sigma^2 f_6 + 1.1$	0.00	8.10	1.74
7	$\omega_7 f_7 + f_7$	1	$\alpha 2\sigma^2 f_7 + 2$	0.00	8.10	1.74
8	$\omega_8 .5f_8 + f_8$	1	$\alpha .5\sigma^2 f_8 + 1.5$	$.2f_8 + 2$	8.10	1.74

of supplies locally. The time target is met in Variant 3 as well. In Variant 2, $v_{R_1} = 18.84$, whereas in Variant 3, $v_{R_1} = 17.41$. We see, as we did in Table 2, that an increase in σ^2 results in fewer relief supplies being delivered in total according to the optimal solution. Hence, relief organizations should try, if at all possible, to reduce the uncertainty associated with their total operational costs in their disaster relief supply chain networks.

In Variants 4 and 5 we explore the impact on the strategies and on the optimal link flows of increasing the risk aversion factor α. Specifically, in Variant 4 we utilize the Variant 1 data in Table 2 but we increase α to 10 and in Variant 5 we increase α even more to 100. We report the input data and results for $\alpha = 10$ and for $\alpha = 100$ in Table 4.

In Variant 4, the optimal path flow pattern is: $x_{p_1}^* = 3.17$ and $x_{p_2}^* = 8.10$, with $v_{R_1} = 11.27$. In Variant 5, the optimal path flow pattern is: $x_{p_1}^* = 0.68$ and $x_{p_2}^* = 1.74$, with $v_{R_1} = 2.46$. As the risk aversion factor α increases, the flows on the paths decrease and, hence, also the total relief supply deliveries at the demand point R_1 decrease. In Variants 4 and 5 the time target is, again, met and, hence, the values of $z_{p_1}^*$, $z_{p_2}^*$ and $\mu_{p_1}^*$ and $\mu_{p_2}^*$ are again all 0.00.

3.4 Example 2 and Variant

Example 2 and its variant consider a realistic, larger scenario setting. The supply chain network topology is as given in Fig. 3. Specifically, with the larger Example 2, and its variant, we focus on Mexico.

According to the United Nations (2011), Mexico is ranked as one of the world's 30 most exposed countries to three or more types of natural disasters, notably, storms, hurricanes, floods as well as earthquakes, and droughts. For example, as reported by The International Bank for Reconstruction and Development/The World Bank (2012), 41 % of Mexico's national territory is exposed to storms, hurricanes, and floods; 27 % to earthquakes, and 29 % to droughts. The hurricanes can come from the Atlantic or Pacific oceans or the Caribbean. As noted by de la Fuente (2011), the single most costly disaster in Mexico were the 1985 earthquakes, followed by the floods in the southern state of Tabasco in 2007, with damages of more than 3.1 billion US dollars.

We consider a humanitarian organization such as the Mexican Red Cross, which is interested in preparing for another possible hurricane, and recall the devastation wrought by Hurricane Manuel and Hurricane Ingrid, which struck Mexico within a 24-h period in September 2013. Ingrid caused 32 deaths, primarily, in eastern Mexico, whereas Manuel resulted in at least 123 deaths, primarily in western Mexico (NOAA 2014). According to Pasch and Zelinsky (2014), the total economic impact of Manuel alone was estimated to be approximately $4.2 billion (US), with the biggest losses occurring in Guerrero. In particular, in Example 2, we assume that the Mexican Red Cross is mainly concerned about the delivery of relief supplies

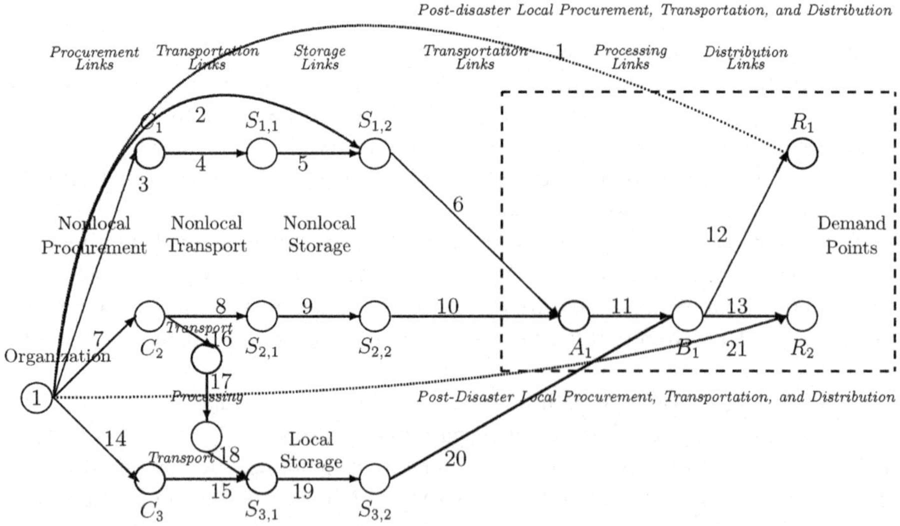

Fig. 3 Disaster relief supply chain network topology for Example 2 and its variant

to the Mexico City area and the Acapulco area. Ingrid affected Mexico City and Manuel affected the Acapulco area and also points northwest.

The Mexican Red Cross represents the organization in Fig. 3 and is denoted by node 1. There are two demand points, R_1 and R_2, for the ultimate delivery of the relief supplies. R_1 is situated closer to Mexico City and R_2 is closer to Acapulco. Nonlocal procurement is done through two locations in Texas, C_1 and C_2. Because of good relationships with the USA and the American Red Cross, there are two nonlocal storage facilities that the Mexican Red Cross can utilize, both located in Texas, and represented by links 5 and 9 emanating from $S_{1,1}$ and $S_{2,1}$, respectively. Local storage, on the other hand, is depicted by the link emanating from node $S_{3,1}$, link 19. The Mexican Red Cross can also procure locally (see C_3). Nonlocal procurement, post-disaster, is depicted by link 2, whereas procurement locally, post-disaster, and direct delivery to R_1 and R_2 are depicted by links 1 and 21, respectively. Link 11 is a processing link to reflect processing of the arriving relief supplies from the USA and we assume one portal A_1, which is in southcentral Mexico. Link 17 is also a processing link but that processing is done prior to storage locally and pre-disaster. Such a link is needed if the goods are procured nonlocally (link 7). The transport is done via road in the disaster relief supply chain network in Fig. 3.

The demand for the relief items at the demand point R_1 (in thousands of units) is assumed to follow a uniform probability distribution on the interval $[20, 40]$. The path flows and the link flows are also in thousands of units. Therefore,

$$P_{R_1}(\sum_{p\in\mathcal{P}_1} x_p) = \frac{\sum_{p\in\mathcal{P}_1} x_p - 20}{40-20} = \frac{\sum_{i=1}^{6} x_{p_i} - 20}{20}.$$

Also, the demand for the relief item at R_2 (in thousands of units) is assumed to follow a uniform probability distribution on the interval $[20, 40]$. Hence,

$$P_{R_2}(\sum_{p\in\mathcal{P}_2} x_p) = \frac{\sum_{p\in\mathcal{P}_1} x_p - 20}{40-20} = \frac{\sum_{i=7}^{12} x_{p_i} - 20}{20}.$$

The time targets for the delivery of supplies at R_1 and R_2, respectively, in hours, are: $T_1 = 48$ and $T_2 = 48$. The penalties at the two demand points for shortages are: $\lambda_1^- = 10{,}000$ and $\lambda_2^- = 10{,}000$ and for surpluses: $\lambda_1^+ = 100$ and $\lambda_2^+ = 100$. The tardiness penalty function $\gamma_{R_1}(z) = 3(\sum_{p\in\mathcal{P}_{R_1}} z_p^2)$ and the tardiness penalty function $\gamma_{R_2}(z) = 3(\sum_{p\in\mathcal{P}_{R_2}} z_p^2)$.

As in Example 1 and its variants, we assume that, for Example 2, the covariance matrix associated with the link total cost functions $\hat{c}_a(f_a, \omega_a)$, $a \in L$, is a 21×21 matrix $\sigma^2 I$. In Example 2, $\sigma^2 = 1$ and the risk aversion factor $\alpha = 10$ since the humanitarian organization is risk averse with respect to its costs associated with its operations.

The additional data for Example 2 are given in Table 5, where we also report the computed optimal link flows via the Euler method, which are calculated from the

Table 5 Link total cost, expected value of random link cost, marginal generalized link total cost, and time completion functions for Example 2 and optimal link flows

Link a	$\hat{c}_a(f_a, \omega_a)$	$E(\omega_a)$	Marginal generalized total link cost $\hat{gc}'_a$	$\tau_a(f_a)$	f_a^*; $\alpha = 10$; $\sigma^2 = 1$
1	$\omega_1 6f_1 + f_1$	2	$\alpha 72\sigma^2 f_1 + 13$	$f_1 + 15$	9.07
2	$\omega_2 3f_2 + f_2$	2	$\alpha 18\sigma^2 f_2 + 7$	$f_2 + 7$	2.54
3	$\omega_3 2f_3 + f_3$	1	$\alpha 8\sigma^2 f_3 + 3$	0.00	2.57
4	$\omega_4 3f_4 + f_4$	1	$\alpha 18\sigma^2 f_4 + 4$	0.00	2.57
5	$\omega_5 2f_5 + f_5$	1	$\alpha 8\sigma^2 f_5 + 3$	0.00	2.57
6	$\omega_6 2f_6 + f_6$	2	$\alpha 8\sigma^2 f_6 + 5$	$2f_6 + 10$	5.11
7	$\omega_7 2f_7 + f_7$	1	$\alpha 8\sigma^2 f_7 + 3$	0.00	8.51
8	$\omega_8 3f_8 + f_8$	1	$\alpha 18\sigma^2 f_8 + 4$	0.00	4.36
9	$\omega_9 2f_9 + f_9$	1	$\alpha 8\sigma^2 f_9 + 3$	0.00	4.36
10	$\omega_{10} 2f_{10} + f_{10}$	1	$\alpha 8\sigma^2 f_{10} + 3$	$2f_{10} + 10$	4.36
11	$\omega_{11} f_{11} + f_{11}$	2	$\alpha 2\sigma^2 f_{11} + 3$	$f_{11} + 2$	9.47
12	$\omega_{12} f_{12} + f_{12}$	2	$\alpha 2\sigma^2 f_{12} + 3$	$f_{12} + 6$	17.78
13	$\omega_{13} f_{13} + f_{13}$	2	$\alpha 2\sigma^2 f_{13} + 3$	$f_{13} + 7$	17.64
14	$\omega_{14} f_{14} + f_{14}$	1	$\alpha 2\sigma^2 f_{14} + 2$	0.00	21.79
15	$\omega_{15} f_{15} + f_{15}$	1	$\alpha 2\sigma^2 f_{15} + 2$	0.00	21.79
16	$\omega_{16} f_{16} + f_{16}$	1	$\alpha 2\sigma^2 f_{16} + 2$	0.00	4.15
17	$\omega_{17} .5f_{17} + f_{17}$	1	$\alpha\sigma^2 .5f_{17} + 1.5$	0.00	4.15
18	$\omega_{18} f_{18} + f_{18}$	1	$\alpha 2\sigma^2 f_{18} + 2$	0.00	4.15
19	$\omega_{19} .5f_{19} + f_{19}$	2	$\alpha\sigma^2 .5f_{19} + 1.5$	0.00	25.94
20	$\omega_{20} f_{20} + f_{20}$	2	$\alpha 2\sigma^2 f_{20} + 2$	$2f_{20} + 5$	25.94
21	$\omega_{21} 6f_{21} + f_{21}$	2	$\alpha 72\sigma^2 f_{21} + 13$	$f_{21} + 14$	9.13

computed path flows reported in Table 6. Note that the time completion functions in Table 5, $\tau_a(f_a)$, $\forall a \in L$, are 0.00 if the links correspond to procurement, transport, and storage, pre-disaster, since such supplies are immediately available for shipment once a disaster strikes.

The definitions of the paths joining node 1 with R_1 and node 1 with R_2, the optimal path flows, optimal path deviations, and the optimal Lagrange multipliers for Example 2 are reported in Table 6. Note that there are 6 paths joining node 1, representing the organization with R_1, and 6 paths joining node 1 with R_2. The paths represent sequences of decisions and activities that must be executed for the relief supplies to reach the destinations.

The largest volumes of relief supplies flow on paths p_1 and p_6 for R_1 and on paths p_{11} and p_{12} for R_2. All these paths correspond to local procurement. Paths p_6 and p_{11} correspond also to local storage. The projected demands are: $v_{R_1} = 26.84$ and $v_{R_2} = 26.76$.

Both pre-positioning and procurement post-disaster strategies are optimal and, hence, used. This makes sense since the organization is interested in risk reduction and, therefore, utilizes a portfolio of strategies. In fact, in Example 2 all paths have positive flow.

Table 6 Path definitions, target times, optimal path flows, optimal path time deviations, and optimal lagrange multipliers for Example 2

	Path definition (links)	x_p^*	z_p^*	μ_p^*
$\mathcal{P}_{R_1}$: set of paths corresponding to demand point R_1	$p_1 = (1)$	9.07	0.00	0.00
	$p_2 = (2, 6, 11, 12)$	1.27	34.75	208.53
	$p_3 = (3, 4, 5, 6, 11, 12)$	1.29	25.26	151.56
	$p_4 = (7, 8, 9, 10, 11, 12)$	2.18	23.78	142.69
	$p_5 = (7, 16, 17, 18, 19, 20, 12)$	2.98	50.48	302.85
	$p_6 = (14, 15, 19, 20, 12)$	10.06	50.48	302.85
$\mathcal{P}_{R_2}$: set of paths corresponding to demand point R_2	$p_7 = (2, 6, 11, 13)$	1.27	35.48	212.88
	$p_8 = (3, 4, 5, 6, 11, 13)$	1.29	25.99	155.91
	$p_9 = (7, 8, 9, 10, 11, 13)$	2.18	24.51	147.04
	$p_{10} = (7, 16, 17, 18, 19, 20, 13)$	1.17	51.20	307.19
	$p_{11} = (14, 15, 19, 20, 13)$	11.74	51.20	307.19
	$p_{12} = (21)$	9.13	0.00	0.00

3.5 *Example 2–Variant 1*

In Variant 1 of Example 2, we kept the data as in Example 2, but now we assumed that the humanitarian organization has a better forecast for the demand at the two demand points. The demand for the relief items at the demand point R_1 again follows a uniform probability distribution but on the interval [30, 40] so that:

$$P_{R_1}(\sum_{p\in\mathcal{P}_1} x_p) = \frac{\sum_{p\in\mathcal{P}_1} x_p - 30}{40 - 30} = \frac{\sum_{i=1}^{6} x_{p_i} - 30}{10}.$$

Also, the demand for the relief item at R_2 follows a uniform probability distribution on the interval [30, 40] so that:

$$P_{R_2}(\sum_{p\in\mathcal{P}_2} x_p) = \frac{\sum_{p\in\mathcal{P}_2} x_p - 30}{40 - 30} = \frac{\sum_{i=7}^{12} x_{p_i} - 30}{10}.$$

The computed path flows are reported in Table 7.

The projected demands are: $v_{R_1} = 31.84$ and $v_{R_2} = 31.79$. The greatest percentage increase in path flow volumes occurs on paths p_1 and p_6 for demand point R_1 and on paths p_{11} and p_{12} for demand point R_2, reinforcing the results obtained for Example 2.

For both Example 2 and its variant the time targets are met for paths p_1 and p_2 since $\mu_{p_1}^*$ and $\mu_{p_2}^* = 0.00$ for both examples. Hence, direct local procurement post-disaster is effective time-wise and cost-wise. Mexico is a large country and this result is quite reasonable.

Table 7 Path definitions, target times, optimal path flows, optimal path time deviations, and optimal lagrange multipliers for Variant 1 of Example 2

	Path definition (links)	x_p^*	z_p^*	μ_p^*
$\mathcal{P}_{R_1}$: set of paths corresponding to demand point R_1	$p_1 = (1)$	11.30	0.00	0.00
	$p_2 = (2, 6, 11, 12)$	1.37	43.13	258.78
	$p_3 = (3, 4, 5, 6, 11, 12)$	1.49	33.42	200.49
	$p_4 = (7, 8, 9, 10, 11, 12)$	2.58	32.28	193.69
	$p_5 = (7, 16, 17, 18, 19, 20, 12)$	2.81	64.37	386.19
	$p_6 = (14, 15, 19, 20, 12)$	12.29	64.37	386.19
$\mathcal{P}_{R_2}$: set of paths corresponding to demand point R_2	$p_7 = (2, 6, 11, 13)$	1.37	43.92	263.49
	$p_8 = (3, 4, 5, 6, 11, 13)$	1.49	34.20	205.20
	$p_9 = (7, 8, 9, 10, 11, 13)$	2.57	33.07	198.40
	$p_{10} = (7, 16, 17, 18, 19, 20, 13)$	1.96	65.15	390.90
	$p_{11} = (14, 15, 19, 20, 13)$	13.04	65.15	390.90
	$p_{12} = (21)$	11.36	0.00	0.00

4 Summary and Conclusions

In this paper, we developed a mean-variance disaster relief supply chain network model for risk reduction with stochastic link costs, uncertain demands for the relief supplies, and time targets associated with the demand points. The humanitarian organization seeks to minimize the expected value of the total operational costs and the weighted variance of these costs in the supply chain network plus the penalized expected shortages and surpluses as well as the deviations from the time targets. Each link has an associated time completion function and the decision-maker determines his risk aversion. This framework handles, in an integrated manner, both the pre-positioning of supplies, which can be local or nonlocal, and the procurement of supplies, both local and nonlocal, post-disaster. The model allows for the investigation of the optimal strategies associated with the paths which are composed of links comprising the necessary activities from procurement to ultimate delivery of the relief supplies to the victims at the demand points.

We presented the optimization model, along with its variational inequality formulation, which enables computation via the Euler method, which, in turn, yields closed form expressions for the path variables at each iteration. Through a series of numerical examples, we illustrated the concepts and computational procedure. Specifically, we presented a series of smaller examples and then construct a set of larger examples, based on a study focusing on Mexico, in the case of hurricanes. We find that, in the case of the Mexico study, although all strategies are used, in that all the path flows are positive, the most highly recommended strategies, in terms of path flow volumes, are those with local procurement, whether with storage pre-disaster, or direct procurement, post-disaster.

The model extends the model of Nagurney et al. (2015) in several dimensions:

1. It considers stochastic link costs, which are relevant given uncertainty in disaster relief supply chain network operations.
2. The objective function includes the minimization of the expected costs as well as the variance with an associated weight for the latter to denote the humanitarian organization's value of risk reduction.
3. The supply chain network topology allows for the procurement and pre-positioning of supplies locally and is more general than that in earlier literature.
4. The generality of the framework allows for numerous sensitivity analysis exercises to evaluate risk aversion, the assessment of the impacts of the size of penalties on shortages and supplies, as well as modifications to the cost and time completion functions.

The framework consolidates decision-making associated with two phases of disaster management: preparedness and response, incorporates uncertainty in costs and demands, and includes the critical time element. Future research may include extending this framework to assess synergies associated with horizontal cooperation among humanitarian organizations in relief operations. In addition, it would be interesting to consider multiple supplies with different associated priorities as well as to include transportation time uncertainty in future work.

Acknowledgements This paper is dedicated to the students in Professor Anna Nagurney's Humanitarian Logistics and Healthcare class in 2015 at the Isenberg School of Management and to all the victims of natural disasters over the centuries as well as to humanitarian professionals.

Professor Anna Nagurney thanks Professor Panos M. Pardalos of the University of Florida and Professor Ilias Kootsireas of Wilfrid Laurier University for the great collaboration on the co-organization of the 2nd International Conference on Dynamics of Disasters in Kalamata, Greece.

The authors also thank the speakers and participants in the conference for comments and stimulating discussions on themes of the conference.

The authors acknowledge helpful comments from two anonymous reviewers on an earlier version of this paper and acknowledge Professor Kotsireas for handling the reviewing process.

References

Balcik, B., Beamon, B.M.: Facility location in humanitarian relief. Int. J. Log. Res. Appl. **11**(2), 101–121 (2005)

Barbarosoglu, G., Arda, Y.: A two-stage stochastic programming framework for transportation planning in disaster response. J. Oper. Res. Soc. **55**(1), 43–53 (2004)

BBC News: Nepal quake: towns near epicentre 'devastated' - Red Cross, 1 May 2015

Beamon, B., Kotleba, S.: Inventory management support systems for emergency humanitarian relief operations in South Sudan. Int. J. Logist. Manag. **17**(2), 187–212 (2006)

Boyles, S.D., Waller, S.T.: A mean-variance model for the minimum cost flow problem with stochastic arc costs. Networks **56**(3), 215–227 (2010)

Bozorgi-Amiri, A., Jabalameli, M.S., Mirzapour Al-e-Hashem, S.M.J.: A multi-objective robust stochastic programming model for disaster relief logistics under uncertainty. OR Spektrum **35**, 905–933 (2013)

CBS News: Typhoon Haiyan slams into northern Vietnam, 10 Nov 2013. Available online at http://www.cbsnews.com/news/typhoon-haiyan-slams-into-northern-vietnam

Chen, F., Federgruen, A.: Mean-variance analysis of basic inventory models. Working Paper, Columbia Business School, Columbia University, New York (2000)

Chicago Tribune: Typhoon Haiyan: Desperate Philippine survivors turn to looting, 13 Nov 2013. Available online at http://http://www.chicagotribune.com/news/chi-philippines-typhoon-haiyan-20131113,0,4099086,full.story

Das, R.: Advancement on uncertainty modeling in humanitarian logistics for earthquake response. PhD dissertation, The Graduate School of Science and Engineering, Department of International Development Engineering, Tokyo Institute of Technology (2014)

de la Fuente, A.: Government expenditures in pre and post-disaster risk management. The World Bank, Washington, DC (2011)

Dong, J., Zhang, D., Nagurney, A.: A supply chain network equilibrium model with random demands. Eur. J. Oper. Res. **156**, 194–212 (2004)

Dupuis, P., Nagurney, A.: Dynamical systems and variational inequalities. Ann. Oper. Res. **44**, 9–42 (1993)

Falasca, M., Zobel, C.W.: A two-stage procurement model for humanitarian relief supply chains. J. Humanitarian Logist. Supply Chain Manag. **1**(2), 151–169 (2011)

Fugate, W.: The state of FEMA – leaning forward: go big, go early, go fast, be smart. A report by the US Department of Homeland Security. Available online at: www.fema.gov/pdf/about/state_of_fema/state_of_fema.pdf (2012)

Hale, T., Moberg, C.: Improving supply chain disaster preparedness: a decision process for secure site location. Int. J. Phys. Distrib. Logist. Manag. **35**(3), 195–207 (2005)

Harris, G.: Nepal's bureaucracy blamed as quake relief supplies pile up. The New York Times, 4 May 2015

Huang, M., Smilowitz, K., Balcik, B.: Models for relief routing: equity, efficiency and efficacy. Transp. Res. E **48**, 2–18 (2012)

Kim, S.H., Cohen, M., Netessine, S.: Performance contracting in after-sales service supply chains. Manag. Sci. **53**, 1843–1858 (2007)

Liu, Z., Nagurney, A.: Risk reduction and cost synergy in mergers and acquisitions via supply chain network integration. J. Financ. Decis. Making **7**(2), 1–18 (2011)

Lorch, D.: Rural Nepal still waiting for relief after devastating earthquake. USA Today, 5 May 2015

Luke, S., McVicker, L.: Local Nepal earthquake survivor: "it was utter chaos." San Diego News, 29 April 2015

Markowitz, H.: Portfolio selection. J. Financ. **56**(6), 2019–2065 (1952)

Markowitz, H.: Portfolio Selection: Efficient Diversification of Investment. Wiley, New York (1959)

Matopoulos, A., Kovacs, G., Hayes, O.: Local resources and procurement practices in humanitarian supply chains: an empirical examination of large-scale house reconstruction projects. Decis. Sci. **25**(4), 621–646 (2014)

Mete, H.O., Zabinsky, Z.B.: Stochastic optimization of medical supply location and distribution in disaster management. Int. J. Prod. Econ. **126**, 76–84 (2010)

Nagurney, A.: Network Economics: A Variational Inequality Approach, second and revised edition. Kluwer Academic, Dordrecht (1999)

Nagurney, A., Masoumi, A.H.: Supply chain network design of a sustainable blood banking system. In: Boone, T., Jayaraman, V., Ganeshan, R. (eds.) Sustainable Supply Chains: Models, Methods and Public Policy Implications, pp. 49–72. Springer, London (2012)

Nagurney, A., Qiang, Q.: Fragile Networks: Identifying Vulnerabilities and Synergies in an Uncertain World. Wiley, Hoboken, NJ (2009)

Nagurney, A., Qiang, Q.: Fragile networks: identifying vulnerabilities and synergies in an uncertain age. Int. Trans. Oper. Res. **19**, 123–160 (2012)

Nagurney, A., Zhang, D.: Projected Dynamical Systems and Variational Inequalities with Applications. Kluwer Academic, Boston, MA (1996)

Nagurney, A., Yu, M., Qiang, Q.: Supply chain network design for critical needs with outsourcing. Pap. Reg. Sci. **90**(1), 123–143 (2011)

Nagurney, A., Yu, M., Qiang, Q.: Multiproduct humanitarian healthcare supply chains: a network modeling and computational framework. In: Proceedings of the 23rd Annual POMS Conference, Chicago, IL (2012a)

Nagurney, A., Masoumi, A.H., Yu, M.: Supply chain network operations management of a blood banking system with cost and risk minimization. Comput. Manag. Sci. **9**(2), 205–231 (2012b)

Nagurney, A., Masoumi, A.H., Yu, M.: An integrated disaster relief supply chain network model with time targets and demand uncertainty. In: Nijkamp, P., Rose, A., Kourtit, K. (eds.) Regional Science Matters: Studies Dedicated to Walter Isard, pp. 287–318. Springer International Publishing, Cham (2015)

NOAA: WMO retires Ingrid and Manuel for Atlantic and Eastern North Pacific basins, 10 April 2014

Ortuño, M.T., Tirado, G., Vitoriano, B.: A lexicographical goal programming based decision support system for logistics of humanitarian aid. TOP **19**(2), 464–479 (2011)

Ortuño, M.T., Cristóbal, P., Ferrer, J.M., Martín-Campo, F.J., Muñoz, S., Tirado, G., Vitoriano, B.: Decision aid models and systems for humanitarian logistics: a survey. In: Vitoriano, B., Montero, J., Ruan, D. (eds.) Decision Aid Models for Disaster Management and Emergencies. Atlantis Computational Intelligence Systems, vol. 7, pp. 17–44. Springer Business + Science Media, New York (2013)

Ozbay, K., Ozguven, E.E.: Stochastic humanitarian inventory control model for disaster planning. In: Proceedings of the Transportation Research Board 86th Annual Meeting, Washington, DC (2007)

Pasch, R.J., Zelinsky, D.A.: Hurricane Manuel. National Hurricane Center, Tropical cyclone report. EP132013, revised 14 April 2014

Qiang, Q., Nagurney, A.: A bi-criteria indicator to assess supply chain network performance for critical needs under capacity and demand disruptions. Transp. Res. A **46**(5), 801–812 (2012)

Rottkemper, B., Fischer, K., Blecken, A.: A transshipment model for distribution and inventory relocation under uncertainty in humanitarian operations. Socio Econ. Plan. Sci. **46**, 98–109 (2012)

Schneeweis, T., Crowder, G., Kazemi, H.: The New Science of Asset Allocation Risk Management in a Multi-Asset World. Wiley, Hoboken, NJ (2010)

The International Bank for Reconstruction and Development/The World Bank: FONDEN: Mexico's natural disaster fund - a review, Washington, DC (2012)

Toyasaki, F., Wakolbinger, T.: Impacts of earmarked private donations for disaster fundraising. Ann. Oper. Res. **221**(1), 427–447 (2014)

Tzeng, G.-H., Cheng, H.-J., Huang, T.: Multi-objective optimal planning for designing relief delivery systems. Transp. Res. E **43**(6), 673–686 (2007)

UNHCR: United Nations High Commissioner for Refugees UNHCR Handbook for Emergencies, 3rd edn. UNHCR, Geneva (2007)

United Nations: Global assessment report on disaster reduction (2011)

USAID: United States Agency for International Development, Office of Foreign Disaster Assistance, Field operations guide for disaster assessment and response, Version 4.0 (2005)

Van Wassenhove, L.N.: Blackett memorial lecture. Humanitarian aid logistics: supply chain management in high gear. J. Oper. Res. Soc. **57**(5), 475–489 (2006)

Vitoriano, B., Ortuño, M., Tirado, G., Montero, M.: A multi-criteria optimization model for humanitarian aid distribution. J. Glob. Optim. **51**, 189–208 (2011)

Vitoriano, B., Montero, J., Ruan, D. (eds.): Decision Aid Models for Disaster Management and Emergencies. Atlantis Computational Intelligence Systems, vol. 7. Springer Business + Science Media, New York (2013)

Walton, R., Mays, R., Haselkorn, M.: Defining "fast": Factors affecting the experience of speed in humanitarian logistics. In: Proceedings of the 8th International ISCRAM Conference, Lisbon, May 2011

A Review of Current Earthquake and Fire Preparedness Campaigns: What Works?

Gabriela Perez-Fuentes, Enrica Verrucci, and Helene Joffe

Abstract Current community preparedness campaigns and interventions for natural hazards are not as effective as they aim to be. Research consistently shows that levels of preparedness for natural hazards are low across cultures, despite increased efforts in public hazard education and outreach. Individuals living in areas at risk of natural disasters are not prepared, despite reporting being aware of such risks. This lack of preparedness increases their probabilities of suffering the cascade of ill consequences that follow a disaster. Most of the natural disaster preparedness campaigns rely mainly on the delivery of information, despite studies consistently showing that simply providing the public with information about risk and safety skills is not sufficient to affect preparedness behaviors. Moreover, many of these campaigns lack evaluation and so their success cannot be proven. Research in the fields of social representations, emergency preparedness, and risk communication indicates that a combination of cognitive, emotional, and cultural factors, as well as messaging style, shapes preparedness behaviors. This paper presents the findings of an online search conducted to identify major earthquake and fire preparedness campaigns. The content, design, and theoretical background of these campaigns are analyzed and the results of this evaluation are discussed. This review serves as a guideline for future interventions and campaigns. It aims to contribute to the field of natural hazard preparedness by extracting the components of existing campaigns that have successfully increased preparedness.

Keywords Natural disaster preparedness • Preparedness campaigns • Earthquake preparedness • Fire preparedness

---

G. Perez-Fuentes (✉) • H. Joffe
Department of Clinical, Educational & Health Psychology, UCL, London, UK
e-mail: g.ruiz@ucl.ac.uk; h.joffe@ucl.ac.uk

E. Verrucci
Department of Civil, Environmental & Geomatic Engineering, UCL, London, UK

I.S. Kotsireas et al. (eds.), *Dynamics of Disasters—Key Concepts, Models, Algorithms, and Insights*, Springer Proceedings in Mathematics & Statistics 185,
DOI 10.1007/978-3-319-43709-5_13

1 Community Preparedness for Natural Disasters

Natural disasters are increasing in frequency and severity across the world for a range of reasons. A key cause is climate change, via changes in temperatures, in rainfall and in sea levels, which magnify the risk of disasters such as severe storms, cyclones, droughts, and earthquakes (Maskrey and Safaie 2015). In addition, increasing social inequality is linked to growing hazard exposure, and rapid urbanization and gentrification are potential factors that can lead to dangerous and unpredictable risk levels with worldwide consequences (Maskrey and Safaie 2015). According to the Global Assessment Report on Disaster Risk Reduction 2015 (GAR15), the global economic cost of natural disasters for the built environment alone is estimated to be US$314 billion, on average, annually (Maskrey and Safaie 2015). Currently, most governments focus solely on disaster management strategies, rather than addressing the triggers of disaster risk such as poverty, climate change, gentrification, poor urban planning, and inadequate (or the lack of) building codes. One way of addressing this problem is by placing more emphasis on developing and improving disaster preparedness, both at the city and community level.

Prepared communities recover faster and more effectively after a disaster (Mileti et al. 1975), which in turn translates into a more resilient public. It is crucial for communities to adopt and maintain preparedness measures at the household level if they are to reduce risk of injury and damage inside the home. Preparedness measures can range from securing heavy objects, structural retrofitting and storing food and water, to having an emergency kit and communication and evacuation plans. Thus, it is in people's best interest to develop and implement safety-related strategies before a disaster occurs; however, the existing literature on natural hazard preparedness consistently shows that most people are not prepared for action in emergency situations (Ballantyne et al. 2000; Lambert et al. 1999; Duval and Mulilis 1999; McClure et al. 1999; Paton 2000; Lindell and Whitney 2000). Even communities living in places at high risk do not prepare sufficiently (Paton et al. 2000; Joffe et al. 2013; Karanci et al. 2005; Rüstemli and Karanci 1999; Garcia 1989). Government authorities and disaster management organizations often attribute the lack of preparedness among communities to the public's lack of information. Thus, the majority of disaster preparedness campaigns and interventions have been designed according to what is sometimes termed the "deficit model" (Bauer et al. 1994; Evans and Durant 1995), which postulates that simply providing information to the public will encourage preparation. However, studies have consistently proven that merely providing communities with risk information is not sufficient to affect preparedness behaviors (Ballantyne et al. 2000; Paton et al. 2000, 2005; Perry and Lindell 2008).

1.1 Community Preparedness for Natural Disasters: Current Problems

There are several factors related to society's deficient levels of preparedness for natural disasters. The first one lies in a lack of firm evidence regarding how to bring about public preparedness. In particular, the interventions that are carried out are often not rigorously evaluated. Secondly, as mentioned above, risk awareness alone is not a sufficient condition to elicit preparedness behaviors (Ballantyne et al. 2000; Lindell and Whitney 2000; Paton et al. 2000, 2005; Perry and Lindell 2008); however, campaigns and interventions tend to focus heavily on this aspect. Furthermore, the literature on natural disaster preparedness suggests that previous experience with a disaster (Mileti and O'Brien 1992) is a poor predictor of preparedness action (Rüstemli and Karanci 1999; Plapp and Werner 2006; Johnston et al. 1999; Palm 1998; Dooley et al. 1992). An example of this is a study that looked at volcanic hazard perception and preparedness in two New Zealand communities, one directly affected by a volcanic eruption (by ash fall) and the other unaffected. Results of measures before and after the volcano eruption showed that the community affected had higher hazard risk perception and lower levels of adjustment measures after the eruption. Preparedness levels of the unaffected community remained unchanged after the eruption of the volcano. Therefore, while risk awareness increased for the affected community after the eruption, its levels of preparedness decreased after the disaster, with respondents stating feeling more confident that they would not sustain any damage in the event of a hazard (Johnston et al. 1999). This study shows how an increase in risk perception does not necessarily translate into action.

What, beyond experience and awareness, might predict preparedness? Knowledge of such factors can be used in devising future preparedness campaigns to successfully increase disaster preparedness.

2 Social Psychological Literature on Natural Hazard Preparedness

Understanding emotive, socio-cultural, and cognitive factors that affect preparedness behaviors is essential if one is to design a preparedness intervention. A study that examined social representations of earthquakes in three cities with high seismic risk—Seattle, USA, Izmir, Turkey, and Osaka, Japan (Joffe et al. 2013)—showed that societies with higher levels of anxiety in relation to earthquakes prepared less than those with lower anxiety. In addition, those cultures with least trust in their societal institutions prepared least, as was the case in Izmir, Turkey, while those with more trust prepared more. Results from this study also showed that higher levels of fatalism correlate with lower preparedness; those with an "I can" attitude tended to prepare more, as was the case for respondents living in Seattle, USA.

The following have been consistently proven to be predictors of preparedness in the existing preparedness studies that have been rigorously evaluated: Perceived self-efficacy, which refers to beliefs regarding personal capacity to act effectively; collective efficacy, which is not the sum of the sense of efficacy of its members but a group-level property of people's shared beliefs in their collective power; empowerment, which is defined by a process of increasing personal, interpersonal, or political power so that individuals can take action to improve their life situations; perceived outcome expectancy, which refers to the perception of whether personal actions will effectively mitigate or reduce a problem; critical awareness, a concept created by Douglas Paton, which is the extent to which people think and talk about a specific hazard within their environment; social cohesion, defined as "a state of affairs concerning both the vertical and the horizontal interactions among members of a society, as characterized by a set of attitudes and norms that include trust, a sense of belonging, and the willingness to participate and help, as well as their behavioral manifestations" (Chen et al. 2006); sense of community, that is, feelings of attachment to people and places; community participation, and trust in authorities. Furthermore, sociocultural variables such as anxiety, trust, and fatalism moderate the relationship between awareness and actual preparedness behaviors, as well as that between awareness and personal responsibility. The empirical literature also shows that being a homeowner and having dependents also increases preparedness.

Another set of factors that play an important role in hazard preparedness are cognitive biases (Kahneman 1979), which are universal cognitive mechanisms that allow individuals to go about their daily lives. They can color risk perception and hinder preparedness. One of them is *optimistic bias*: most people believe that they are less at risk of being affected by a danger than similar others (Helweg-Larsen 1999; Spittal et al. 2005). Another example of such mechanisms is *normalization bias*: when people experience little or no harm in a natural disaster they become less likely to heed future disaster warnings (Mileti and O'Brien 1992; Johnston et al. 1999). Thus, it seems that a combination of personal (emotional and cognitive), social, and cultural factors influence people's interpretation of risks and their preparedness behaviors. On the basis of this social psychological literature, studies have developed models to predict the adoption of natural hazard preparedness with good results (Paton and Commission 2001; Paton et al. 2005). However, such models are often not implemented in the design of hazard preparedness interventions, such as campaigns at the large scale level, or other type of interventions such as community trainings or drills on a smaller scale. This paper focuses on a review of three of the major campaigns on earthquake and/or home fire preparedness.

3 A Review of the Key Campaigns on Earthquake and Fire Preparedness

Few of the existing campaigns on earthquake and home fire preparedness have been systematically evaluated. There are a wide range of mass media and internet-based natural disaster preparedness sites but documentation and evaluation of them is rare. When they are documented and/or evaluated, results are described in a way that is vague. An online Google and Google Scholar search in English of "earthquake preparedness campaigns," "home fire safety campaigns," and "natural disaster preparedness campaigns" was conducted. Those English language campaigns that targeted earthquake and/or fire preparedness (or both), and those targeting the largest number of participants (e.g., national and county level) were selected.

On this basis, three campaigns were chosen for review, all of them North American: (1) "The Great ShakeOut" for earthquake drills, (2) "The American Red Cross for Home Fire Preparedness Campaign," and (3) "Make it Through," a multihazard preparedness campaign that includes home fire and earthquake preparedness.

3.1 The Great ShakeOut

The Great ShakeOut is an annual community earthquake drill that began in November 2008 in California with the goal of providing people with an opportunity to learn what to do before, during, and after an earthquake. This drill is based on a hypothetical earthquake of magnitude 7.8, called the ShakeOut Scenario that could occur along the southern portion of the San Andreas Fault. Almost five and a half million Californians participated in the 2008 ShakeOut, the largest earthquake drill in North American history at that time. Since then, the Great ShakeOut has expanded nationally across the entire USA, as well as to other countries, such as New Zealand, Japan and Southern Italy. The main component of this preparedness campaign is participation in a community national drill. Disaster drills (e.g., fire, flood, earthquakes) play an important role in emergency preparedness in schools, organizations, businesses, and communities. Furthermore, drills have been demonstrated to increase preparedness among participants (Simpson 2002). Likewise, community-based drills have been shown to have a deep and motivating effect on the participating individuals. By demonstrating proficiency in a simulated drill, individuals involved will perform more efficiently in real-life situations because the situation will not be new (Simpson 2002). For marketing, standard "media relations" tools such as press conferences and news releases, as well as social media were used, rather than paid advertising. This generated many news articles promoting ShakeOut, allowing it to become known rapidly nationally as well as at an international level.

The ShakeOut's success in reaching national and international communities resulted from the consistency of communications, such as websites and distribution materials, as well as the extensive advertising and media outreach. The ShakeOut website is the main channel for delivering and receiving consistent information about the drill. To prepare for the drill, participants can engage with the various games and media tools that the website offers to promote experiential learning and skill rehearsal, both of which are effective techniques for behavior change (Michie et al. 2008). In addition, hands-on practice and goal setting were some of the other behavior change techniques used to promote preparedness behaviors.

Despite reaching a large number of participants and becoming widely known, the Great ShakeOut lacks ongoing evaluation. Three evaluations were carried out after the first drill in 2008 but after that, the organization reported lack of funding for further monitoring. Subsequently, the Southern California Earthquake Center formed a Research and Evaluation Committee to develop and implement evaluation activities. The Committee conducted a household preparedness survey of the participants, which was implemented each year from 2009 to 2012; however, the data have not been analyzed yet. However, major findings from the initial evaluation report (Blakley et al. 2009) are the following: (a) Increased individual and organizational awareness about earthquakes (b) Enhanced understanding of what to expect during a high magnitude earthquake and how to respond (c) The drill prompted preparedness behaviors such as getting an emergency kit, developing evacuation and sheltering plans and sharing knowledge with family, friends, and neighbors (Wood and Glik 2013). Results from the 2008 survey (Wood and Glik 2013; Blakley et al. 2009) showed that 79 % of those who registered on the ShakeOut website reported having physically participated in the drill. Thus participation in the drill appears to have led to the further behavior of registering on the website. Nearly all of these people (97 %) said they would continue to participate in an annual earthquake drill. Those who participated in the drill were more likely to recruit others to participate (84 % v. 70 %), to practice other aspects of their disaster plan (49 % v. 27 %) and to assist others in their earthquake preparations (46 % v. 18 %) than those who did not participate. The creators of the ShakeOut kept it simple and concise and asked people to perform one simple activity (a drill), using proven effective behavior change techniques such as interactive learning, hands-on practice, and goal setting, which are all scientifically based ingredients likely to result in adoption of preparedness measures. Consistent with additional proven techniques to enhance preparedness behaviors, they targeted empowerment and community cohesion.

Despite its success, the Great ShakeOut has some limitations. Firstly, the changes are based upon self-reports. There does not appear to be material verification that the behavior changes reported actually occurred or that participants really participated in the drill. Secondly, consistent and ongoing assessment is needed to evaluate its impact longitudinally.

3.2 American Red Cross Home Fire Preparedness Campaign

The American Red Cross launched a national campaign to reduce deaths and injuries from home fires by 25 % over the following five years in October 2014. Its goal is to increase the use of smoke alarms in neighborhoods with higher numbers of home fires and to encourage Americans to practice their fire evacuation plans. Along with installing smoke alarms, the Red Cross joins with fire departments, churches, businesses, schools, public health departments, social services, neighborhood leaders and other community groups nationwide to canvass neighborhoods and teach people about fire safety.

Red Cross volunteers go door-to-door installing smoke alarms. They ask people to do two things—check their smoke alarms and develop an evacuation plan. This campaign is ongoing and no results are available yet. A survey is being implemented where Red Cross volunteers report the number of alarms installed, batteries installed and if the resident/family had both an evacuation plan and other hazard plan during their home visits.

Consistent with the literature on home fire intervention studies, the Red Cross Home Fire Campaign uses techniques proven to be among the most effective methods to improve fire preparedness behaviors, as well as reducing fire related deaths and injuries, which are smoke alarm canvassing and smoke alarm installation (Mallonee et al. 1996; Ta et al. 2006). The presence of fire service personnel appears to be the most effective method of distributing smoke alarms (Douglas et al. 1998) and face-to-face interaction is proven to be an effective technique for improving home fire preparedness (Miller et al. 2014), both of which are employed by the Red Cross in this campaign. Hands-on training is also among the most effective techniques in improving preparedness responses for home fires (Miller et al. 2014).

This campaign uses observational assessment of the preparedness measures taken, in addition to a brief survey, thereby overcoming any putative problems with self-report measures used to evaluate the ShakeOut. However, there are no results yet and thus its effectiveness in improving preparedness behaviors is as yet unknown.

3.3 Make It Through

This is a multi-hazard campaign in the state of Washington (USA) that focuses on a range of hazards (including home fires and earthquakes). It was launched in April 2012. Its aim is to raise awareness and encourage individuals to prepare for catastrophes by making a plan, building a kit with supplies for 7–10 days, and helping each other. It focuses on three preparedness behaviors: make a plan, build, a kit and help each other.

The only medium of communication is the "Make it Through" website, which provides a brochure, an educational toolkit for public educators and several videos of earthquake survivors, a trusted source to increase motivation to prepare. A research agency was hired to conduct the initial evaluation. The research included: interviews with program managers, reviewing 16 regional and national preparedness campaigns, evaluating demographics in the region and conducting a series of focus groups.

The agency evaluated social media engagement, website users and used a telephone pre- and post-campaign survey. This survey measured campaign awareness, source from which awareness came, perceived risk, level of preparedness, likelihood to prepare if not already prepared and demographics. Results of the evaluation indicated that reading about preparedness (13 % from 1 %) and needing to be prepared in general (15 % from 4 %) increased from pre- to post-survey. Respondents reported having stored food and water for a median of 7 days, an increase from the pre-campaign survey's median of 3 or 5 days. The website received nearly 16,000 visits in the first 90 days. However, the majority of survey respondents (69 %) were not aware of an existing website for catastrophe preparedness and only 1 % had heard of the "Make it Through," suggesting that respondents received the information and motivation to prepare from other sources.

This campaign contains several limitations. Its only way of communication is via its website. There is no face-to-face interaction, no hands-on training, no social interaction, nor interactive learning, no feedback, and no technique targeting empowerment or social cohesion (although it attempts to measure them). Furthermore, the campaign targets several complex behaviors (make a kit, make a plan, help each other), which are not clearly defined; since they are vague they are not likely to result in behavior change, according to the importance of implementation intentions (Gollwitzer and Sheeran 2006). No results regarding adoption of preparedness measures have been reported yet.

4 Guidelines for Designing an Intervention on Earthquake and Fire Preparedness: UCL Challenging RISK Study

Based on the three campaigns reviewed above, as well as interventions studies previously reviewed targeting earthquake and home fire preparedness (Perez-Fuentes and Joffe 2015), interventions on natural hazard preparedness lack assessment and when evaluated most are poorly described. In addition, the majority use self-report measures alone. Two of the above reviewed campaigns were designed in accordance with the evidence in the field (Great ShakeOut and Red Cross); however, there is still a need for more explicit and better designed natural hazard preparedness interventions and campaigns so that they can be replicated and improved to engage publics in interventions based on strong empirical evidence.

To date, there is no published, evaluated cross-cultural, longitudinal intervention study combining earthquake and home fire safety preparedness measures. A group of researchers from UCL's EPICentre, headed by a psychologist (Helene Joffe), developed an intervention on earthquake and home fire preparedness within the Challenging RISK EPSRC-funded study. A manuscript describing the study protocol of this cross-cultural, longitudinal intervention study is being published by Natural Hazards (Joffe et al., in press). Study objectives are: (1) to increase household preparedness measures for earthquakes and home fires in lay people; (2) to evaluate changes in lay people's levels of motivation, self-efficacy, perceived outcome expectancy, trust, empowerment, anxiety, and social cohesion, as well as levels of adjustment, before and after the intervention. In addition, the researchers will conduct follow-up assessments three and 12 months after the intervention to measure long-term effects; and (3) to compare results in two highly seismic regions: Izmir in Turkey and Seattle, in North America.

The intervention itself consists of interactive, face-to face, hands-on practice workshops led by two researchers, experts in earthquake emergency preparedness in the different cities. The study was conducted in Seattle in September 2015, with the 3-month follow-up assessments being completed in January 2016. The intervention was conducted in Izmir between May and June 2016. The intervention includes home visits to evaluate home preparedness, a group-based, face-to-face 6-hour workshop (over two days), review photos of participants' safety measures at home, as well as pre- and post-intervention assessments. By including home fire and earthquake preparedness, the more everyday risk gets paired with the longer return period risk. This may facilitate a routinization of adopting disaster preparedness measures at home.

The pre- and post-intervention surveys, in a control and the intervention group, measure the following: individual measures (self-efficacy, outcome expectancy, earthquake anxiety, critical awareness, fatalism), interpersonal measures (trust, levels of corruption, responsibility), social measures (collective efficacy, social cohesion, social empowerment), demographic measures. In addition, participants in both groups receive home visits from a trained member of a recruitment agency to observe their earthquake and home fire adjustment behaviors in their households. A checklist is used to assess their baseline level of preparedness, their preparedness one week after the intervention and three and 12 months after the intervention. This study is novel in that it combines an observational measure of behavior (the home checklist) with participants' self-reports on preparedness levels and individual, interpersonal, social, and demographic factors.

5 Concluding Remarks

There is a need for better designed campaigns and interventions in order to successfully change preparedness behaviors. There is also a need for a multihazard approach to emergency preparedness interventions. A public that is better prepared for multiple hazards is better prepared for specific and unpredictable hazards and is therefore more resilient.

This intervention was carefully designed by a team of multidisciplinary researchers and is tailored to match the social representations of each location, obtained in a previous study (Joffe et al., 2013). This study, which combines observational and self-report evaluations, has significant implications for the field of natural disaster preparedness at an international level, as well as for the area of interventions on natural hazard preparedness, as it will allow for replication, improvement, and therefore development of the field.

References

Ballantyne, M., Paton, D., Johnston, D., Kozuch, M., Daly, M.: Information on Volcanic and Earthquake Hazards: The Impact on Awareness and Preparation. Institute of Geological and Nuclear Sciences Limited Science Report, Wellington (2000)

Bauer, M., Durant, J., Evans, G.: European public perceptions of science. Int. J. Public Opin. Res. **6**(2), 163–186 (1994)

Blakley, J., Chen, N., Kaplan, M.: An Evaluation of the First Great Southern California ShakeOut. The Norman Lear Center/University of Southern California, Los Angeles (2009)

Chen, L., Liu, Y., Chan, K.: Integrated community-based disaster management program in Taiwan: a case study of Shang-An village. Nat. Hazards **37**(1–2), 209–223 (2006)

Dooley, D., Catalano, R., Mishra, S., Serxner, S.: Earthquake preparedness: predictors in a community survey1. J. Appl. Soc. Psychol. **22**(6), 451–470 (1992)

Douglas, M.R., Mallonee, S., Istre, G.R.: Comparison of community based smoke detector distribution methods in an urban community. Inj. Prev. **4**(1), 28–32 (1998)

Duval, T., Mulilis, J.P.: A person-relative-to-event (PrE) approach to negative threat appeals and earthquake preparedness: a field study1. J. Appl. Soc. Psychol. **29**(3), 495–516 (1999)

Evans, G., Durant, J.: The relationship between knowledge and attitudes in the public understanding of science in Britain. Public Underst. Sci. **4**(1), 57–74 (1995)

Garcia, E.M.: Earthquake Preparedness in California: A Survey of Irvine Residents. National Emergency Training Center (1989)

Gollwitzer, P.M., Sheeran, P.: Implementation intentions and goal achievement: a meta-analysis of effects and processes. Adv. Exp. Soc. Psychol. **38**, 69–119 (2006)

Helweg-Larsen, M.: (The lack of) optimistic biases in response to the 1994 Northridge earthquake: the role of personal experience. Basic Appl. Soc. Psychol. **21**(2), 119–129 (1999)

Joffe, H., Rossetto, T., Solberg, C., O'Connor, C.: Social representations of earthquakes: a study of people living in three highly seismic areas. Earthquake Spectra **29**(2), 367–397 (2013)

Joffe, H., Perez-Fuentes, G., Potts, H., Rossetto, T.: How to increase earthquake and home fire preparedness: the fix-it intervention. J. Nat. Hazards (in press)

Johnston, D.M., Bebbington Chin-Diew Lai, M.S., Houghton, B.F., Paton, D.: Volcanic hazard perceptions: comparative shifts in knowledge and risk. Disaster Prev. Manag. **8**(2), 118–126 (1999)

Kahneman, D., Tversky, A.: Prospect theory: an analysis of decision under risk. Econometrica J. Econ. Soc. 263–291 (1979)

Karanci, A.N., Aksit, B., Dirik, G.: Impact of a community disaster awareness training program in Turkey: does it influence hazard-related cognitions and preparedness behaviors. Soc. Behav. Personal. Int. J. **33**(3), 243–258 (2005)

Lambert, A.J., Burroughs, T., Nguyen, T.: Perceptions of risk and the buffering hypothesis: the role of just world beliefs and right-wing authoritarianism. Personal. Soc. Psychol. Bull. **25**(6), 643–665 (1999)

Lindell, M.K., Whitney, D.J.: Correlates of household seismic hazard adjustment adoption. Risk Anal. **20**(1), 13–26 (2000). doi:10.1111/0272-4332.00002

Mallonee, S., Istre, G.R., Rosenberg, M., Reddish-Douglas, M., Jordan, F., Silverstein, P., et al.: Surveillance and prevention of residential-fire injuries. N. Engl. J. Med. **335**(1), 27–31 (1996)

Maskrey, A., Safaie, S.: GAR global risk assessment. In: EGU General Assembly Conference Abstracts, 2015, vol. 17, p. 6494 (2015)

McClure, J., Walkey, F., Allen, M.: When earthquake damage is seen as preventable: attributions, locus of control and attitudes to risk. Appl. Psychol. **48**(2), 239–256 (1999). doi:10.1111/j.1464-0597.1999.tb00060.x

Michie, S., Johnston, M., Francis, J., Hardeman, W., Eccles, M.: From theory to intervention: mapping theoretically derived behavioral determinants to behavior change techniques. Appl. Psychol. **57**(4), 660–680 (2008)

Mileti, D.S., O'Brien, P.W.: Warnings during disaster: normalizing communicated risk. Soc. Probl. **39**(1), 40–57 (1992)

Mileti, D.S., Drabek, T.E., Haas, J.E.: Human Systems in Extreme Environments: A Sociological Perspective, vol. 21. Institute of Behavioral Science, University of Colorado, Boulder (1975)

Miller, T.R., Bergen, G., Ballesteros, M.F., Bhattacharya, S., Gielen, A.C., Sheppard, M.S.: Increasing smoke alarm operability through theory-based health education: a randomised trial. J. Epidemiol. Community Health **68**(12), 1168–1174 (2014)

Palm, R.: Urban earthquake hazards: the impacts of culture on perceived risk and response in the USA and Japan. Appl. Geogr. **18**(1), 35–46 (1998)

Paton, D.: Emergency planning: integrating community development, community resilience and hazard mitigation. J. Am. Soc. Prof. Emerg. Manage. **7**, 109–118 (2000)

Paton, D., Smith, L.M., Johnston, D.M.: Volcanic hazards: risk perception and preparedness. N. Z. J. Psychol. **29**(2), 86–91 (2000)

Paton, D., Commission, N.Z.E.: Developing a Model to Predict the Adoption of Natural Hazard Risk Reduction and Preparatory Adjustments. Earthquake Commission (2001)

Paton, D., Smith, L., Johnston, D.: When good intentions turn bad: promoting natural hazard preparedness. Aust. J. Emerg. Manage. **20**(1), 25–30 (2005)

Perez-Fuentes, G., Joffe, H.: An intervention to increase earthquake and fire preparedness. Paper presented at the Sustainable Development Conference Proceedings, 2015

Perry, R.W., Lindell, M.K.: Volcanic risk perception and adjustment in a multi-hazard environment. J. Volcanol. Geotherm. Res. **172**(3), 170–178 (2008)

Plapp, T., Werner, U.: Understanding risk perception from natural hazards: examples from Germany. Risk **21**, 101–108 (2006)

Rüstemli, A., Karanci, A.N.: Correlates of earthquake cognitions and preparedness behavior in a victimized population. J. Soc. Psychol. **139**(1), 91–101 (1999)

Simpson, D.M.: Earthquake drills and simulations in community-based training and preparedness programmes. Disasters **26**(1), 55–69 (2002)

Spittal, M.J., McClure, J., Siegert, R.J., Walkey, F.H.: Optimistic bias in relation to preparedness for earthquakes. Australas. J. Disaster Trauma Stud. **1**, 1–10 (2005)

Ta, V.M., Frattaroli, S., Bergen, G., Gielen, A.C.: Evaluated community fire safety interventions in the United States: a review of current literature. J. Community Health **31**(3), 176–197 (2006)

Wood, M.M., Glik, D.: Engaging Californians in a Shared Vision for Resiliency: Practical Lessons Learned from the Great California Shakeout. The Alfred E. Alquist Seismic Safety Commission, Sacramento (2013)

The Impact of the Syria Crisis on Lebanon

Denise Sumpf, Vladimir Isaila, and Kristine Najjar

Abstract Despite popular expectations of political transition and economic development associated with the 2010 Arab Spring movement, the Arab region sees unrelenting conflict in many of its parts (for example, Libya, Syria, and Yemen). Against the backdrop of regional challenges, the present paper will discuss the impact of the Syria Crisis on its neighboring country, the Republic of Lebanon.

The Syria Crisis just entered its sixth year and the destruction of historical cities such as Aleppo and Palmyra reach unprecedented levels as international actors (e.g., Russia) enter the conflict. Streams of refugees leave the camps in neighboring Jordan, Turkey, and Lebanon to embark on the very risky journey across the Mediterranean Sea to Europe. Without any durable political solution anticipated to emerge anytime soon, the conflict takes uglier turns every day, for example: (1) increasing disrespect for the protection of civilians (e.g., Geneva Convention) results in a rising direct and indirect civilian death toll also affecting the few crucial health workers that still remain in country; (2) continuing geographical expansion of Islamic State militants and unsavory shifting coalitions at the national and regional level—be they political or military; and so on.

Lebanon—being an immediate neighbor of Syria—and its population have felt the brunt of the Syria Crisis. Hosting more than 1 million refugees, the economic burden is assumed to have accumulated to 7.5 billion USD. Contributing to these "costs of conflict" are the economic, social, and environmental impacts of the Syria Crisis. Setting the stage for the discussion, the present paper provides a brief history of the special relationship between the two countries in the Levant. It highlights the close political, economic, and social ties over centuries, often neither peaceful nor constructive. With an understanding for the relationship between Lebanon and Syria, the paper details the economic, social (health, education), and environmental (agriculture, food security) costs for Lebanon that emanate from the continuing Syria Crisis.

D. Sumpf (✉) • V. Isaila • K. Najjar
Economic Governance and Planning Section, Economic Development and Integration Division, United Nations Economic and Social Commission for Western Asia, Beirut, Lebanon
e-mail: sumpf@un.org

I.S. Kotsireas et al. (eds.), *Dynamics of Disasters—Key Concepts, Models, Algorithms, and Insights*, Springer Proceedings in Mathematics & Statistics 185,
DOI 10.1007/978-3-319-43709-5_14

The resilience of Lebanon and its people is remarkable, though without support the impacts of the Syria Crisis could prove insurmountable. In February 2016, the London Conference brought together stakeholders that committed to substantial humanitarian and development support for Lebanon. They emphasised the importance of greater integration and cooperation among public, private, and civil society institutions to improve economic (e.g., labor market access, subsidy reform, competition, and market access), social (e.g., safety nets), and environmental (e.g., upgrade irrigation practices) policies in Lebanon with a view towards enhancing resilience and returning to a peaceful, sustainable, and inclusive development growth path.

Keywords Lebanon • Syria crisis • Economic costs • Social costs • Environmental impact • Refugees • Humanitarian crisis • Development challenge • Neighbours

1 Introduction: The Context of the Syria Crisis and Current Status Quo

In one of his first statements as UN High Commissioner for Refugees, Filippo Grandi sums up the human costs of humanitarian crises: "we live in a world of emergencies, of disasters, of 60 million people in exile, displaced or stateless and in a world where wars are so many and so difficult to resolve." The Middle East has suffered from long-standing lines of conflicts, to which emerging ones add heavy burdens in terms of political, economic, and social challenges. For example, the so-called Arab Spring triggered pro-democracy uprisings in vast parts of the Middle East. The demonstrations oscillated between peaceful and forceful, while government responses ran the full gamut from violent military response to appeasement by socio-economic measures (e.g., subsidies, cash transfers). Some countries—for example, Morocco and Tunisia—maintained relative stability during the political, economic, and social transitions since the Arab Spring, others—Libya and Syria—remain mired in protracted conflict with contested proposals for political solutions brokered by the international community.

In March 2016, the Syrian civil war entered its sixth year. The resulting humanitarian crisis has spiralled well beyond the limits of the country to its direct neighbors, the Arab region and to Europe. As of October 2015 and based on local information, Human Rights Watch[1] estimated the total death toll at over 250,000 people,[2] of which at least 100,000 were civilians. According to UNHCR[3] and as of

[1]Human Rights Watch (2016).

[2]The Syrian Center for Policy Research (http://scpr-syria.org/publications/policy-reports/scpr-alienation-and-violence-report-2014-2/) calculated the direct or indirectly war related death toll at 470,000 people.

[3]Continuously updated figures available from the UN inter-agency information sharing portal (UNHCR 2016).

February 2016, the total number of persons of concern stood at 4,598,594 displaced Syrians seeking shelter in Egypt, Iraq, Jordan, Lebanon, and Turkey as well as other North African countries. This is in addition to the estimated 8 million internally displaced Syrians.

Lebanon, a small Mediterranean country, currently hosts more than 1 million Syrian refugees, which adds approximately 25 percent to the existing Lebanese population of 4.5 million people[4] (in comparison and with a population of around 81 million people, Germany shelters 573,828 persons of concern, including asylum seekers, refugees, and stateless persons). The political, economic, environmental, and social strain is palpable across Lebanon and tests an already fragile equilibrium. To give an overview on the dynamics of the Syria Crisis and its impact on Lebanon, one section of this paper analyses the context in Lebanon with respect to the historic relationship to Syria and another part of the paper addresses the economic, social, and environmental costs of the Syria War and the resulting migrant crisis.

Thus far, humanitarian prioritised to address immediate needs, but for that to be effective in cushioning the impact of the continuing conflict the integration of a development perspective is necessary to find a medium to long-term solution.[5] Prior to the Syria Crisis, both Lebanon and Syria were on a positive track towards achievement of internationally agreed development objectives (i.e. the Millennium Development Goals). However, development progress, if not all lost has slowed significantly, and the transition towards achievement of the Sustainable Development Goals, which requires strong country ownership and leadership, is a major challenge. Taking into consideration the complexity of the political, economic, social, and environmental conditions in the Arab region, the present paper will limit its focus on the dynamics and spillovers of the war in Syria on its immediate neighbor, Lebanon. Nonetheless, the consequences of the Syria conflict affect the region and the rest of the world.

2 Historical Relationship Between Syria and Lebanon: From Sykes-Picot to Today[6]

The geo-political region of Syria, throughout history, has included various ethnic and religious groups. To this day, Syria encompasses populations of Syrian Arab, Aramean, Armenian, Turkmen, Greek, and Kurdish descent. They adhere to numerous religious denominations adding a varied but important presence in a geographical area where Muslims (both Sunni and Shi'a) form the majority, while Christians and Druze represent a strong minority. Both the modern day Republic of Lebanon and the Syrian Arab Republic were once part of the same country, thus sharing ethnic, cultural, and religious affinities given their hundreds of

[4]See World Bank Development Indicators: World Bank (2016a).

[5]Lebanon Center for Policy Studies (2016).

[6]For a summary timeline, please refer to Annex.

years of mutual political, economic, and cultural experience. Their shared historic experience links these two countries, for better or for worse, and the relationship has defined each other's politics, economies, and societies throughout the twentieth century, and continues to do so to this day.

2.1 Sykes-Picot and the French Mandate in Syria and Lebanon

The Sykes-Picot Agreement, officially known as the Asia Minor Treaty, marks the beginning of the development and eventual establishment of the modern states of Lebanon and Syria as separate entities.[7] At the San Remo conference of 1920, the Allied powers agreed that based upon the Asia Minor Treaty, the former Ottoman territories and vassals in the Middle East would be extended eventual independence under the *Mandate* of the French Republic and the United Kingdom. This was agreed upon by the Council of the League of Nations in 1922.[8] The French Mandate included current day Syria and Lebanon, while the British Mandate included present day Israel, Palestine, Jordan, and Iraq.[9] French ambitions for the region were manifested when the recently appointed King of Syria, Faisal I was ousted by French forces and a regional administration was set up in his place to directly oversee Syrian affairs of state.[10]

The Syrian region subsequently divided into smaller statelets centered on the religious communities present. Thus, the new Syrian Federation constituted the State of Aleppo, Damascus, Jabal al-Druze, Alawite, and Greater Lebanon.[11] Under the French Mandate, local elites were brought into political and economic functions (mostly) from the various religious minorities in the region, in an effort to curry favor among the local populations that were not prone to insurrection or ideas of pan-Arabism or pan-Islamism. The strategy of separating the Syrian region into smaller statelets while championing religious minorities and local elites was done by French colonial authorities to ensure that the region would not and could not rise up against the French administration. Indeed, General Henri Gouraud, who oversaw the French Mandate from 1919 to 1923 advocated for the separation of Syria into

[7]The Agreement, named after its authors Mark Sykes and François Georges-Picot, signed on the 16th of May 1916 was a secret agreement between the French Republic and the United Kingdom on the partition of the Middle-Eastern holdings of the Ottoman Empire contingent on the defeat of the Central Powers. See: Chaitani (2007), Cleveland and Bunton (2013), Fildis (2011), Khoury (1987) and others.

[8]Kamrava (2005).

[9]Shambrook (1998).

[10]Faisal I bin Hussein bin Ali al-Hashimi, an ally of British forces during the First World War, was declared King of Syria in 1920, and subsequently expelled from Syria following the French occupation of Damascus. He would later on be crowned King of Iraq in 1921 with the support of the British under their mandate of Iraq. See: Fildis (2011), pp. 40–43.

[11]Shambrook (1998).

regional statelets, stating that "It will be easy to maintain a balance among three or four [Syrian] states that will be large enough to achieve self-sufficiency and, if need be, pit one against the other."[12]

2.2 Greater Lebanon and the 6:5 Rule

The territories of the respective provinces of the Syrian Federation were relatively equally distributed in order to remove the dominance of one single region. The Lebanese State was therefore given territories North, South and East of it into an expanded "Greater Lebanon." However, the demographic map changed with the creation of Greater Lebanon. Where the previously minor Muslim population constituted a very low percentage of the total population, as of the 1932 census, they constituted almost half, with some estimates placing the Muslim minority at 49 percent of the resident population.[13] The 1926 Lebanese Constitution was formulated in an effort to keep the balance of power with Lebanon's Maronite community, given the demographic shift.[14] The 6:5 ratio meant that seats in Parliament were distributed according to a six to five Christian/Muslim ratio. Furthermore, the president, who retained veto power over the legislative institutions, was to be Christian. This effectively guaranteed that the constitution could not be changed in favor of Muslim representation, and that Parliament always held a majority Christian assembly.[15] This constitutional framework, which applies to this day, together with future demographic and economic discrepancies in the region were some of the primary reasons for the outbreak of the Lebanese Civil War as well as for modern day political crises.

2.3 The Syrian and Lebanese Republics

The region did rise up against the French administration, when in 1925 the *Great Syrian Revolt* began first in the Jabal al-Druze state and quickly spread over the entire region, culminating in the creation of the Republic of Syria in 1930, including

[12]Henry Gouraud, quoted in Traboulsi (2007), p. 86.

[13]There is a lot of disagreement on the methods used for this census. The authorities agreed to include the Armenian community as part of the Lebanese citizenry, while some Muslim communities in the North and South of Lebanon were excluded in an effort to minimize the existing Muslim population. Khoury and Jaulin (2012).

[14]Khoury (1987).

[15]Khoury (1981).

all former Syrian regions, except for Greater Lebanon.[16] The French occupation and administration were seen as liberators and preservers of the Maronite Christians in Lebanon, and as such refused to join the Syrian Republic. This was aided by the rise in Lebanese nationalism in the years prior to the Revolt. This nationalism was based on a distinct Christian character and the relative rejection of pan-Arabism so often found in political and economic elites in the region at the time.[17]

In 1936, the Syrian government drafted the Franco-Syrian Treaty of Independence and sought to diplomatically end French imperial ambitions in Syria. However, the Treaty was outright refused by the French government leading to tensions with the Syrian elites and sporadic skirmishes with local militias.[18] Tensions continued throughout the next decade, until 1941 when an occupied France was unable to respond to an increased push from nationalist elements. This forced France to accept the independence of Syria under a renewed momentum for self-determination in Syria and the region. Despite this, the French government stated that it maintained responsibility in the region through its Mandate. In 1945, French army personnel bombed Damascus and tried to arrest members of the democratically elected government. Similarly, Lebanese authorities declared independence and following French intervention, the majority of the government was thrown into prison, indicating a French unwillingness to end its occupation and involvement in the region under the pretext of its Mandate over Syria and Lebanon.[19] Due to militarily and economically untenable conditions and with rising pressure from Britain, the International Community as well as Syrian and Lebanese nationalist groups, the French were forced to evacuate and in mid-1946, Syria and Lebanon gained de facto independence.[20]

[16]The revolt against the French Mandate in Syria was mainly aimed at popular dissatisfaction against the French Administration viz-a-viz economic and political appropriation. For example, the French officers and administrative attaches were meant to provide training and expertise in governance so that they may in the future be able to rule independently; however, in quick order, the French authorities placed themselves in direct control of all governmental institutions. This disdain against Syrian self-rule as well as the economic disempowerment of established Syrian elites and tribal Shaykhs are the main causes of the Great Syrian Revolt. Even though the French forces soundly put down the rebellion by 1927, French authorities were forced to compromise to maintain order. This compromise involved the appointment of Syrian urban elites and rural Shaykhs into political and economic positions, as well as the unification of Syria under a central government. See: Fildis (2011), Khoury (1981, 1982).

[17]Lebanse Nationalists prominent among Lebanese intellectuals before and during the French Mandate supported the idea of an Independent Lebanon, distinct from Syria. Its pioneers include Adib, al-Sawada, K.T. Khaiarallah and Bulus Nujaym among others. See: Fildis (2011), Traboulsi (2007).

[18]Cleveland and Bunton (2013).

[19]Ibid.

[20]Chaitani (2007).

2.4 Independence and Continued Political and Sectarian Conflict: 1946–1990

2.4.1 Lebanon

Following independence, Maronite President Bishara Al-Khuri and Sunni Prime Minister Riad El-Solh agreed to an informal "National Pact," apart from the formal constitution. The National Pact was meant to "supplement and correct the constitution on essential questions of the country's identity, its Arab and international relations and the incorporation of Muslim communities in the power structure." [21] The Pact was meant as a political compromise among the sectarian political entities and allowed for political stability throughout the 1950s and 1960s.

Lebanese elites consolidated their influence over all aspects of political life and economic functioning. Under the presidency of Kamil Sham'un in the 1950s, political and economic power was concentrated among 30 ruling families, which firmly controlled the financial, banking, construction, trade, and tourism industries. The stakeholders of these industries were also heavily involved in political life; therefore economic policy would develop to support the service and financial sector, at the detriment of the Lebanese production and manufacturing industry. Despite this, Beirut quickly became the gateway to the Arab world in all aspects of banking, finance, and trade, flourishing due to the vast amounts of capital and oil flowing from the Gulf to the West and vice-versa.[22] The Lebanese average annual growth rate from 1950 to 1974 was estimated at a steady 7 percent, while low inflation, remittances, and financial and monetary stability nurtured the burgeoning service economy.[23]

At the same time, the various Israeli–Arab conflicts caused massive influxes of Palestinian and other refugees throughout the 1940s, 1950s, and 1960s into Lebanon, usually along its southern border with Israel and Syria. Drawn in by either conflict or economic reasons, refugees, and migrants in Lebanon contributed to a gradual demographic shift in the country. Furthermore, the economic prosperity witnessed in Lebanon was concentrated mainly in the capital and surrounding regions, and largely within the Maronite community. Large economic and political discrepancies formed between the capital and the rest of the country, and between Christian and Muslim communities, fuelling sectarian resentment.

2.4.2 Syria

Political instability and economic woes plagued the nascent Syrian government and following Syria's defeat in the 1948 Arab-Israeli war, a military coup d'état overthrew the civilian democratic government in 1949 and assumed control of

[21] Traboulsi (2007).

[22] Ibid.

[23] Makdisi and Sadaka (2003).

governance.[24] Syrian political instability (20 cabinets and four constitutions in a span of a decade) as well as various conflicts with the new Israeli state caused its economy to lag well behind its neighbor. During the 1946–1970 period, Syria underwent multiple phases of economic development. Following its independence, Syria was primarily agrarian, and continued to be so until the mid-1960s; however, economic growth in industrial sectors grew rapidly following the nationalization of several sectors of the economy during the 1960s.[25] Syria's cooperation with the USSR during this period led to the adoption of Soviet style economic planning and industrialisation. Given its national economic planning and ineffective and corrupt bureaucracy, Syrian industrial growth was not sustained beyond the initial industrialization phase. This meant Syria's economic growth was much lower than that of Lebanon's during this time period, whose economy was moving towards the service and financial sectors.[26]

Ideas of pan-Arabism flourished throughout the late 1940s and 1950s. Following Egypt's Abdal Nasser's leadership in the Suez Canal Crisis in 1956 against the British, French, and Israelis, the Arab nations throughout the Middle-East began discussions of forming an Arab federation. These discussions and deliberations led to the formation of the United Arab Republic in 1958 between Egypt and Syria. However, dissatisfaction among Syrian military elites with the new federal arrangement saw another military coup in late 1961, officially re-establishing the Syrian Arab Republic.[27] A worsening economic situation, popular dissatisfaction with the military leadership and established elite led the military and civilian arm of the Arab Socialist Resurrection Party (Ba'ath Party) to stage a coup and take over executive and legislative authority in the country.[28] In 1970, disputes among the ruling Ba'ath Party members over support for the Palestinian Liberation Organization led to divisions among its members. In the same year, Minister Hafez al-Asad staged a bloodless coup and assumed the role of President, organizing around himself a loyal military and civilian elite, mainly from his own Alawite community.[29] With strong backing from the Soviet Union he reformed many aspects of the Syrian economy and increased efforts to continue the industrialization process started in the 1950s and 1960s. However, "the failure of a state-led economic development project; bureaucratic corruption; rising foreign debt, inflation and unemployment; and high levels of domestic repression"[30] all plagued the Hafez

[24] Heydemann (1999), Krókowska (2011).

[25] Ibid.

[26] Ibid.

[27] Syrian political and military elites felt underrepresented and their interests sidelined under the new federal arrangement. While Nasr's popularity was waning, Syrian dissatisfaction was increasing resulting in the collapse of the UAR. For more detailed information on the UAR, see Kamrava (2005).

[28] Heydemann (1999).

[29] Ibid.

[30] Ibid, p. 6.

Al-Asad regime. With the fall of the Soviet Block in 1991, Syria began looking to the West and the USA for support, undergoing economic liberalisation efforts and maintaining a gradual but strong economic growth rate.[31]

Hafez Al-Asad's authoritarian regime was extremely resilient, having tackled a worsening economic situation with increasing dissent in his country. In 1982, for example, the fundamentalist Sunni Muslim Brotherhood staged a rebellion against the regime in Hama, central Syria; Asad's heavy-handed response all but wiped out Muslim Brotherhood resistance in Syria. This issued a warning to future dissidents that the regime was willing to do whatever it takes to preserve itself.[32] These events proved to be a chilling foreshadowing of what would happen three decades later under Hafez Al-Asad's son and heir, Bashar Al-Asad.

2.5 The Lebanese Civil War and Syrian Intervention: A Return to the Fold?

In the late 1960s and early 1970s Muslim Lebanese and the political left in the country were calling for the dismantling of the 6:5 rule and the formulation of a new constitution to reflect the changed demographic and religious makeup of Lebanon. Conflicts arose from sectarian tensions derived from dissatisfaction among Muslim Lebanese over political and economic representation. This developed into a full blown Civil War by 1975.[33] Conflicts between Christian and Muslim militia throughout Lebanon led to political division in the governance structures as well as the military. As fighting intensified, the various Christian militias and army groups began losing ground to the Palestinian, Sunni, Leftist and Druze militias and Lebanese splinter army cells.[34] The Syrian army was invited into Lebanon at the behest of the Christian political elite in order to fight off the advancing Palestinian Liberation Organization (PLO) and its allies.[35] Over the next decade, internal fighting as well as Syrian and Israeli intervention resulted in the death of over 100,000 Lebanese and the displacement of a further 900,000 Lebanese citizens.

---

[31] Between 1990 and 2010, Syria had an average GDP growth of 4.98 percent, while immediately after liberalization during the 1990s, Syria's average GDP growth was 5.39 percent. In the meantime, unemployment in the country decreased by half from 1997 until 2010, from 16.761 to 8.613 (data on unemployment not available prior to 1997). International Monetary Fund (2013).

[32] Cleveland and Bunton (2013).

[33] There are many more nuanced reasons for the Lebanese Civil War and the sectarian conflicts in Lebanon. Beyond economic and political discrepancies, issues of identity, Lebanon's place in the Cold War, relations with its neighbors, especially Syria and Israel formed many of the problems dividing Lebanese opinion. For more details on these issues and for a more comprehensive analysis on the Lebanese Civil war, see: Chaitani (2007), Makdisi and Sadaka (2003), Rowayheb (2011), Traboulsi (2007) among others.

[34] For more detail on the factions and their interaction, see: Rowayheb (2011).

[35] Rowayheb (2011).

The conflict officially ended with the Tai'f Agreement in 1989, which distributed Cabinet and government positions 50/50 among Christians and Muslims and placed an emphasis on "coexistence."[36] The Syrian intervention, however, did not end, and its troops remained in Lebanon until 2005. This had an enormous impact on domestic affairs as exemplified by the importation of between 400,000 and 800,000 Syrian workers during the 1980s of which 200,000 were given Lebanese citizenship as a result of Syrian political pressure.[37]

2.6 The Brotherhood Treaty: The Legal Grounds for Increasing Political and Economic Convergence Between Lebanon and Syria

Signed in 1992, the Brotherhood Treaty is a broad (both in scope and meaning) treaty aimed at giving legal and normative grounds for Syrian-Lebanese cooperation on political, economic, and cultural issues, based on historic proximity between the two nations. This has been seen by many as an attempt to set the stage for the return of Lebanon into the Syrian fold and was reluctantly signed by the Lebanese legislative assembly at the (not so subtle) behest of the occupying Syrian forces. The Treaty also legitimised the Syrian presence in Lebanon, with Article 5 and 6(5) providing for the mutual "support [...] in matters pertaining to its security and national interests in accordance with the provisions of the present Treaty."[38] Although the emphasis was on cooperation and the increase of communal economic and social linkages, very little was done in that respect. Indeed, for many, this treaty served only to legimitise Syrian political and military intervention in Lebanon. Syrian occupation ended in 2005, following the alleged involvement of Syria in the assassination of ex-Premier Rafiq Hariri, and the subsequent public uprising against the Syrian regime and occupation.[39]

2.7 In Recent History ...

Following Hafez Al-Assad death, his son took over the reins of the Syrian Republic in 2000. Bashar Al-Assad began by instituting economic reforms and allowing civil society groups to form, resulting in what was called the "Damascus

[36] Makdisi and Sadaka (2003).

[37] Balanche (2005).

[38] "Treaty of Brotherhood, Cooperation and Coordination Between the Syrian Arab Republic and the Lebanese Republic" (1992).

[39] Osoegawa (2013).

Spring."[40] However, over time political, economic, and social reform processes slowed while the security apparatus grew and increased control over people and country. The 2011 "Arab Spring" affected domestic Syrian support for the regime, and resulted in violent clashes (beginning in the city of Dar'a and Homs) between the regime and the various protest or dissident groups. The heavy-handed response of the Government enabled the conflict to grow into today's civil war with all its consequences.

3 The Economic Costs of the Syria Crisis

Community resilience refers mainly to the ability of communities to prepare for and cope with natural disasters. Nonetheless, communities can also prove resilient in times of armed conflict. While Lebanon is not party to the conflict in Syria, despite the involvement of Hezbollah[41] (a non-state actor—political and military organization—following Shi'a Islam and being based in Lebanon), Lebanese people carry the burden of the spillover of the conflict and have so far shown remarkable resilience despite national political stalemate (e.g., absence of a President for over a year).

In his speech addressing the so-called London Conference in February 2016 (see Box 1), the Lebanese Prime Minister drew attention to the stress on the Lebanese society, economy, and environment. As few other countries in the Arab region, Lebanon suffers from limited availability of renewable and non-renewable resources. Even prior to the impact of the conflict in Syria, the general political, security level of infrastructure and public services in Lebanon has not been sufficient for the country's population, largely due to substantial claims of corruption.[42] For example, there are daily electricity cuts in urban centers for at least three hours, while rural areas suffer from prolonged power outages; the unresolved waste management approach of the Government led not only to widespread popular demonstrations in August 2015 but also to rampant pollution and random garbage disposal on empty plots of land across the country. In its statement at the London Conference, the Lebanese Government made clear that "Lebanon has incurred losses of $13.1 billion USD since 2012 out of which $5.6 billion USD in 2015 alone (over 11 percent of GDP), as well as massive impacts on public services including education, health, energy, water, waste collection and treatment and infrastructure, all of which had been strained even before the crisis."[43] Furthermore, they asked the

[40] Alvarez-Ossorio (2012).

[41] For details on Hezbollah's objectives, structure, and emergence as well as involvement in Syria, please refer to Masters and Laub (2014), Norton (2014), Sullivan (2014).

[42] Lebanon ranks 123 among 168 countries in the Transparency International "Corruption Perceptions Index 2015" (Transparency International 2015).

[43] Government of Lebanon (2016).

international community for support in coping with the impact of the Syria Crisis in form of grants and concessional loans over the 5 years, which amount to about 11 billion USD.[44]

Box 1
Excerpt of the Speech of Lebanon's Prime Minister Tammam Salam at the Syria Donors' Conference on 4 February 2016 in London:

"The tragedy striking the people of Syria, who sought refuge away from the carnage in their country, has dramatically afflicted other people in other countries. Poverty, misery, disease, have swelled and indiscriminately plagued the Lebanese and the one million and a half Syrians that now constitute almost a third of the local population, grafted, as they are, on a limited pool of national resources and services that were largely insufficient for Lebanon's own inhabitants.

Recognizing that there is categorically nothing acceptable, other than a temporary presence, and despite the difficulties, they face, both the Government and the people of Lebanon have interacted with the Syrians in ways that go far beyond what is required by international conventions, human rights and moral principles. This has not gone without a price. Unemployment and poverty increased considerably, growth fell to zero and shortages became more critical in everything, from space in schools and hospitals to water and energy supply, all at a prohibitive cost to Lebanon's economy and to its environment."

Source: Government of Lebanon (2016)

3.1 Official Estimations of the Costs of Conflict

Calculations by researchers affiliated to the Government of Lebanon quantify the financial burden and refer to Lebanese and Syrians in the country as becoming "partners in deprivation"[45]:

- "Since March 2011, the economic losses due to the crisis in Syria have exceeded $7.5 Billion.
- [...]
- Cost of Syrian Refugees on public infrastructure: $589 Million in 2014.
- Cost of Syrian Refugees on electricity 500-580 $Million between 2012-2014.
- Cost of Syrian Refugees on education sector: $194 Million between 2012-2014.
- Lebanon lost 894 thousand tourists (around 41.5 percent of total tourists) between 2011-2013."

---

[44]Ibid.

[45]Gebara (2015), all currency references are in USD

Even though such numbers give an idea of the monetary scale, they have to be treated cautiously due to the limited availability of data and the difficulty to establish cause-and-effect relationships that allow differentiation of costs either due to the Syria Crisis or due to the general political, security and socio-economic environment of Lebanon in itself. To elicit the context of these numbers and to shed some light on how such substantial sums emerge, the remainder of the paper discusses the impact of the Syria Crisis in terms of the economic, social, and environmental impact.

3.2 Selected Characteristics of the Lebanese Economy and Their Implications

Lebanon is an upper middle-income country, though the Lebanese economy is small, services based (e.g., banking) and currently without natural resources rents. Vulnerability to political risks (e.g., recognizable deterioration in relations with countries of Gulf Cooperation council) and high debt-to-GDP ratio (increase from 130 percent to about 143 percent in 2015)[46] constrain the Lebanese economy. Recent offshore exploration gives hope for the country to be able to benefit from revenues from hydrocarbon resources. The macro-economic opportunities and governance challenges related to revenue from natural resources are part of the current debate[47]: with hydrocarbon exploration in the Levantine Basin at a very early stage and in light of the existing overall political and economic governance challenges (e.g., concentration of political power, sectarianism), think tanks discuss the potential negative consequences ("resource curse") for Lebanon. Objective assessments (e.g., actual size of reserves), expectation management concerning generated income and prudent policy decisions (e.g., fiscal planning) together with political vision concerning the contentious location of hydrocarbon resources (proximity to Israel) are crucial to ensure broad-based benefit of natural resource revenues.

The annual growth rate of gross domestic product (GDP) for 2015 is projected to be 2.5 percent[48] and with the exception in 2013 (0.9 percent) more or less consistent with the growth rate in previous years (about 2 percent in 2011, 2012 and 2014).[49] During the "golden years" between 2007 and 2010, Lebanon saw unprecedented GDP growth between 8 and 10.3 percent and with the economic growth target formulated in the internationally agreed development "Agenda 2030," these rates set the aspirational goal. Nonetheless, growth projections linked to the developments in Syria show a return to more realistic growth rates (see Fig. 1), if the Syria Crisis ends

[46]Banque du Liban (2015).

[47]Chabaan (2016), Chabaan and Harb (2015) and Leenders (2016).

[48]Investment Development Authority of Lebanon (2015).

[49]World Bank (2016a).

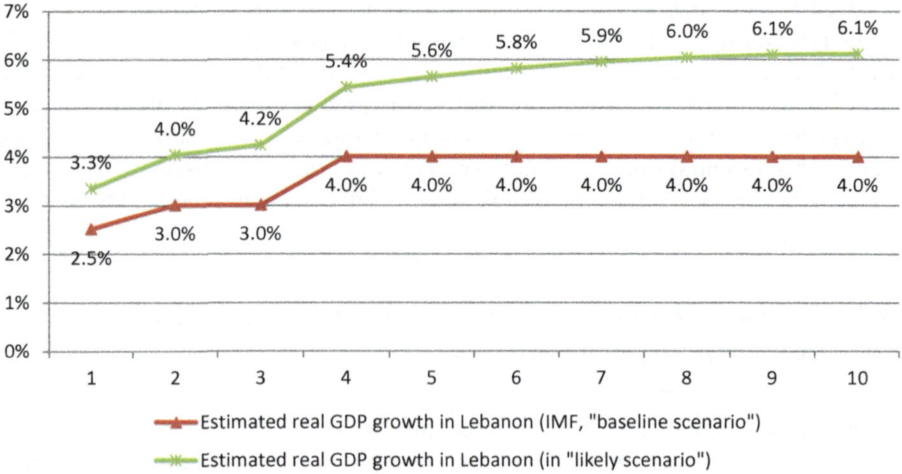

Fig. 1 GDP growth in Lebanon as influenced by the situation in Syria: Scenario 1—IMF baseline calculation (*brown*) vs Scenario 2—"ESCWA likely scenario" (*Green*). *Source*: Calculations by ESCWA/National Agenda for the Future of Syria Project

in a scenario that sees Syria become mainly peaceful, though with continuation of armed conflict or emergence of other conflicts (so-called ESCWA likely scenario[50]).

Figure 2 illustrates the correlation between the Syrian and Lebanese GDP growth rates: the divergence in the early 1990s is owed to the end of the Lebanese Civil War triggering a period of tremendous economic recovery. On the other hand, the difference in 2005 can be explained by the period of political instability and volatile security resulting from the assassination of Rafiq Hariri on 14 February 2005. Despite the period of strong economic growth between 2007 and 2010 in Lebanon and coinciding with the first phase of the Syria Crisis, the government of Rafiq Hariri's son, Saad Hariri, collapsed in 2011 and marked the beginning of yet another period of political and institutional insecurity in the country that continues until today.

In terms of composition of value added to the economy and with the beginning of the Syria Crisis, the contribution of industry and manufacturing has increased in Lebanon, while the contribution of services declined (see Fig. 3). Several interpretations are possible: The influx of Syrian refugees and their willingness to work could mean some services are now delivered under informal arrangements and thus difficult to caption. Increasing contribution of industry and manufacturing may be attributable to increased availability of cheap unskilled labor, which reduces, for

[50]Under the project "National Agenda for the Future of Syria" different political scenarios (e.g., "business as usual," "ESCWA likely") represent the basis for modeling socio-economic requirements for reconstruction and stabilization (United Nations Economic and Social Commission for Western Asia 2016b).

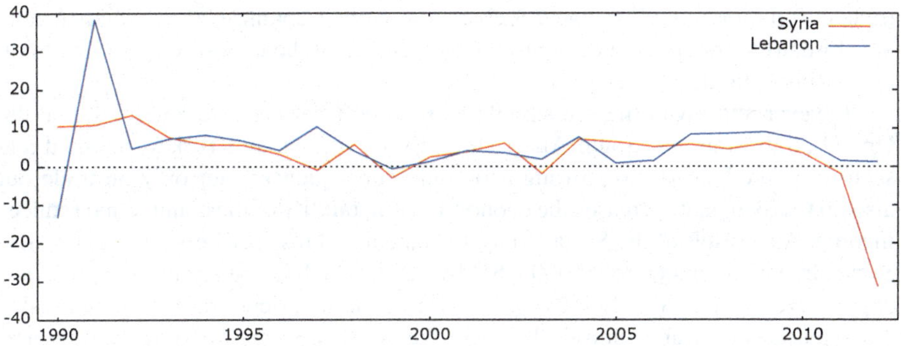

Fig. 2 Correlation of growth rates (Syria—*red*, Lebanon—*blue*). *Source*: Calculations by ESCWA/National Agenda for the Future of Syria Project

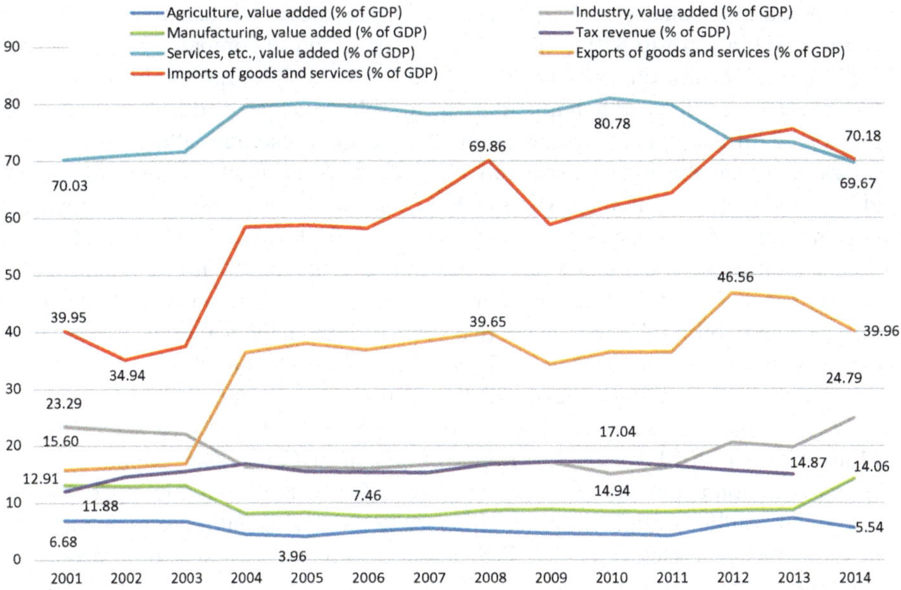

Fig. 3 Value added (per cent of GDP). *Source*: Based on World Bank (2016)

example, construction costs. Tax revenues constitute about 14.87 percent of the GDP (latest available information from 2013) and with a view to implementing required fiscal reforms such contribution could increase in the future.

The Levant has a strong merchant history and therefore, the important role exports and imports play comes as no surprise: within the Arab region the high trade correlation index[51] between Lebanon and other countries (e.g., Saudi Arabia and Jordan) highlights the economic dependence and the exposure to political risk emanating from the Syria Crisis.

However and given that Lebanon's main export partners are Saudi Arabia, the United Arab Emirates, South Africa, Iraq, Syria, Jordan, and Turkey, political and security-related shifts have strong economic consequences, not only on trade but also on tourism, that increases the economic vulnerability of the country and reduces income. As a result of the Syria Crisis, the calculated loss per Lebanese exporter to Syria already amounted to 90,000 USD by 2012, which is a quarter of the pre-crisis export level.[52] A small silver lining is that Lebanese agricultural exports now also reach markets in Arab countries that were previously covered by Syrian agricultural exports.

As a middle-income country, Lebanon is normally not on the list of countries to have donor priority. With the Syria Crisis, however, capital inflow in form of official development assistance (ODA) continues to increase over the last years (from 621 million USD in 2013 to 820 million USD in 2014) and in comparison to pre-crisis levels (e.g., 448 million USD in 2010) (see Table 1). Unsurprisingly the bulk of ODA received is allocated to humanitarian aid, followed by the education sector and other social infrastructure (together about 34 percent of resources allocated). Crucial sectors such as health and infrastructure receive relatively small amounts of financial aid. In this context, it is interesting to note that the level of official development assistance in 2014 corresponds almost exactly to the level of ODA in 2006 when a major conflict between Israel and Lebanon flared up for about one month (July) and caused significant destruction estimated to have cost Lebanon between 1 billion USD[53] and 3.6. billion USD.[54] In the two years following the war ODA for Lebanon peaked with 979 million USD in 2007 and 1.07 billion USD in 2008.

Contrary to increasing official aid flows, private investment from abroad has severely declined between 2013 and 2014 (net private flows were negative 38.2 million USD in 2014 versus 320 million USD in 2013). Private capital actually leaving Lebanon is indicative of the regional strife and the economic repercussions of political conflict and the different alliances in the Syria conflict (e.g., Saudi Arabia vs. Qatar). The lack of private investment also means that structural economic challenges (e.g., implementation of legislative frameworks that are conducive to an improvement of the business climate) and employment creation do not take place. As an example, the implementation and enforcement of an effective competition

---

[51]Merchandise trade correlation index (annual, 1995–2012)—Lebanon vis-à-vis: (1) Jordan: 0.3347, (2) Saudi Arabia: 0.2730; (3) Syria: 0.2437; (4) Palestine: 0.2438; (5) Bahrain: 0.1977; (6) Oman: 0.1844; (7) Turkey: 0.1486; (8) United Arab Emirates: 0.1432; (10) Kuwait: 0.1015; (11) Qatar: 0.0734; (12) Yemen: 0.0173; and (14) Iraq: −0.0039 (United Nations Conference on Trade and Development 2016).

[52]Calì et al. (2015).

[53]Darwish et al. (2009).

[54]Heinrich Boell Foundation (2006).

Table 1 Development assistance receipts and their allocation (average in 2013–2014)

Receipts for Lebanon (in million USD)	*2012*	*2013*	*2014*
Net Official Development Assistance (ODA)	711.5	621	819.6
Average over 2 years		720.3	
Net private flows	206.7	320.4	−38.2
Bilateral ODA by Sector for Lebanon (2013–2014 average)			
	in per cent	*in million USD*	
Humanitarian aid	56.87	409.63	
Education	18.52	133.40	
Other social infrastructure	16.24	116.98	
Multisector	2.51	18.08	
Production	2.01	14.48	
Other/unspecified	1.55	11.16	
Health and population	1.15	8.28	
Economic infrastructure	0.67	4.83	
Programme assistance	0.48	3.46	
Total	100	720.3	

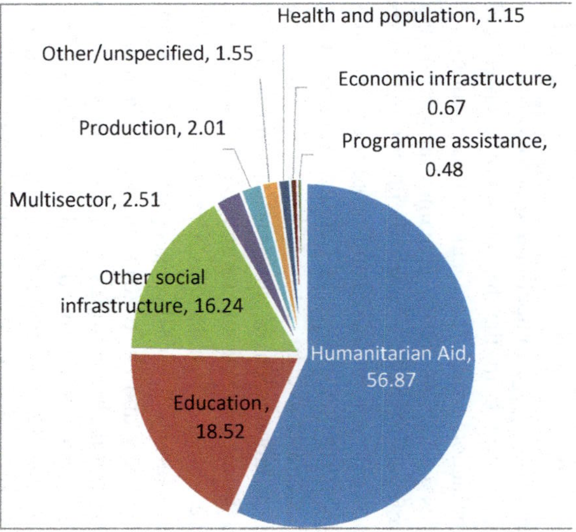

Source: Based on OECD (2016)

legislation is pending, even though—in comparison to other models in the Arab region—the Lebanese draft law would be one of the more comprehensive legislative tools, with a scope of application extending on both private and public sector under the leadership of an overarching competition authority.[55]

Low employment growth in the formal labor market and displacement to the informal sector accompanies the weak macro-economic context. Officially, unemployment rates in Lebanon have remained around 11 percent prior to the influx of Syrian refugees. However, non-official estimates expect it to be more likely around 20 percent with an increasing proportion of youth being affected.[56] With the Syria Crisis, the labor force in Lebanon increased significantly leading to decreasing wages and crowding out of Lebanese laborers, especially in sectors requiring low-skilled labor, such as construction and agriculture. Labor market contractions result in an increase of poverty rates among both in the Lebanese population and the refugee population. Already in 2004, 28.6 percent of the Lebanese population (about 1.5. million people) lived below the national poverty line of 2.40 USD per day and an in 2015 this figure increased by approximately 4 percent.[57] The structural shortcomings of the Lebanese labor market (including a mismatch between supply and demand) and a high degree of informality are not results of the Syria Crisis. They are rather long-term features that need to be urgently reformed to increase resilience, because a current study reveals approximately half of Lebanese households and 75 percent of refugee households experience personal impact in the form of increased cost of living, higher food prices and unaffordability of appropriate housing (see Table 2).

Table 2 Vulnerability to different shock among Lebanese and Syrian refugees

	Economic shocks	Shocks to cost of living	Demographic shocks	Other shocks
Non-refugees	−0.08***	−0.28***	−0.08***	−0.06***
	(0.02)	(0.03)	(0.02)	(0.01)
Constant	0.18***	0.78***	0.18***	0.09***
	(0.02)	(0.03)	(0.02)	(0.01)
Observations	2449	2449	2449	2449

Standard errors in parentheses ***$p<0.01$, **$p<0.05$, *$p<0.1$
Source: World Bank (2016b)
Note: Economic shocks = Loss of jobs, working hours, non-payment or delay in wages; Cost of living shocks = Increase in price of food, cost of housing, loss of assets or forced eviction; Demographic shocks = Breakup of family, serious illness or injury, death of a household member; Other shocks = Decrease in remittances, loss of livestock, crops, or agricultural assets, reduced or suspended assistance

[55] United Nations Economic and Social Commission for Western Asia (2015).

[56] Calì et al. (2015), International Monetary Fund (2014), World Bank (2015, 2016a), and Yaacoub and Badre (2011).

[57] Central Administration of Statistics and World Bank (2015), World Bank (2016a).

4 The Social Costs of the Syria Crisis

Since the beginning of the Syrian conflict, Lebanon has maintained an open borders policy towards Syrian refugees. Given the already large Palestinian refugee population of some 450,000, Lebanon's strained social infrastructure is beginning to show its fragility.[58] In 2013, the World Bank raised the alarm on the negative GDP and its consequences on wages, profits, the ability to invest, save or consume, which in turn would increase poverty rates and unemployment.[59] Beyond the macroeconomic forecast, state provision of basic health and education services are failing to cover the needs of the large refugee population. The growing number of refugees and the relative lack of resources means deteriorating circumstances for refugees, both Syrian and Palestinian who make up over a third of the population of Lebanon and an increasing strain on Lebanese citizens.

4.1 Health

The Lebanese Healthcare sector is characterised by a private–public system. In the last decades, the health system has become increasingly reliant on privately run institutions. Public hospitals, which generally cater to the poorest in Lebanese society, who cannot afford private healthcare, are chronically understaffed, lack finances, and offer lower quality care.[60] Given the extensive medical needs of the Syrian refugees, many of whom are suffering from diseases related to malnourishment, lack of proper hygienic measures and injuries from the conflict itself are overwhelming the already overburdened public health system. As chronic insufficiencies prevail across Lebanon in healthcare provision, many fear the compounding of health issues plaguing the refugee population will lead to a decline in their quality of life, and a breakdown of the Lebanese public health system.[61]

4.1.1 Challenges

A World Bank report argues that the conflict in Syria is impacting the Lebanese health system through increased demand for health care services, which far outweighs the system's operational capacity.[62] Furthermore, the Lebanese Ministry of

[58] UNRWA (2014).
[59] World Bank (2013).
[60] MoE/EU/UNDP (2016).
[61] Lyles and Doocy (2015), UNHCR (2014b), and UNICEF (2014).
[62] World Bank (2013).

Public Health (MOPH) has consistently failed to deliver on payments to contracted hospitals given the lack of finances, while at the same time health care worker shortages are acute.[63]

Given both the number and scattered nature of the Syrian refugees, many are not covered by even basic medical infrastructure and services. Since 2014, over 514,000 consultations for primary health care (PHC) services and over 29,000 consultations for secondary health care (SHC) were conducted.[64] Whilst local and international efforts have been made to address the lack of medical coverage for the most vulnerable, 67.1 percent of refugee households and 48.4 percent of Lebanese households stated that they are not able to get medical care when it is needed in recent survey.[65] There are currently no field hospitals in the regions of Lebanon with the largest population of Syrian refugees; as such, the clinics and hospitals already established in these regions have to deal with the massive increase in people needing urgent medical treatments. This has put an enormous strain on Lebanon's ability to cater to both the refugee's and Lebanese citizen's PHC and SHC needs. Some successes can be reported however: over 100 public healthcare centers, supported by numerous mobile clinics, funded by private, public, and international funds have been established, catering specifically to Syrian refugees and vulnerable Lebanese.[66] In 2014, "20.2 per cent of Syrian refugee households reported one or more hospitalizations of a household member in Lebanon for reasons other than childbirth."[67] Syrian refugees spend on average 105 USD on health services per month, constituting a third of the mean Syrian refugee monthly household income of 316 USD.[68] By comparison, Lebanese host community households spend an average of 217 USD on monthly health services, constituting roughly a fifth of household income of 953 USD during the same time frame.[69] With one third of the mean household income for Syrian refugees going to pay for medical services, many are forced to work illegally or go into debt; the average Syrian refugee household monthly expenditure amounts to 718 USD, more than double the mean monthly income.[70] This indicates that the Lebanese health service provision infrastructure is unable to provide adequate and affordable healthcare for its most vulnerable citizens and the majority of Syrian refugees.

Immunisation, vaccination, and pre-emptive health programmes were run by the Lebanese Ministry of Public Health with support from the WHO, UNICEF, UNHCR, and NGO Beyond Association, targeting over a million children in

[63] World Bank (2013).

[64] UNHCR (2014b).

[65] Lyles and Doocy (2015).

[66] International Medical Corps (2013).

[67] Lyles and Doocy (2015), p. 33.

[68] Ibid.

[69] Ibid.

[70] Ibid.

Lebanon, both Lebanese citizens and Syrian refugees.[71] "In 2014, two million doses of vaccines were procured by UNICEF for the March and April nation-wide campaigns for the Ministry of Public Health," aiming to prevent polio, measles, and rubella outbreaks.[72] Furthermore, the WHO, UNHCR and other NGOs have supplemented the lack of PHC and SHC services through provision of equipment, medical supplies, and medication for chronic diseases.[73] The costs and scarcity issues in the health services are somewhat buoyed by external aid, usually administered through UN institutions such as the WHO and UNHCR as well as many smaller NGO's operating in Lebanon. However, the assistance provided is still far from achieving full and equitable access for all vulnerable Lebanese citizens and refugees.

Thus, the Lebanese health service system is straining, and is fast reaching a breaking point. The main issues plaguing the healthcare infrastructure and system are accessibility to health services, that needs exceed available resources, and the high cost for both the host communities and the refugees.[74] The negative effects of these issues are attenuated to some extent by a large influx of funds and programs from the international community. However, although buoyed by international efforts to cater to the needs of the Lebanese host communities and the refugee populations, the costs are mounting; the funds, subsidies, and programs provided either directly to vulnerable communities or through the Lebanese government and international institutions are not enough to redress the chronic institutional and infrastructure deficiencies. The quality, quantity, and access to medical and health services are quickly degenerating, putting massive strains on Lebanese citizens and worsening the already disastrous situation of the refugee population. As basic medical and health services are not available to large swaths of the population, both Lebanese and Syrian, diseases which have long been thought extinct in the region are now returning and threatening the general livelihood of these communities; polio, measles, and others have resurged in the last year, with thousands of cases reported. Similarly, other communicable diseases, which are otherwise easily preventable with access to basic sanitary measures, are quickly becoming commonplace.

4.1.2 Potential Solutions

Significant efforts have been made by the Lebanese Government and the international community in redressing some of the key issue areas in healthcare provision and coverage, especially in areas of preventative medicine such as immunization campaigns. There remains, however, a thorough shortage of medical coverage for

[71] UNHCR (2014b).

[72] Ibid.

[73] Ibid.

[74] UNHCR (2015).

vulnerable Lebanese and Syrian refugees, as well as a stark decline in the quality and quantity of healthcare in the public system.

Lebanese and international stakeholders must increase efforts to strengthen existing governmental primary health infrastructure and public healthcare systems. An enlargement of resources earmarked for the Ministry of Public Health (MoPH) can increase the Lebanese capacity to monitor and provide early warning in respect to communicable diseases, while increasing response capabilities, and limiting the circulation of otherwise easily treated diseases, which affect a large percentage of the refugee population.[75] Furthermore, operational capabilities need to increase in order to deal with the increased demand of primary, secondary and tertiary health care. This includes further investment in public clinics and hospitals, procurement of medical products and addressing the lack of qualified medical practitioners.

Given that 76 percent of the refugee population are women and children, reproductive, neo-natal and child specific healthcare should be targeted in order to reach the most vulnerable and numerous members of the refugee population.[76]

Greater emphasis on preventative and rehabilitative low-cost measures such as immunisation and nutrition programs, targeting vulnerable Lebanese and Syrian communities can increase the long-term resiliency of these communities, while ensuring that the development of children is not hampered. Although over 2 million doses of vaccines were procured and distributed by UNICEF and UNHCR in 2014, gaps remain in other areas of preventative medicine such as sanitation, nutrition, and routine medical check-ups.[77] Therefore, the procurement and distribution of medical products as well as pharmaceutical products which are necessary in preventative medicine for the MoPH should be a priority.

Underpinning the institutional and infrastructural deficiencies is a gross lack of awareness among vulnerable Lebanese and Syrian refugee communities about their options in accessing medical care of any kind. Therefore, the Government and international stakeholders should conduct awareness programs to address these issues.

4.2 Education

According to UNHCR, half the Syrian refugees in Lebanon are children: as of late 2014, there are over 400,000 Syrian school-aged children in Lebanon, outweighing the number of Lebanese children in public schools.[78] Slightly less than half the school-aged children are not enrolled in any form of education program, be it the public school system, private or non-formal education.[79] According to the UNHCR

[75] World Health Organization (2014).

[76] Ibid.

[77] UNHCR (2014b).

[78] UNHCR and REACH (2014).

[79] UNHCR (2014a).

education update for Syrian refugees report for 2013–2014, most Syrian children have been out of school for more than two years, while efforts to promote education and enrollment have been dismal given the lack of educational institutions (formal and informal, public or private) and resources available.

4.2.1 Challenges

This "Lost Generation" of Syrian children, deprived of safety and opportunity will contribute to large social and economic issues in the future. This includes increasing social inequalities and decreasing access to economic opportunities. The future of the Syrian refugee youth is in question; if the war persists, many of the Syrian refugees growing up under these conditions will add to the worsening economic conditions of Lebanon, where the low wage labor sectors will not be able to ensure equitable employability (construction, farming, industry, etc.), which will contribute to unemployment rates, poverty, crime, and security issues. This is not abstract, it is 5 years since the start of the Syrian civil war, children of 12 or 13 years of age when the war broke are now entering adulthood, many deprived of basic education and future opportunities, exacerbating the already dire social and economic situation of Lebanon. In short, the provision of education to Syrian refugees is of paramount importance, as it can contribute to the medium and long-term resilience of the refugee population.

Various agencies and the Government of Lebanon have requested as much as 1.68 billion USD in order to implement educational programs, develop the educational infrastructure and enroll the vast majority of the refugee children. However, only 38 percent or 646 million USD of that sum has been received as of late 2014.[80] With a total lack of resources, financial, institutional and infrastructural, the Lebanese government and other agencies are at a loss in regard to how to stretch the already thin means. The issues highlighted by the UNHCR in a recent survey indicate two main barriers to accessing education by Syrian refugees: a "Hard Barrier" referring to financial, space, and awareness constraints, as well as a "Soft Barrier" meaning causes, which do not directly affect enrollment but lead to drop-out.

4.2.2 Barriers to Accessing Education

"Hard barriers" to accessing education revolve around the lack of awareness of education opportunities. Most Syrian households polled throughout Lebanon cite "lack of opportunities" or that "they did not know they had the right to education" as the main reasons for non-enrollment of their children.[81] Furthermore, many more families cited "cost of education" as a main barrier; as in the previous

[80] Ibid.

[81] UNHCR and REACH (2014).

case, no household surveyed knew they had access to financial aid for tuition fees. This implication is far reaching; even if the Government of Lebanon and various agencies manage to implement educational programs able to cater to the vast majority of school-aged children, it seems the main issue is the lack of awareness of these opportunities. These opportunities include available resources for educational purposes, the rights of Syrian refugee households to attend and enroll their children in educational programs (either public, private, formal or informal), financial assistance and aid for educational purposes and the procedures of enrollment.

"Soft barriers" to accessing education opportunities include issues of transportation, the necessity of some children to support the family income, language barriers, bullying and security. The same UNHCR survey found that the high costs of programs aimed at transporting children to schooling or educational centers are not sustainable, but have been proven to contribute heavily to high attendance rates.[82] Thus in many cases, especially in regions where Syrian refugee communities are dispersed, one of the main barriers remains lack of access to transport and logistics. Furthermore, many children are forced to contribute to the household income given the high ratio of income to debt the average Syrian family incurs, as enumerated in the above chapters. Lastly, issues of language barriers,[83] bullying and security persist, with many Syrian parents stating that although they would want to enroll their child in schools, one or more of the "soft barriers" prevent them from doing so.

4.2.3 Capacity

The Lebanese Educational System (LES) is at a breaking point. More than 400,000 Syrian school-aged children need educational services, effectively more than doubling the school-aged children in the country. There already are around 275,000 Lebanese children in public schools, within which there is a maximum institutional capacity for 300,000 students, one UNICEF report finds.[84] Where there are large concentrations of Syrian refugees in the already saturated LES, such as the greater Beirut area, there are issues of overcrowding and decline in teaching quality.[85] However, in many areas of Lebanon, where Lebanese and Syrian populations are dispersed, issues of educational service capacity are less pressing. Instead, issues of transport, logistics and awareness prevail (Bekaa region, Tripoli, South Lebanon, etc.). In both cases, however, the costs of education are prohibitive to many Syrian refugee households. Even with subsidies and financial assistance from the Ministry

[82]Ibid.

[83]Many Lebanese children are taught in either French or English, with Arabic as a separate class; Syrian children were taught wholly in Arabic, presenting an impasse to performance and integration of many Syrian children who are in the LES.

[84]UNICEF (2014).

[85]UNHCR and REACH (2014).

of Education and Higher Education (MEHE), the negative opportunity costs of education represented by the loss of income potential of young adults and children is a large deterrent for many families.

These issues are somewhat attenuated by large humanitarian assistance, both financial and institutional, provided by NGO's and private organizations. The UNHCR Yearly Report on Education for Syrian Refugees indicates that of the 219,000 children who have had access to education, more than half have done so through non-formal classes in the 2013–2014 period. The non-formal classes take "place in a diverse range of environments and situations outside the established formal system. These [...] may be temporary and intend to serve identifiable groups of children and learning objectives such as preparing children to be enrolled in school."[86] Financial assistance and subsidies provided by the MEHE in partnership with the UNHCR, UNICEF and other international donors and NGO's is extended to all refugee households (registered or unregistered), and are the main contributor to educational programs and learning centers outside the LES.

4.2.4 Potential Solutions

Although the efforts of the Lebanese government and international organisations in rehabilitating and sustaining educational programs and services for vulnerable Syrian, Palestinian, and Lebanese children are Herculean, similar issues to the health sector persist. The institutional incapacity of the LES to absorb a doubling in the population of school age children, the inability of ad hoc or informal education to counterweigh the deficiencies of the LES, lack of resources, high costs of education relative to the income of Syrian refugees and other hard and soft barriers in accessing education. Overall, this contributes to a negative impact in the medium and long-term resiliency of Syrian refugees, while negatively impacting the Lebanese capacity for providing economic and social security for its citizens.

The decision of the Ministry of Education to introduce a two-shift public education system in order to gradually enroll the Syrian school-aged children is a very positive outcome. Though, the quality and accessibility of that education remains somewhat in question given the lack of resources available. Therefore, concerned stakeholders should aim to strengthen the LES and direct resources for the capacity building of current programs.

The dispersed geographical area which Syrian refugees occupy make accessibility to educational centers a difficulty for some communities and remains one of the main reasons for non-enrollment, especially in remote areas of Lebanon. The construction or rehabilitation of more schools across Lebanon and state sponsored commuter programs and public transport could attenuate the overcrowding and distance problems.

[86] UNHCR (2014a), p. 2.

International agencies such as the UNHCR and other stakeholders, including the LES subsidise Syrian households for educational purposes. Unfortunately, many refugee parents are not aware of the financial assistance programs in place or of their right to access the LES. Awareness programs at a grassroots level could help inform refugee households of the education possibilities and reduce perceived barriers to entry.

5 The Environmental Costs of the Syria Crisis

The agricultural sector, food management and waste management are some of the environmental aspects that the conflict has negatively impacted in Lebanon. There is evidence of increasing competition between Syrian refugees and host communities for scarce natural resources and jobs, as well as social crowding, particularly in disadvantaged and vulnerable governorates, harming both groups in attaining food security. From an environmental perspective the influx of refugees has placed increasing pressures on an already challenged natural resource base (particularly in terms of water and land), impacting negatively on carrying capacity and ecosystem services.

5.1 Agriculture

The agricultural sector has been decreasing in relative terms in the Lebanese economy, making up 4 percent of GDP in 2011,[87] however, it is an important aspect of the rural economy since it has a big impact on the rural population's livelihoods. When focusing on the geographical areas that are most influenced by the Syrian civil war, Northern Lebanon and the Bekaa region will be key, due to their high concentration of Syrian refugees. In these areas agriculture is the most important sector and makes up about 80 percent of local GDP.[88] This places the agricultural sector at the epicenter of the discussion of the impact of the Syrian crisis on the environment in Lebanon. Currently, there is a negative impact on the agricultural sector in Lebanon where 42 percent of Lebanese crops are yielding less than 50 percent of what they used to 2 years ago.[89]

[87] Central Administration of Statistics (2013).

[88] Food and Agriculture Organization (2014).

[89] Food and Agriculture Organization (2015).

5.1.1 Challenges

The agricultural sector in Lebanon has long had close ties with Syria, Lebanese farmers benefited from cheaper, subsided agricultural goods from Syria and had strong trading routes with its neighbor. Since the crisis, Lebanese farmers have been unable to access these trade routes or the agricultural products (fertilizers and pesticides), both of which have increased production costs and had a grave effect on the agricultural sector.[90] The prices of certain agricultural inputs are increasing, 67 percent of farmers claim that prices increased between 2012 and 2014, of whom 47 percent claim traders are manipulating prices upwards.[91] The farmers closer to the border with Syria have at times been unable to access their lands for irrigation or harvest due to its proximity with the border, which has greatly decreased farmers' production.[92] The impact of refugees on natural resources specifically water and land is also having a detrimental impact on the agricultural sector. Water and land pollution as well as competition for natural resources and an increase in demand for water has limited farmers' access to natural resources for irrigation. Indeed the influx of refugees has increased the demand for water by 8–12 percent, which places a stress on water resources as evidenced by the decrease in water access in groundwater systems and river sources. For example, from 2013 to 2014 groundwater levels in wells decreased by 1–20 m in different areas of Lebanon, while three water sources of the Litani River Basin have decreased their water volume in 2012–2013, in comparison to water volume in 2011–2012.[93] According to an FAO study, farmers attribute 58 percent of their reduction in crop yield to a reduction in water.[94] There has also been a high increase in wastewater of 8–14 percent, some of which is untreated and reaches water bodies, open land and soil, which affects agricultural crops. Water disposal in dumpsites also increases the land mass of these dumpsites, having increased by 109,075 m^2, which could potentially infringe on agricultural lands.[95]

In the meantime, there is also a drastic increase in Informal Tented Settlements (ITS), highly concentrated in the Northern Lebanon and Bekaa region, which are horizontal by nature and must comply with inter-tent spacing specifications, further expanding their surface area. The ITSs, which increased in number from 250 in 2011 to 1,224 in 2014 with a largest concentration in Lebanon's main agricultural regions, will directly infringe on agricultural lands, prohibiting farmers take advantage of

[90] Food and Agriculture Organization (2014).
[91] Food and Agriculture Organization (2015).
[92] Food and Agriculture Organization (2014).
[93] MoE/EU/UNDP (2014).
[94] Food and Agriculture Organization (2015).
[95] MoE/EU/UNDP (2014).

what would usually be productive lands.[96] In 2015 the number of ITS reached an alarming 5,082 (out of which approximately 1,303 are inactive sites and 1,700 contain less than four tents) (see maps in Annex).[97]

5.1.2 Potential Solutions

The Government of Lebanon should do all within its means to eliminate the negative environmental impacts of the Syrian conflict and its detrimental effect on the agricultural sector, especially in the areas of the highest concentration of Syrian refugees. It should increase the capacity of the Ministry of Agriculture and other stakeholders to manage disasters and shocks, as well as enhance environmental governance to manage and monitor natural resources to maintain ecosystems services in support of vulnerable populations and to reduce potential for conflict over natural resources.[98]

To appease the competition over water for irrigation, the Government of Lebanon can take advantage of its water endowment in the region and improve its storing methods, as it only stores 6 percent of its total resources, in comparison to 85 percent in the MENA region on average. It can also implement the National Water Sector Strategy (NWSS) from 2012. [99] In relation to the increase in the cost of production, the Government of Lebanon should ensure balanced prices for inputs for farmers by enhancing the capacity of regulatory bodies on price fixing and enhancing agricultural services such as trainings and provision of agricultural extensions.[100]

In order to decrease the contamination of land, soil, and water in the medium term attributed to the refugees, the Government of Lebanon is advised to build the necessary infrastructure, such as disposal and collection equipment, sanitary landfills, and solid waste treatment plants, firstly in impacted areas so that the solid waste attributed to refugees can be correctly discarded. With the necessary infrastructure, priority dumps of high refugee concentration that have a negative impact on water pollution should then be closed.[101] When it comes to the reduction in availability of water, the Government of Lebanon should control the use of water and develop an emergency action plan. Furthermore, the Government of Lebanon should raise awareness of water management and conservation and develop the water infrastructure storage and distribution systems, specifically for affected areas.[102] To detain the expanse of ITSs on limited agricultural lands, which

---

[96]Ibid.

[97]MoE/EU/UNDP (2016).

[98]Food and Agriculture Organization (2014).

[99]United Nations Economic and Social Commission for Western Asia (2016a).

[100]Food and Agriculture Organization (2015).

[101]MoE/EU/UNDP (2014).

[102]Ibid.

could potentially become toxic and unproductive due to uncontrolled solid waste, the Government of Lebanon should contain their encroachment and potentially designate agricultural lands as exclusion zones. [103]

5.2 *Food Security*

Food security covers four dimensions: availability, access, utilisation, and stability. Today and in conjunction with a high public-debt ratio, Lebanon can maintain a reasonably sufficient supply of food, even though it is import dependent with respect to numerous essential food items, such as cereals. Despite fertile plains, the agriculture sector contributes only 4 percent to the country's GDP, compared to 23 percent at the end of the civil war in the 1990s. Notwithstanding, the increase in demand for food due to the influx of Syrian refugees affects not only the availability of food, but also the economic accessibility due to an increasing vulnerability to price shocks. The impact of food security[104] differs across populations (see Fig. 4). This is arguably a result of different economic opportunities and challenges across the groups, which implies that the food security and nutrition challenges differ among the Lebanese population and the refugee population, be they Palestinian or Syrian. In this context, it is alarming that 49 percent of Lebanese are reportedly worried about their ability to source enough food and 31 percent stipulate that they were unable to eat healthy and nutritious food over the course of a year. Syrian refugees, on the other hand, have a more immediate and urgent situation when it comes to food security as 89 percent of them are considered food insecure

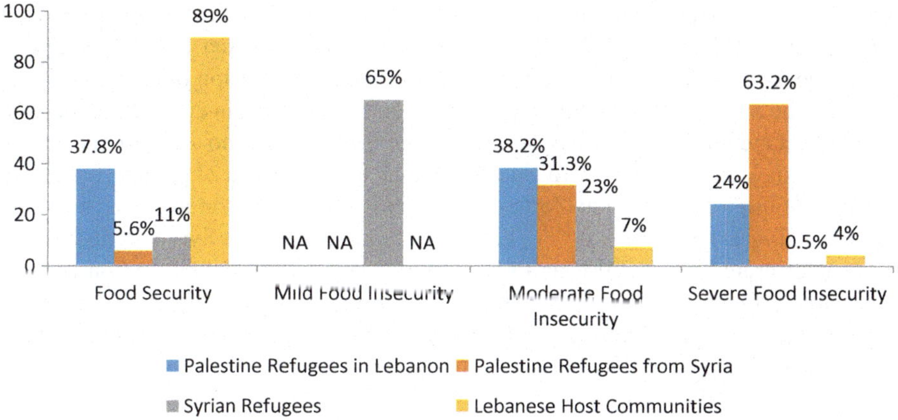

Fig. 4 Food security in Lebanon. *Source*: Food and Agriculture Organization (2015), Rose (2012), United Nations (2015) and WFP/UNHCR/UNICEF (2015)

[103] Ibid.

[104] United Nations Economic and Social Commission for Western Asia (2016a).

(65 percent of them as mildly food insecure and 23 percent as moderately food insecure) and impressively almost 50 percent of their monthly expenditure is on food on average.[105] Furthermore, changes in the value chain of food production and in dietary preferences lead to new challenges related to micronutrient deficiencies and increasing levels of obesity.

Related to the agricultural sector, the food security in Lebanon has been greatly impacted by the Syrian civil war. As the farmers face obstacles such as lack of natural resources and increasing cost of production, their production has undoubtedly dropped, which influences food availability.[106] Although food security is not an immediate problem for Lebanese host communities at a national level, certain local areas lack a physical access to food. Probably the biggest problem when it comes to food security is the issue of economic vulnerability which makes access to food more restrained and could increase risk of food insecurity in the future.[107] When it comes to Syrian refugees, all aspects of food security are at a high risk and more gravely have shown deterioration over time.[108]

5.2.1 Challenges

Availability: When there is an increase of around 33 percent in the population in a few years due to the influx of refugees, the demand for food inevitably shoots up. Lebanese farmers have not been able to keep up with supply for a number of reasons aforementioned above and food imports from Syria have drastically decreased. This results in a gap for food availability, mainly due to the performance of the agricultural sector in Lebanon.

Utilisation: Overall, food utilization in Lebanon varies depending on the region. There is a reduction in meals eaten per day by Lebanese host communities when it comes to those areas that have a higher proportion of Syrian refugees. 13 percent of respondents had eaten two meals in the last day in Bekaa and 18 percent of respondents in Akkar, two of the regions with highest proportion of refugees.[109]

With the Syrian refugees, number of meals eaten has fallen in 2015 in comparison to 2014. One in three households consumed just one or no cooked meals the previous day, in comparison to one in four in 2014. About 27 percent of Syrian refugee households were unable to cook at least once a day on average, the main reason being a lack of food.[110] The nutrients consumed by Syrian refugees has also deteriorated since 2014, as the households who were unable to consume of vegetables doubled to 60 percent, those unable to consume vitamin A rich food groups increased to 33 percent and sugary products were consumed more, being consumed almost on a daily basis.[111]

---

[105] WFP/UNHCR/UNICEF (2015).

[106] Food and Agriculture Organization (2015).

[107] Food and Agriculture Organization (2015), United Nations Economic and Social Commission for Western Asia (2016a).

[108] WFP/UNHCR/UNICEF (2015).

[109] Food and Agriculture Organization (2015).

[110] WFP/UNHCR/UNICEF (2015).

[111] Ibid.

Access: What seems to be the biggest problem is the non-physical access to food, mainly finance measures. There has been an increase of 200,000 workers in the labor supply in the agricultural sector. This increase in supply of labor by the Syrian refugees, which usually offer lower more competitive wages than Lebanese citizens, lead to unemployment and/or a reduction in their income.[112] This phenomenon is mostly seen in the regions that are most dependent on agriculture and with highest concentration of Syrian refugees. Syrian refugees allowed to work in the labor market are suffering from this competition as well, facing a decline in their income, while also facing limitations from the Lebanese government to access the Lebanese labor market.[113]

Lebanese communities face higher food prices which decrease their purchasing power. This has led to an increase in usage of debt and savings and a decrease in other needs, in order to maintain the same food consumption as before. If that is not done, then there is a reduction in the number and size of meals eaten per day.[114] Unemployment and reduction of income related to the influx of refugees also weakens food access in Lebanon. According to the Development Management International report on Lebanese hosting households 19 percent of respondents in Bekaa were borrowing food or purchasing food on credit and 4 percent of them were relying on humanitarian actors for food, while in the north 9 percent were borrowing food or purchasing food on credit and 13 percent of them were relying on humanitarian actors for food.[115] The refugees have mostly relied on food vouchers by organizations for food access, as 48 percent of the food they obtained was through these vouchers. Syrian refugees have also increased their credit and formal and informal borrowing in order to attain food, as witnessed in Akkar and in Bekaa where borrowing increased by 24 percent and 25 percent, respectively.[116] In Lebanon, Syrian refugees are 15 percent less likely to buy food with their own economic resources than they were one year ago, and this can be seen in the increase in household dependency on food vouchers.[117] The future of non-physical access to food for Syrian refugees and Lebanese host communities seems grim as food market trends exhibit an increment in food commodity prices.[118]

Stability: Food instability is a threat that is growing, since in 2014, there has been a meteorological drought, increased unemployment in the areas of higher refugee presence, increase in food prices, and a decrease in overall security. The crisis in Syria, Lebanon's historically, economic, and politically connected neighbor, and

[112] Food and Agriculture Organization (2015), United Nations Economic and Social Commission for Western Asia (2016a).

[113] WFP/UNHCR/UNICEF (2015).

[114] Food and Agriculture Organization (2015).

[115] Development Management International (2012).

[116] WFP/UNHCR/UNICEF (2015).

[117] Ibid.

[118] United Nations Economic and Social Commission for Western Asia (2016a).

the influx of refugees compromise the overall political and economic stability in Lebanon, gravely affecting food stability.

5.2.2 Potential Solutions

In order to enhance food security, the Government of Lebanon should focus on its more pressing issue of availability of and access to food. The Government of Lebanon should focus on increasing its self-sufficiency of food, as it is heavily dependent on imported food, mainly from Syria which has decreased its food production thus decreasing Syria's food exports to Lebanon.[119] This requires support from the Government of Lebanon through different projects for a sustainable agricultural sector. Supporting the sustainable development of the agricultural sector, apart from increasing the food availability, will also ensure a steady and potentially increasing income for those in the areas that are most affected by the Syrian crisis. The Government of Lebanon should implement the complete Ministry of Agriculture 2015–2019 strategy that would improve the food security in Lebanon. Ensuring an overall stable security situation, where road blocks do not happen, would also avoid food instability. When it comes to non-physical access to food the government should provide support for consumers through income-generating labor intensive programs in order to make them more resilient and less susceptible to shocks. Lebanon has the potential to increase its agricultural sector as they have comparative advantage in fruits and vegetables and produce enough products to export a select number.[120] A food security surveillance system could also monitor the status of Lebanese households in order to create appropriate responses to shocks that would threaten food security.

5.3 Waste Management

Waste management can come in two main forms through wastewater and solid waste. The impact the Syrian crisis has brought on waste management is through the influx of refugees, which increases the amount of waste and puts pressure on the current and weakened waste management infrastructure. The amount of municipal solid waste (MSW) attributed to refugees is expected to increment to 15.7 percent in comparison to before the crisis, since the main areas that witnessed this increase in MSW contain a high presence of refugees. When it comes to wastewater, there has been an increase of 8–14 percent by the end of 2014.[121]

[119] UN High Commissioner for Refugees (2014).

[120] United Nations Economic and Social Commission for Western Asia (2016a).

[121] MoE/EU/UNDP (2014).

5.3.1 Challenges

The increase in solid waste, attributed to refugees, has been overstressing the existing solid waste management (SWM) infrastructure, as only 48 percent of the increase is capable of being correctly disposed of by the existing infrastructure. This overstress can be witnessed by the increase in municipal expenditure on SWM, increasing 11 percent from 2011 to 2012 and 40 percent from 2012 to 2013. The remaining 52 percent that cannot be sustained by the current SWM infrastructure is being disposed of in open dumps, further contaminating the environment on land, soil, and groundwater.[122] The increase in solid waste has also worsened the health and safety conditions around dumpsites, as they become insect and rodent breeding locations that could spread diseases. There has also been an increase of infectious waste, of which 18 percent of it is being improperly managed without any treatment, which leads to grave environmental problems when it comes to air, water, and land pollution.[123] The increase in wastewater attributed to refugees is estimated to have created an increase of 40,000 tonnes of Biological Oxygen Demand (BOD_5) per year,[124] or 34 percent increase at the national level, which greatly increases the organic biodegradable load in the environment. The areas of highest presence of refugees have one of the highest pressures of BOD_5.[125] The impacts of untreated wastewater, which is not being managed by existing infrastructure, on water bodies and open lands, negatively affects fish and wildlife populations as well as recreational water use and drinking water.

5.3.2 Potential Solutions

When it comes to waste, in general, the Government of Lebanon should focus on increasing the capacities of the municipalities that are overstrained and its infrastructure overstressed to be able to manage the increase in waste. Specifically for solid waste, the Government of Lebanon should provide additional waste collection bins and trucks to host communities, promote and implement recycling activities, collect and treat health care waste, reduce the burdens for financial pressure on municipalities, build necessary infrastructure in key areas and close identified priority dumps in areas of high refugee concentration.[126]

---

[122] Ibid.

[123] Ibid.

[124] BOD_5 calculates the quantity of oxygen required or used for the microbiological decomposition (oxidation) of organic material in water. It is frequently used by academics and environmental agencies to measure the pollution load in water.

[125] Ibid.

[126] Ibid.

In relation to wastewater management the Government of Lebanon should ensure the supervision of sludge disposal and should implement wastewater collection and treatment infrastructure in the impacted areas.[127]

6 Where Do We Go Now (لـ ويـ ن وهلّا)...

To date (April 2016) a fragile and partial ceasefire brokered by the USA and Russia does not appear to hold and the UN-led peace process may falter as parties to the conflict prepare yet again for an increase in armed confrontation (e.g., in and around Aleppo). In a parallel and national process, parliamentary elections were held on 13 April 2016 though depending on who is asked about legitimacy and validity of the elections, they are considered a success[128] or a sham.[129] Without a durable political situation in place (or even in sight), an end to the war in Syria and the Syria Crisis affecting the country itself, the neighbors regionally and globally appears not within a foreseeable future.

The present paper described the impact of the civil war in Syria on its neighboring country Lebanon. It highlighted the historical relationship between the two countries and illustrated economic, social, and environmental costs. Given the constraints of the chapter, the discussion provided only an overview and can be analyzed in much further detail (e.g., health impact beyond the physical extending to addressing mental health concerns). While Lebanon has had political, economic, and social challenges prior to the Syria Crisis, the costs resulting from the Syria war for Lebanon are undoubtedly high and not abstract monetary sums. Each day the crisis affects people from both the host community and the refugees in Lebanon in very tangible ways. As a consequence, stress—continuing and at varying degrees of intensity—subjects the resilience of the individual, as well as that of society, to the test. A sentiment of people partnering in deprivation is taking over the rhetoric of politicians, researchers, and other stakeholders.

Nevertheless, seeing how people live alongside and try to make the most out of their situation means that hope for a better future is still there. Together with other countries around the world, Lebanon adopted "Transforming our World: The 2030 Agenda for Sustainable Development" in September 2015, which means a commitment towards the implementation of development targets subsumed under the 17 Sustainable Development Goals (SDGs) by the year 2030. In addition to calling for the formulation of a clear, comprehensive, multi-stakeholder "National Development Strategy" as an integrated response plan for Lebanon, the present paper would like to make a strong case for greater integration and cooperation among many public, private, and civil society institutions to improve economic (e.g., labor market access, subsidy reform, competition, and market access), social (e.g., safety nets) and environmental (e.g., upgrade irrigation practices) policies in

[127] Ibid.

[128] MoE/EU/UNDP (2016).

[129] World Health Organization (2014).

Lebanon with a view towards enhancing resilience and returning to a peaceful, sustainable and inclusive development growth path.

Acknowledgments The authors thank the reviewers for their time and valuable comments. We also appreciate the editorial assistance from Ms. Rebecca Crompton.

A.1 Annex

Summary timeline of events from the French Mandate onwards

1916	Asia Minor Treaty (Sykes-Picot Agreement) between the French Republic and the UK on the partition of spheres of influence in the Middle East holdings of the Ottoman Empire
1920	League of Nations sanctioned Mandate of the Middle East under French and British supervision and governance. French Mandate included present day Syria and Lebanon
1920–1922	Syrian Kingdom formed under Faisal I bin Hussein bin Ali al-Hashimi. French Mandate forces ousted Faisal and established a Mandatory French Administration
1925–1930	Great Syrian Revolt began in Jabal Druze province and quickly spread over the entire Syrian Federation with the purpose of achieving Syrian independence and unification. The Syrian region was unified into the Republic of Syria without Lebanon
1936–1946	Franco-Syrian Treaty of Independence, not ratified by the French Government, continuing the French Mandate in the region. In 1946, under pressure from the UN, the British and Syrian and Lebanese Nationalists, French forces evacuated the region and Syria and Lebanon gained de facto independence
1948	Arab-Israeli War with the newly created State of Israel over Arab dissatisfaction with the UN sponsored partition plan
1949	Military Coup d'état led by Hussni al-Zaimy overthrowing the civilian democratic government in Syria
1956	Suez Canal Crisis involving Egypt, Britain, France and Israel
1958–1961	Formation of United Arab Republic under Egypt's Gamal Abdal Nasser's leadership between Egypt and Syria
1961–1963	Members of the Arab Socialist Resurrection Party (Ba'ath Party) took over all executive and legislative authority in the country
1970	Hafez al-Assad, member of the Ba'ath party and Minister of Defense took power in a coup and assumed leadership of the Syrian Arab Republic as President
1973	Yom Kippur War—over Sinai and the Golan Heights
1975–1990	Lebanese Civil War. Syrian forces begin occupation of Lebanon in 1976
1990–1995	End of the Lebanese Civil War in 1990 with the signing of the Syrian sponsored Tai'f Agreement and the subsequent Brotherhood Treaty between Lebanon and Syria in 1992
2000–2011	Syrian forces leave Lebanon in 2005. Hafez al-Assad dies in 2000 and is succeeded by his son, Bashar al-Assad
2011-present	Syrian Civil War

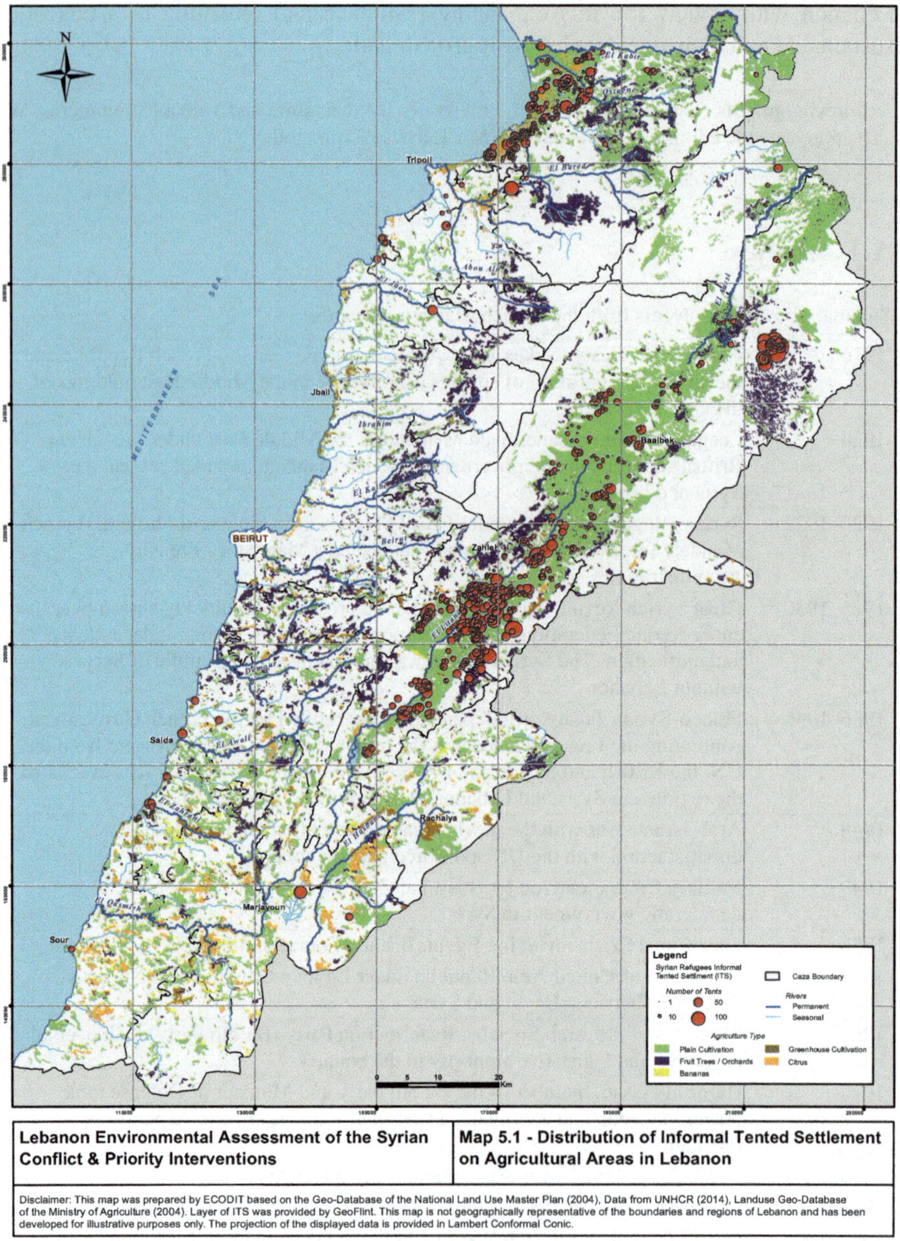

Map 1 Lebanese environmental assessment of the Syrian conflict and priority intervention—distribution of informal and tented settlements on agricultural areas. *Source*: MoE/EU/UNDP (2016)

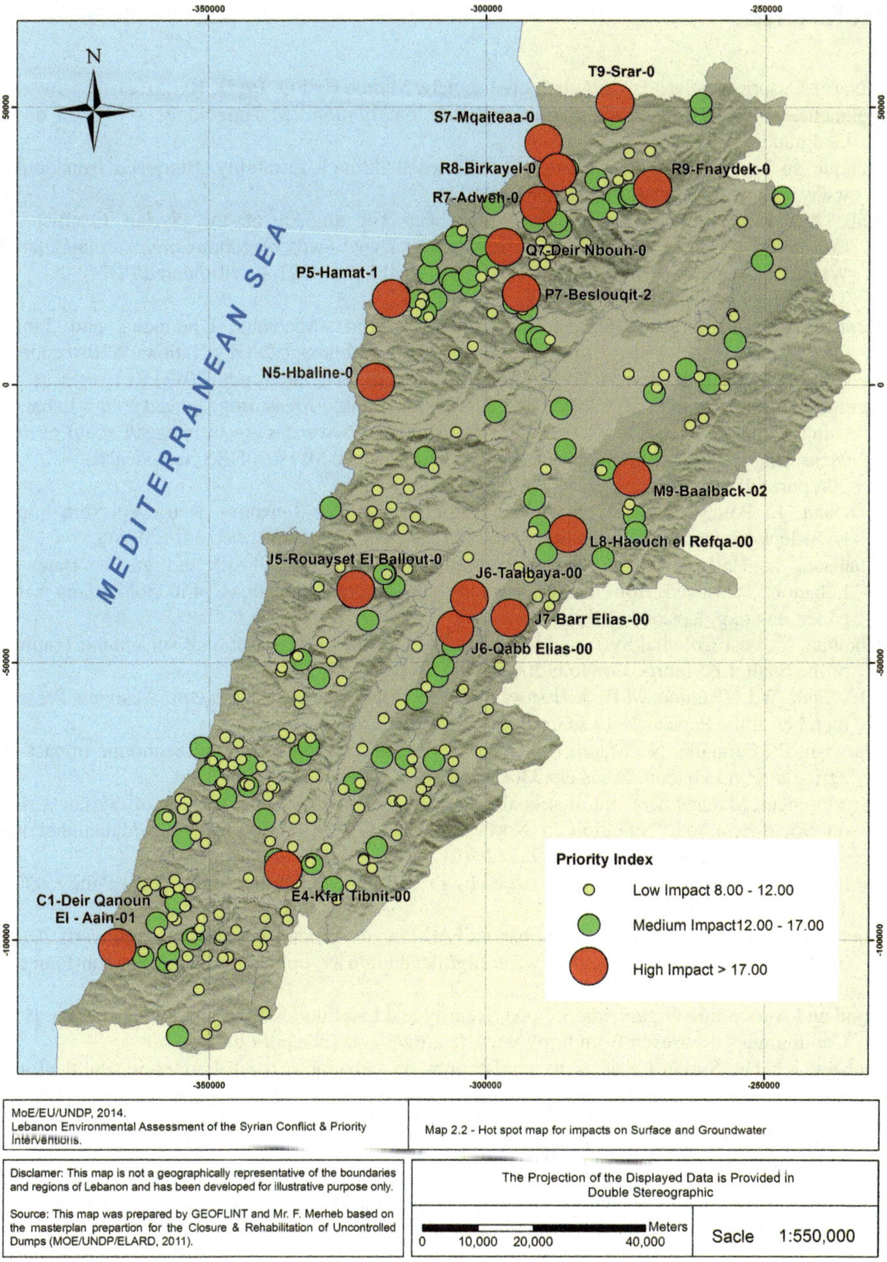

Map 2 Hot spot map for MSW impacts on surface and ground water. *Source*: MoE/EU/UNDP (2014)

References

Alvarez-Ossorio, I.: Syria's struggling civil society. Middle East Q. **19**(2), 10 (2012)

Balanche, F.: Syrie-Liban: intégration régionale ou dilution? M@ppemonde - Maison de la Geographie, **13** (2005)

Banque du Liban: Lebanon's Resilience will see it through instability. Retrieved from http://www.bdl.gov.lb/news/more/8/182/316 (2015)

Calì, M., Harake, W., Hassan, F., Struck, C.: The impact of the Syrian Conflict on Lebanese Trade. Retrieved from http://www-wds.worldbank.org/external/default/WDSContentServer/WDSP/IB/2015/04/28/090224b082e14021/1_0/Rendered/PDF/The0impact0of00ct0on0Lebanese0trade.pdf (2015)

Central Administration of Statistics: Lebanese National Accounts: Comments and Tables 2004-2011. Retrieved from http://www.cas.gov.lb/images/PDFs/National%20Accounts/CAS_Lebanon_National_Accounts_2011_Comments_and_tables.pdf (2013)

Central Administration of Statistics, & World Bank: Measuring poverty in Lebanon using 2011 HBS. Retrieved from http://www.cas.gov.lb/images/Excel/Poverty/Measuring%20poverty%20in%20Lebanon%20using%202011%20HBS_technical%20report.pdf (2015)

Chabaan, J.: Policy Brief: Avoiding the Resource Curse in Lebanon. Retrieved from http://www.lcps-lebanon.org/publications/1459424105-policy_brief_18_web.pdf (2016)

Chabaan, J., Harb, J.: Macroeconomic implications of windfall oil and gas revenues in Lebanon. Retrieved from http://www.lcps-lebanon.org/publications/1450350528-jana_harb-paper_eng.qxp_lcps.pdf (2015)

Chaitani, Y.: Post-Colonial Syria and Lebanon: The Decline of Arab Nationalism and the Triumph of the State. I.B. Tauris, London (2007)

Cleveland, W.L., Bunton, M.P.: A History of the Modern Middle East, 5th edn. Westview Press, a member of the Perseus Books Group, Boulder (2013)

Darwish, R., Farajalla, N., Masri, R.: The 2006 war and its inter-temporal economic impact on agriculture in Lebanon. Disasters **33**(4), 629–644 (2009)

Development Management International: Rapid Assessment of the Impact of Syrian Crisis on Socio-Economic Situation in North and Bekaa. Retrieved from https://data.unhcr.org/syrianrefugees/download.php?id=957 (2012)

Fildis, A.T.: The troubles in Syria: spawned by French divide and rule. Middle East Policy **18**(2), 10 (2011)

Food and Agriculture Organization: Lebanon: FAO Plan of Action for Resilient Livelihoods 2014–2018. Retrieved from http://www.fao.org/fileadmin/user_upload/rne/docs/Lebanon-Plan.pdf (2014)

Food and Agriculture Organization: Food Security and Livelihoods Assessment of Lebanese Host Communities. Retrieved from http://www.fao.org/3/a-az720e.pdf (2015)

Gebara, K.: The Syrian Crisis & its implications on Lebanon (Social, Economic, Political and Security Challenges an Potential Solutions). Retrieved from http://www.sciences-po.usj.edu.lb/pdf/The%20Syrian%20Crisis%20%20its%20Implications%20on%20Lebanon%20-%20Khalil%20Gebara.pdf (2015)

Government of Lebanon: London Conference – Lebanon Statement of Intent (Presented by the Republic of Lebanon). Retrieved from http://www.businessnews.com.lb/download/LondonConferenceLebanonStatementOfIntent4Feb2016.pdf (2016)

Heinrich Boell Foundation: Environmental Impact of the 2006 Lebanon War. Retrieved from http://www.rebuildlebanon.gov.lb/images_Gallery/Heinrich%20Boll%20Foundation-161006.pdf (2006)

Heydemann, S.: Authoritarianism in Syria: Institutions and Social Conflict, 1946–1970. Cornell University Press, Ithaca (1999)

Human Rights Watch: Syria. Retrieved 3 February 2016, from https://www.hrw.org/middle-east/n-africa/syria (2016)

International Medical Corps: International Medical Corps Lebanon Syrian Refugee Response. International Medical Corps (2013)
International Monetary Fund: IMF Report for Selected Countries and Subjects: Syria. Retrieved 15 February 2016, from http://www.imf.org/external/pubs/ft/weo/2013/01/weodata/weorept.aspx?sy=1990&ey=2010&scsm=1&ssd=1&sort=country&ds=.&br=1&pr1.x=45&pr1.y=6&c=463&s=NGDP_R%2CNGDP_RPCH%2CNGDPD%2CLUR&grp=0&a= (2013)
International Monetary Fund: Lebanon - Selected Issues Report. IMF Country Report No. 14/238. Retrieved from https://www.imf.org/external/pubs/ft/scr/2014/cr14238.pdf (2014)
Investment Development Authority of Lebanon: Economic Performance. Retrieved from http://investinlebanon.gov.lb/en/lebanon_at_a_glance/economic_profile/economic_performance (2015)
Kamrava, M.: The Modern Middle East - A Political History Since the First World War. University of California Press, Berkley/Los Angeles (2005)
Khoury, M., Jaulin, T.: EUDO Citizenship Observatory - Country Report Lebanon. Retrieved from http://eudo-citizenship.eu/docs/CountryReports/Lebanon.pdf (2012)
Khoury, P.S.: Factionalism among Syrian nationalists during the French Mandate. Int. J. Middle East. Stud. **13**(4), 441–469 (1981)
Khoury, P.S.: The Tribal Shaykh, French Tribal Policy, and the Nationalist Movement in Syria between two World Wars. Middle East. Stud. **18**(2), 13 (1982)
Khoury, P.S.: Syria and the French Mandate: The Politics of Arab Nationalism 1920–1945. Princeton University Press, Princeton (1987)
Krókowska, K.: The fall of democracy in Syria. Perceptions **16**(2), 17 (2011)
Lebanon Center for Policy Studies: The repercussions of the Syrian refugee crisis on Lebanon: The challenges providing services and creating jobs. Retrieved from http://www.lcps-lebanon.org/publication.php?id=282&category=900&year=2016 (2016)
Leenders, R.: Policy Brief: Safeguarding against Corruption Risks in Lebanon's Offshore Petroleum Sector. Retrieved from http://www.lcps-lebanon.org/publications/1452173264-policy_brief_16_web.pdf (2016)
Lyles, E., Doocy, S.: Syrian refugee and Affected Host Population Health Access Survey in Lebanon. Retrieved from https://data.unhcr.org/syrianrefugees/download.php?id=9550 (2015)
Makdisi, S., Sadaka, R.: The Lebanese Civil War, 1975–1990. American University of Beirut Institute of Financial Economics - Lecture and Working Paper Series, 3 (2003)
Masters, J., Laub, Z.: Hezbollah (a.k.a. Hizbollah, Hizbu'llah). Retrieved from http://www.cfr.org/lebanon/hezbollah-k-hizbollah-hizbullah/p9155 (2014)
MoE/EU/UNDP: Lebanon Environmental Assessment of the Syrian Conflict & Priority Interventions. Retrieved from http://www.moe.gov.lb/getattachment/6c5fbe66-e28d-4fca-a0ce-3ce92add3f83/Lebanon-Environmental-Assessment-of-the-Syrian-Con.aspx (2014)
MoE/EU/UNDP: Lebanon Environmental Assessment of the Syrian Conflict & Priority Interventions - Updated Factsheet (December 2015). Retrieved from http://www.moe.gov.lb/getattachment/dd4e6574-a959-47c1-b2ed-1fc1847b7515/Updated-Fact-Sheet-December-2015-Environmental-Ass.aspx (2016)
Norton, A.R.: Hezbollah: A Short History (Princeton Studies in Muslim Politics). Princeton University Press, Princeton (2014)
Osoegawa, T.: Syria and Lebanon: International Relations and Diplomacy in the Middle East. I.B.Tauris, London (2013)
Rose, D.: Assessing Food Security at WFP: Towards a Unified Approach. Retrieved from https://resources.vam.wfp.org/sites/default/files/WFP%20Fd%20Sec%20Assessment_Design%20Phase%201%20Report.pdf (2012)
Rowayheb, M.G.: Political change and the outbreak of civil war: the case of Lebanon. Civil War **13**(4), 22 (2011)
Shambrook, P.A.: French Imperialism in Syria, 1927–1936. Ithaca Press, Reading (1998)
Sullivan, M.: Hezbollah in Syria. Retrieved from http://www.understandingwar.org/sites/default/files/Hezbollah_Sullivan_FINAL.pdf (2014)
Traboulsi, F.: A History of Modern Lebanon. Pluto Press, London/Ann Arbor (2007)

Transparency International: Table of results: Corruption Perceptions Index 2015. https://www.transparency.org/cpi2015#results-table (2015)
Treaty of Brotherhood, Cooperation and Coordination Between the Syrian Arab Republic and the Lebanese Republic United Nations - Treaty Series (1992)
UN High Commissioner for Refugees: Inter-agency Multi-Sector Needs Assessment (MSNA) Phase One Report: Secondary Data Review and Analysis. Retrieved from https://data.unhcr.org/syrianrefugees/download.php?id=6241 (2014)
UNHCR: Syrian Refugee Response in Lebanon: Education Update. UNHCR, Lebanon (2014a)
UNHCR: Syrian Refugee Response: Health Within the Lebanon Refugee Context. UNHCR, Lebanon (2014b)
UNHCR: 2015 UNHCR country operations profile - Lebanon. Retrieved 15 March 2016, from http://www.unhcr.org/pages/49e486676.html (2015)
UNHCR: Syria Regional Refugee Response. Retrieved 3 February 2016, from http://data.unhcr.org/syrianrefugees/regional.php (2016)
UNHCR, REACH: Barriers to Education for Syrian Refugee Children in Lebanon: UNHCR, REACH (2014)
UNICEF: UNICEF Lebanon: Syria Crisis UNICEF (2014)
United Nations: Lebanon Crisis Response Plan 2015–2016: Food Security Sector Plan. Retrieved from http://www.un.org.lb/library/assets/FoodSecurity-SectorPlan-065649.pdf (2015)
United Nations Conference on Trade and Development: Merchandise trade correlation index, annual, 1995–2012 (2016)
United Nations Economic and Social Commission for Western Asia: Economic Governance Series: Competition and Regulation in the Arab Region. Retrieved from https://www.unescwa.org/sites/www.unescwa.org/files/publications/files/e_escwa_edid_15_5_e.pdf (2015)
United Nations Economic and Social Commission for Western Asia: Strategic Review of Food and Nutrition Security in Lebanon. Retrieved from https://www.unescwa.org/publications/food-nutrition-security-lebanon (2016a)
United Nations Economic and Social Commission for Western Asia: UNESCWA National Agenda for the Future of Syria Programme. Beirut (2016b)
UNRWA: Where we work. Retrieved 15 March 2016, from http://www.unrwa.org/where-we-work/lebanon (2014)
WFP/UNHCR/UNICEF: Vulnerability Assessment of Syrian Refugees in Lebanon. Retrieved from https://data.unhcr.org/syrianrefugees/download.php?id=10006 (2015)
World Bank: Lebanon: Economic and Social Impact Assessment of the Syrian Conflict, Vol. Special Report, p. 7. World Bank (2013)
World Bank: Lebanon: Economic and Social Impact Assessment of the Syrian Conflict. Retrieved from http://www.worldbank.org/content/dam/Worldbank/document/MNA/LBN-ESIA%20of%20Syrian%20Conflict-%20EX%20SUMMARY%20ENGLISH.pdf (2015)
World Bank: World Development Indicators for Lebanon. Retrieved 3 February 2016, from http://data.worldbank.org/country/lebanon (2016a)
World Bank: Findings from a 2015 Survey of Syrian Refugees and Host Communities in Lebanon (2016b)
World Health Organization: WHO Donor Snapshot - Lebanon. WHO Donor Snapshot. Retrieved from http://www.who.int/hac/donorinfo/syria_lebanon_donor_snapshot_1july2014.pdf (2014)
Yaacoub, N., Badre, L.: The labour market in Lebanon, Statistics In Focus (SIF), Central Administration of Statistics, Lebanon, Issue number 1, October 2011. Retrieved from http://www.cas.gov.lb/images/pdfs/sif/cas_labour_market_in_lebanon_sif1.pdf (2011)

Absenteeism Impact on Local Economy During a Pandemic via Hybrid SIR Dynamics

E.W. Thommes, M.G. Cojocaru, and Safia Athar

Abstract In this paper we study the cost of absenteeism and presenteeism (going to work while sick) during a pandemic in a local economy with several geographically distinct locations, and with work force populations consisting of individuals who live and work in the same city, and individuals who live and work in different locations (daily commuters).

We run simulations to study the effects of the fear factor and of the severity of disease on the number of missed work days in the region, which we translate into loss of productivity costs. We find that higher values of the fear parameter lead to high absenteeism and lower infection levels. However, we also show that for severe pandemics (such as the number of secondary infections is higher) there are scenarios where there exists a unique value of the fear parameter which leads to minimum economic costs for the regional economy. This indicates that "staying at home" policies during an epidemic could be implemented for the work force, without reaching a state of emergency.

Keywords Projected SIR dynamics • Cost of absenteeism in pandemic • Local economies and pandemics

1 Introduction

Modelling in the field of epidemiology has its roots in the early twentieth century. Modern epidemiology has its theoretical roots founded on modelling the spread of a disease and showing that if certain conditions are met, then a disease will go extinct. Some of the earliest disease models were developed by Kermack and McKendrick (Capasso and Serio, 1978). Longini in Longini (1988) studied that the history of infectious agents follow a specific pattern that can be developed in any epidemic model. Epidemiologic models are currently both well-known and widely used (see, for instance, Diekmann and Heesterbeek 2000; Hethcote 2000;

E.W. Thommes • M.G. Cojocaru (✉) • S. Athar
University of Guelph, Guelph, ON, Canada
e-mail: ethommes@uoguelph.ca; mcojocar@uoguelph.ca; sathar@uoguelph.ca

I.S. Kotsireas et al. (eds.), *Dynamics of Disasters—Key Concepts, Models, Algorithms, and Insights*, Springer Proceedings in Mathematics & Statistics 185,
DOI 10.1007/978-3-319-43709-5_15

Thommes et al. 2014; Heesterbeek 2002; Longini 1988 and the references therein). Among these, compartmental models of Susceptible-Infected-Recovered (SIR) type and their many variations are used to also model pandemics and they are concerned with finding ways to estimate the number of secondary infections by one infected individual [the number of secondary infections produced by one individual in completely susceptible populations is denoted by R_0, whereas the same number is denoted by R in partially susceptible populations (Heesterbeek 2002; Chowella et al. 2006; Balcan et al. 2009), etc.; completely susceptible populations here means that all individuals are equally susceptible to the disease; partially susceptible populations, on the other hand, contain some individuals who have immunity or partial immunity to the disease].

In general, compartmental models encompass a population as a whole, often heterogenized by age groups and sometimes other demographic factors (Thommes et al., 2014; Arino and Van den Driessche, 2003; Hyman and LaForce, 2003; Longini, 1988; Nichol et al., 2010). There are also works that regard the population from a different point of view, such as the ones in Sattenspiel and Dietz (1995), which presented a migration model of a population for travelling between different communities. They investigated the transmission of measles in the Caribbean Island of Dominica. Sattenspiel and Herring (1998) worked on the same model but applied it to the Canadian subarctic region to look into population movements in closed regions. Arino and Van den Driessche (2003) adapted the model described by Sattenspiel and Dietz (1995) and introduced the time evolution of a disease using a Susceptible-Infective-Susceptible (SIS) compartmental model and gave accurate derivation of the reproduction number introducing bounds to it. In these works, the models are dynamic, and time evolution shows the evolution of numbers of individuals in each compartment.

In this work, we take the point of view of modelling a population of workers, rather than a biologically complete population. Our working populations live in geographically distinct locations ("cities"), but each location's work force contains a percentage of daily commuters to the other locations, reflecting the current type of workforce in many local economies (Greenwood 1985; Greenwood and Gary 1989). The mathematical framework we employ here is that of a hybrid dynamical systems, specifically we combine a non-standard continuous time infection model (SIR) and a discrete transition state (DTS) matrix encapsulating individuals' daily decision to go to or miss work that day.

In our model the compartments are still S, I, and R; however, the birth rate here signifies the people starting work for first time per unit time, and death rate would be workers retiring per unit time. To account for behavior change in the presence of a pandemic event, we consider that a healthy susceptible individual (S) has a probability of missing work proportional to their city's overall infection level. The scaling factor, which we call "fear" factor, represents the sensitivity of the work force in one location to the level of infection in that location. An infected individual (I) has a probability of missing work dependent on the severity of the disease. Susceptible individuals contract the disease during the work day, at the respective work locations. The infected individuals who choose to miss work are considered to be staying home and are thus removed from the transmission dynamics.

Furthermore, our work was inspired by several interesting facts reported in the press at large regarding loss of productivity in the economy due to both absenteeism and presenteeism [see The Globe and Mail (Wency 2011)]. It seems to be the case that there is a growing opinion of a majority of employers that ill employees cost double in output losses as compared to healthy employees. Moreover it seems that "83 percent of U.S. adult survey participants continued to attend work or school while experiencing flu-like symptoms." The spread of influenza (commonly known as "*flu*") during every winter season is common in most countries around the world, and while it accounts for some deaths in the population, it also benefits from a prophylactic treatments (vaccines) (Thommes et al., 2014). In general, the working population is accustomed to the presence of influenza, and while there are people missing work during one season, we choose to model here a pandemic infection [such as the H1N1 (Balcan et al. 2009), for instance], where the entire population is susceptible and where there are no prophylactic treatments. In such a case, a "stay home from work" policy could help bring down the numbers of infected workers and the duration of the infection peak, of course at the expense of economic loss from absenteeism.

Given the above considerations and our discussion of the working population we like to model, we modify a classic SIR model by imposing population conservation and flow constraints that reflect daily commuters trips. This gives rise to a projected dynamics [see, for instance, Cojocaru et al. (2007, 2009), Dupuis and Nagurney (1993) for similar details] which allows for solutions to our model to be respecting the nonnegativity and flow constraints at every time t.

Projected dynamical systems (PDS) are used in operations research, economic theory, finance and network analysis (see, for example, Aubin and Cellina 1984; Dupuis and Nagurney 1993; Nagurney 1998; Cojocaru and Greenhalgh 2012; Cojocaru et al. 2009, 2005, 2007; Cojocaru 2007; Nagurney and Zhang 2012 and the references therein). In general, a PDS is a dynamical system whose flow is constrained to evolve on a closed and convex subset, generically denoted by K, of the ambient space. In this model we consider the ambient space to be the Euclidean space $\mathbb{R}^n$ and we consider the constraint set K to be described by the usual population constraints of an SIR model, together with in- and outflows of population to and from fixed geographic locations. The results present in the PDS literature (both on Euclidean spaces and on more general Hilbert spaces) are based on nonlinear and convex analysis and differential inclusions (see, for example, Dupuis and Nagurney 1993; Cojocaru and Jonker 2004). In our use here, all theoretical conditions for existence and uniqueness of solutions to such a system are satisfied.

Further, the projected SIR model is blended with a DTS matrix induced by the presence of a pandemic, resulting in a hybrid dynamical system with two time scales (see Van Der Schaft 2004 and the references therein). The continuous time scale is that of the SIR model, and the discrete events are given by individuals who decide each morning whether to go to work or stay home that day, based on their previous day's state. Studies and current uses of hybrid systems can be found in, for instance, Van Der Schaft (2004), whereas projected dynamics and hybrid systems models are present in Cojocaru and Greenhalgh (2012) and Cojocaru et al. (2009).

The aim of this research is to investigate the behavior of the model starting from the baseline model of the SIR with a fixed DTS matrix in the absence of disease. The day-to-day changes in the population's decisions to go to work or not are then studied under the influence of a pandemic disease with various forces of infection and severity. We highlight ranges of the "fear" parameter that affect the amount of missing work as well as the proportion of infected workers who still go to work. Specifically, we quantify the amount of productivity loss in the local economy due to absenteeism and we run scenarios on a two-city economy as an example of how our model can be used as a scenario test bed.

Last but not least, we need to remark at this point that migration of populations specifically understood as families moving to other geographic locations in search of a better life exist in the economic and operations research literature; of particular interest to us were the early works (Nagurney et al., 1992; Nagurney, 1989, 1990) (as well as Cojocaru 2007) where the dynamics of the migration process is looked at from a network equilibrium viewpoint. Our model here shares similarities in outflow constraints on the population, however, we impose that "migrants" return at end of work day to their locations. It differs from the mentioned approaches because we are not looking for an equilibrium state of the migratory process.

The structure of the paper is as follows: in Sect. 2 we build the hybrid model of the work population for a local economy with daily commuters and n locations. In Sect. 3 we quantify the costs of absenteeism and presenteeism, and we show the data we used from existing literature in our scenarios. In Sect. 4, we present simulations of our model applied to a two-city local economy with overall constant population, and we conduct analyses of the pandemic wave and the cost to local economy as impacted by changing values of model parameters. We close with some conclusions and ideas for future investigations.

2 Constrained SIR and Discrete Transition States

In the current study we develop a new hybrid system model by combining an SIR model and discrete probability transition matrices. We present this model below for the purpose of studying the dynamic changes in the movement of a working population over a certain period of time under the presence of a pandemic, while incorporating the everyday decision of workers of going to work or missing work.

2.1 Intercity Commuter Model

In the model presented in Arino and Van den Driessche (2003) intercity travelling has been taken into account, with outgoing and incoming flow rates that were time-dependent. We take a modified view of the ideas in these papers, as we are interested to model the working population in an area, i.e., intercity travelling is only done for work, individuals come home at end of work day, and the outgoing and incoming flows are removed from being continuously time-dependent; instead they are incorporated into probability transition matrices for each of the workers in an S, I, or R compartment, for each location (city). Each individual in a compartment decides between two states (go to work "W" or stay home "H"), based on their state on the previous day.

In general our simulations span a physical time period of 120–300 work days of 10 hrs each; at end of each interval of 10 hrs, we simulate the next morning state transitions for all population compartments, and include these numbers as initial condition for the continuous time dynamics of the new 10 hrs intervals.

We assume that the number of locations (usually thought of as urban areas) is represented by $n > 1$. The residents of a city $i \in \{1, \ldots, n\}$ are the individuals who work and live in that city. Daily commuters are those who are travelling between cities for work and come home at end of work day. In our model, we cannot have outflows larger than the maximum number of daily commuters to work. We also assume a constant overall population, thus it is supposed that the "birth" and "death" rates are equal (roughly speaking, the number of people starting work for the first time in a day equals the number retiring from the work force in a day).

We let the number of residents of city i working in a city $j \neq i$, $j \in \{1, \ldots, n\}$ at a particular time t during the work day to be denoted by N_{ij}. Thus N_{ii} are people who live and work in city i. We denote by $N_i(t)$ the population of city i at time t (workers who live in i and workers who live in i and work in j for all $j \neq i$), thus:

$$N_i(t) = \sum_{j=1}^{n} N_{ij}(t) \tag{1}$$

Assuming μ is our birth and death rate, then we can simply write that:

$$\begin{cases} \dfrac{dN_{ii}}{dt} = \mu(N_i^r - N_{ii}), i \in \{1, 2, \ldots, n\} \\ \\ \dfrac{dN_{ij}}{dt} = -\mu N_{ij}, \ \forall i \neq j \in \{1, 2, \ldots, n\} \\ \text{s.t. } N_i(t) = \textit{constant} \text{ and } N_{ij}(t) \geq 0, i, j \in \{1, \ldots, n\} \end{cases} \tag{2}$$

Assuming a proportion α_i of the population N_i is living in i but not working in i, we need to add the constraints (this ensures the nondepletion of city i with more than the total proportion of commuters):

$$\sum_{j \neq i} N_{ij}(t) \leq \alpha_i N_i(t), \ \forall i \tag{3}$$

2.2 Discrete Transition Matrix

Every morning, a person in city i faces the same choice: to either *go to work* or *to miss work*, given that they either went to work or stayed home the previous day. Therefore we define a transition probability matrix between the two states every day: *Go to work (W)* or *Miss work (H)*. The transition matrix from the states (W, H) to the two states (W, H) is then defined as :

$$T_i = \begin{matrix} & W & H \\ W & p_{11} & p_{12} \\ H & p_{21} & p_{22} \end{matrix} \text{ with } p_{11} + p_{12} = 1, p_{21} + p_{22} = 1.$$

For simplicity, we will assume that in the absence of disease, the matrices are: $T_i := \begin{pmatrix} 1 & 0 \\ 1 & 0 \end{pmatrix}$, i.e., everyone goes to work. To highlight how the constrained SIR and the transition matrices are combined, we first think of the population of city i as made of two types of individuals:

$$\begin{pmatrix} X_i = \text{people who went to work yesterday} \\ Y_i = \text{people who missed work yesterday} \end{pmatrix}$$

We assume initially everyone went to work yesterday in i, i.e.,

$$\begin{pmatrix} X_i(0) \\ Y_i(0) \end{pmatrix} = \begin{pmatrix} N_i \\ 0 \end{pmatrix}$$

Applying the transition matrix T_i we get the initial data for the population dynamics model in system (2) at the beginning of the first simulated work day:

$$(N_i, 0)\, T_i = (X_i(1), Y_i(1))$$

Using our earlier assumption that a maximum of $\alpha_i N_i$ people of city i are commuters, we can further conclude that:

- $(1 - \alpha_i)X_i(1)$ represents workers from i who go to work in city i today; it represents the initial conditions for the equations of N_{ii} in (2);
- $\alpha_i X_i(1)$ represents workers from i who commute to work elsewhere today; it represents the initial condition for the equations of N_{ij} in (2).

2.3 Presence of Disease

When a disease is present we consider each city i's population ($i \in \{1, 2, \ldots, n\}$) divided into three traditional compartments: *Susceptible* (S), *Infectious* (I), and *Recovered* (R). Further, each compartment divides into two groups: people who live and work in city i, and people who live in i but work elsewhere.

In order to highlight the different decisions in different compartments, we introduce two new parameters; C_i the sensitivity to infected fractions in the local (each city's) population, and p_{inc} the probability of being incapacitated by infection, thus resulting in not going to work. Here the term p_{inc} applies to the I compartments and is the same in all cities, while the fear factor C_i is city dependent.

Every compartment of city i has then a fixed transition matrix defined as:

$$T_i^S = \begin{array}{c} \\ \text{W} \\ \text{H} \end{array} \begin{array}{c} \begin{array}{cc} \text{W} & \text{H} \end{array} \\ \begin{pmatrix} 1 - (p_{12} + C_i \frac{\sum_{j=1}^n I_{ij}}{N_i}) & p_{12} + C_i \frac{\sum_{j=1}^n I_{ij}}{N_i} \\ 1 - (p_{22} + C_i \frac{\sum_{j=1}^n I_{ij}}{N_i}) & p_{22} + C_i \frac{\sum_{j=1}^n I_{ij}}{N_i} \end{pmatrix} \end{array} \tag{4}$$

$$T_i^I = \begin{array}{c} \\ \text{W} \\ \text{H} \end{array} \begin{array}{c} \begin{array}{cc} \text{W} & \text{H} \end{array} \\ \begin{pmatrix} 1 - (p_{12} + p_{inc}) & p_{12} + p_{inc} \\ 1 - (p_{22} + p_{inc}) & p_{22} + p_{inc} \end{pmatrix} \end{array},$$

where we choose values of C_i so that $C_i \geq 0$, $p_{12} + C_i \frac{\sum_{j=1}^n I_{ij}}{N_i} \leq 1$ and $p_{22} + C_i \frac{\sum_{j=1}^n I_{ij}}{N_i} \leq 1$. This implies that in general, the parameter C_i has the following range:

$$C_i \in \left[0, C_i^{max}\right], \text{ where } C_i^{max} := \max \left\{ \frac{(1 - p_{12})N_i}{\sum_j I_{ij}}, \frac{(1 - p_{22})N_i}{\sum_j I_{ij}} \right\} \Big]$$

and

$$p_{inc} \in \left[0, \max\{1 - p_{12}, 1 - p_{22}\}\right].$$

Remark 1. 1. Note that if C_i reaches its maximal value for city i, then the transition matrix will be $T_i^S := \begin{pmatrix} 0 & 0 \\ 1 & 1 \end{pmatrix}$, which is the state where all workers stay home the next day (state of emergency for local economy).

2. If p_{inc} equals its maximal value for a given disease, then every infected person misses work the next day.

Last but not least, $T_i^R = T$ represents the recovered workers who revert to the transition matrix before infection.

We apply these transition matrices to each compartment of city i in order to derive the initial conditions of the SIR system below, given by the system (2), where now we keep track of each compartment of the city's population, as well as of the infection impact terms, which include contact rates k_i, transmission probabilities $\beta_i, \beta_j, i \neq j \in \{1, 2, \ldots, n\}$ and recovery rate γ:

$$City_i := \begin{cases} \frac{dS_{ii}}{dt} = -\sum_{k=1}^{n} \kappa_i \beta_i \frac{S_{ii} I_{ki}}{N_i} + \mu(N_i - S_{ii}) \\ \frac{dS_{ij}}{dt} = -\sum_{k=1}^{n} \kappa_i \beta_j \frac{S_{jj} I_{kj}}{N_j} + \mu(N_j - S_{jj}) \\ \frac{dI_{ii}}{dt} = \sum_{k=1}^{n} \kappa_i \beta_i \frac{S_{ii} I_{ki}}{N_i} - (\gamma + \mu) I_{ii} \\ \frac{dI_{ij}}{dt} = \sum_{k=1}^{n} \kappa_i \beta_j \frac{S_{ij} I_{ki}}{N_j} - (\gamma + \mu) I_{jj} \\ \frac{dR_{ii}}{dt} = \gamma I_{ii} - \mu R_{ii} \\ \\ \frac{dR_{ij}}{dt} = \gamma I_{ij} - \mu R_{ij} \end{cases} \tag{5}$$

Under the following constraints:

$$\begin{cases} \sum_{j=1}^{n} \left(S_{ij}(t) + I_{ij}(t) + R_{ij}(t) \right) = N_i - (Y_i^S(t) + Y_i^I(t) + Y_i^R(t)), \quad \forall i \in \{1, \ldots, n\} \\ S_{ij}(t) \geq 0, I_{ij}(t) \geq 0, R_{ij}(t) \geq 0, \quad \forall i, j \in \{1, \ldots, n\} \\ \sum_{j \neq i} S_{ij}(t) \leq \alpha_i (X_i^S(0)); \sum_{j \neq i} I_{ij}(t) \leq \alpha_i (X_i^I(0)) \ \sum_{j \neq i} R_{ij}(t) \leq \alpha_i (X_i^R(0)), \end{cases} \tag{6}$$

where $(X_i^q(0), Y_i^q(0))$ are given by the use of transition matrices above as:

$$\left(q_i(0), 0 \right) T_i^q = \left(X_i^q(0), Y_i^q(0) \right), \text{ where } q \in \{S, I, R\}, \ i \in \{1, ..., n\} \tag{7}$$

There are important remarks to be made at this point, regarding system (5):

1. We do not model the time interval from the moment that workers come home to the next morning's decision in this model.
2. We do not model the workers who stay home during a day, as in general they are assumed to keep themselves at home, rather than out in the cities.
3. We assume that $\alpha_i X_i^S(0)$ are likely to work in city i per day, where 0 signifies the beginning of each work day;
4. $\alpha_i X_i^S(0)$ is used as the initial condition $S_{ii}(0)$ of Eq. (1) in system (5) above.
5. Furthermore, $(1-\alpha) X_i^S(0)$ are likely to commute from city i to work in a different city per day;
6. $(1-\alpha) X_i^S(0)$ is used as $S_{ij}(0), j \neq i \in \{1, 2, \ldots, n\}$ in Eq. (2) of system (5) above.
7. The same explanation as in items 5, 6 above applies to the last two flow constraints of (6).

The assumptions above are simplifications of the overall transmission of disease process throughout the region. Specifically, the fact that we do not model the progression of infection between the end of work day to next morning is such a simplification. This can be addressed in an extension of the model where one can either reformulate the SIR dynamics while workers are at home, or assume an overall progression of the infection in the entire population (workers and non-workers, students, children, seniors, etc.) and apply that progression in the infected numbers every morning.

Assumptions 4 and 5 are realistic for the working population we focus on; the numbers of commuters from one city to another tend to be fairly constant within a seasonal bout of influenza (October to March/April). Evidently, if the population modelled is changed to include not only commuters, but also occasional travellers, then the constraints imposed on the SIR in formulae (6) will change.

3 Cost of Absenteeism in the Local Economy

The questions we would like to investigate in this paper, based on the above presented model, follow two main ideas. The first regards investigating the hybrid model in two hypothetic cases of pandemics, drawn from R_0 values of Table 1, i.e., the case of H1N1-like pandemic ($R_0 = 1.63$) and the case of the Spanish flu pandemic ($R_0 = 4$).

We used (Nichol et al., 2010) to derive values of contact rates between workers (k_1, k_2) and we used an infectiousness period of 8 days to derive the recovery rate γ. Finally, we used $\beta_i = \frac{R_0\gamma}{k_i}$ for the transmission probabilities.

Specifically, we conduct sensitivity analyses on the parameters p_{inc} and C_1, C_2 in the above pandemic scenarios. We then discuss the effects of varying values of $C := C_1 = C_2$ on the level of infection in the populations of each city.

The second idea is to estimate the loss of productivity per work day (*LPD*), thus we divide the annual average salary of the local economy over the number of work days in a year (253 days out of 365). Given that people who go to work while infected are known to have a reduced productivity [by a factor of half (Davis et al., 2005; Mitchell and Bates, 2011)], we can further infer the losses in productivity (which is otherwise known as the cost of presenteeism) for the local economy.

Further, we derive, in all instances, values of C_1, C_2 which achieve the minimal loss of productivity in the local economy, given a pandemic type and specific values of p_{inc}.

Table 1 Pandemic values used for R_0

Infection type	R_0	γ	k_1	k_2
Spanish flu (Chowella et al., 2006; Sattenspiel and Herring, 2003)	[2.4, 4.3]	1/8	10	17
H1N1 [2009, Canada (Balcan et al., 2009)]	1.6–1.7	1/8	10	17

4 Two Cities Local Economy Scenario

We illustrate our model on a two-city example, where the population data is given below, and where the disease data is as in Sect. 3:

Remark 2. We developed the code for the scenarios below in Matlab 2012 and ran it on desktops with Intel-Core I5 4-quad processors. Each scenario was resolved in under 5 min on one machine. We implemented a projection-type algorithm to simulate the solutions of the projected SIR system in continuous time [other possibilities to code this part can be found in Nagurney et al. (1992)]. The code can be extended to simulate any number of cities/locations; however, we focused here on two locations for the ease of presentation. We have not investigated the possibility of parallelizing our code.

Experiment 1. In order to illustrate our first experiment, we run a simulation in our two cities local economy as a function of time $[0, 120\, work\, days]$, where we take $p_{inc} = 0$ (i.e. no one is incapacitated enough by infection to motivate missing work), and where the susceptible population is not influenced at all by the presence of infection ($C_1 = C_2 = 0$). We also assume that the pandemic in this instance has $R_0 = 1.63$.

The results of the simulations are presented in Fig. 1 below. We see that there is no absenteeism due to infection; however, there are infected people going to work every day, thus leading to a cumulative cost of presenteeism in the local economy of approximately $7.5 \cdot 10^7$ \$ (lower right panel).

When $R_0 = 4$, a more dramatic cost of presenteeism takes place, as shown in Fig. 2 (approx. $10 \cdot 10^7$ \$), where in general there are more infected individuals, which leads to a higher loss of productivity, while no absenteeism takes place.

Experiment 2. Next we look at the effect of changes in the probability of becoming incapacitated by infection $p_{inc} \in [0, 1]$. We present two distinct scenarios, $p_{inc} = 0.25$ and $p_{inc} = 0.75$ in a population where susceptibles are not influenced by the presence of infection, i.e., $C_1 = C_2 = 0$. We see that larger values of p_{inc} result in a delay of the infectious wave (see Fig. 3 below with $R_0 = 1.63$—left panels, respectively $R_0 = 4$ for right panels) and where the number of work days for a simulation run is increased to 300, to cover the actual peak of infection.

Clearly the absenteeism here is due to infected individuals being incapacitated to the point of staying home. The effects of an increase in p_{inc} in a pandemic similar to H1N1 will lead to further loss of productivity in the economy (from absenteeism) adding to the cost of presenteeism already highlighted in Experiment 1, coupled with a longer (delayed) pandemic wave.

An increase in the values of p_{inc} for a more severe pandemic such as the Spanish flu (Fig. 3 right panels) leads to higher costs, but an interesting infection peak profile, where there is an initial larger peak, and a subsequent much smaller one.

In Fig. 4 we see that for values of $p_{inc} = 0.75$, the pandemic wave is very delayed, so much so that it takes 300 work days to reach its peak (left panels, $R_0 = 1.63$), while for $R_0 = 4$ case (right panels) the costs increase, but the peak is reached

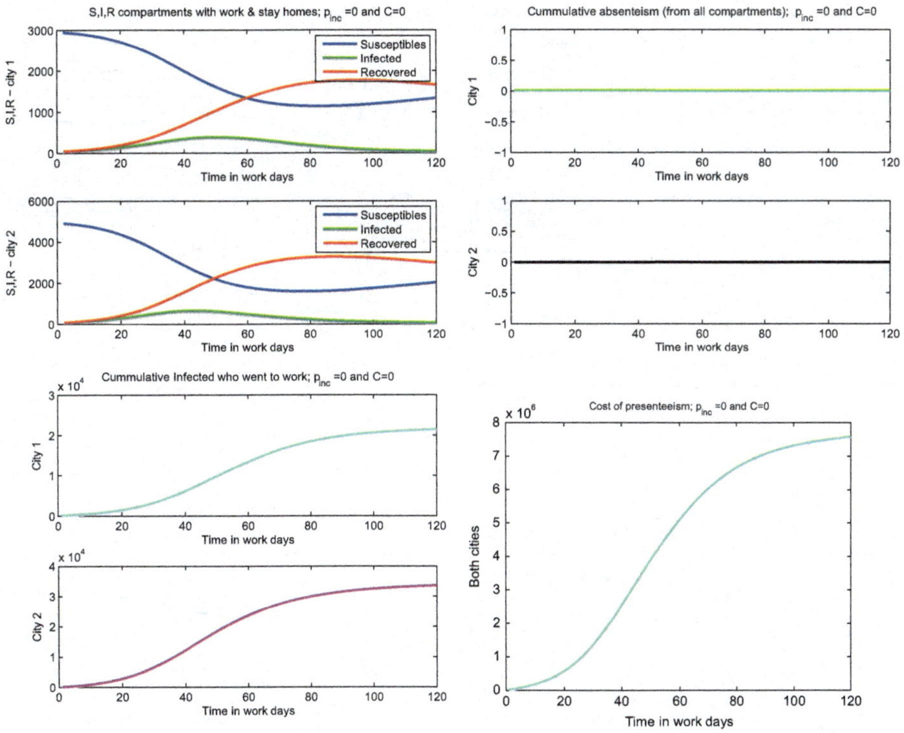

Fig. 1 This is the SIR progression during 120 work days in the while all workers go to work and no sensitivity to infection is deterring going to work. Here we assume $R_0 = 1.6$

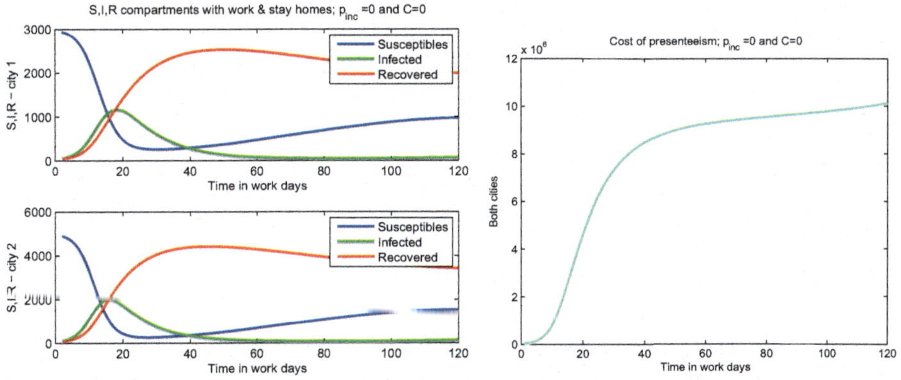

Fig. 2 This is the SIR progression during 120 work days in the while all workers go to work and no sensitivity to infection is deterring going to work. Here we assume $R_0 = 4$. We see a much higher economic cost in the case of this more severe infection

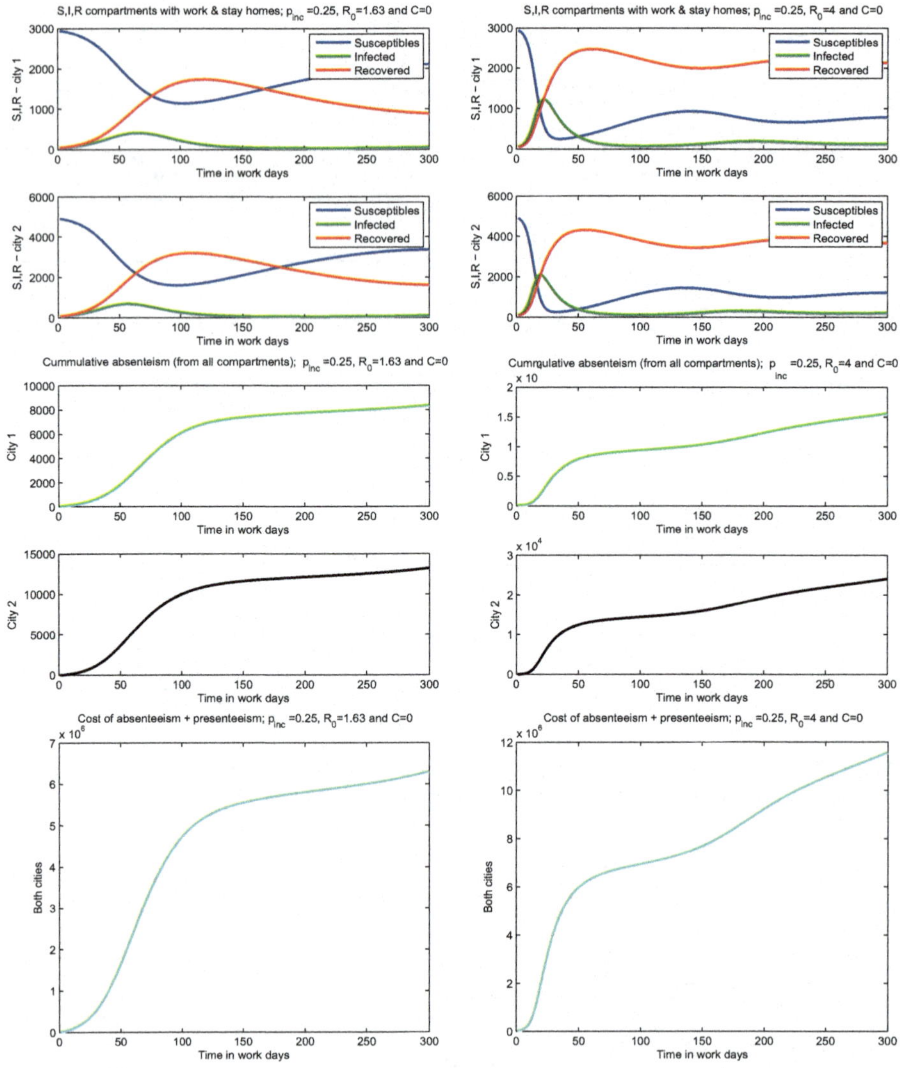

Fig. 3 $p_{inc} = 0.25$

much earlier. These results take place due to the fact that a lower R_0 leads to a slower growth of the I compartment, thus the pandemic wave is slower. However, not allowing any susceptible to miss work leads to high infected numbers and high economic costs in all cases in this experiment.

Experiment 3. Clearly the scenarios simulated so far do not allow absenteeism among susceptibles. Therefore next we investigate the effects of the sensitivity to infection levels parameter C, considering it first the same in all locations: $C_1 = C_2 \in [0, C^{max}]$, where in general we think of $C > 0$ as modelling a population with

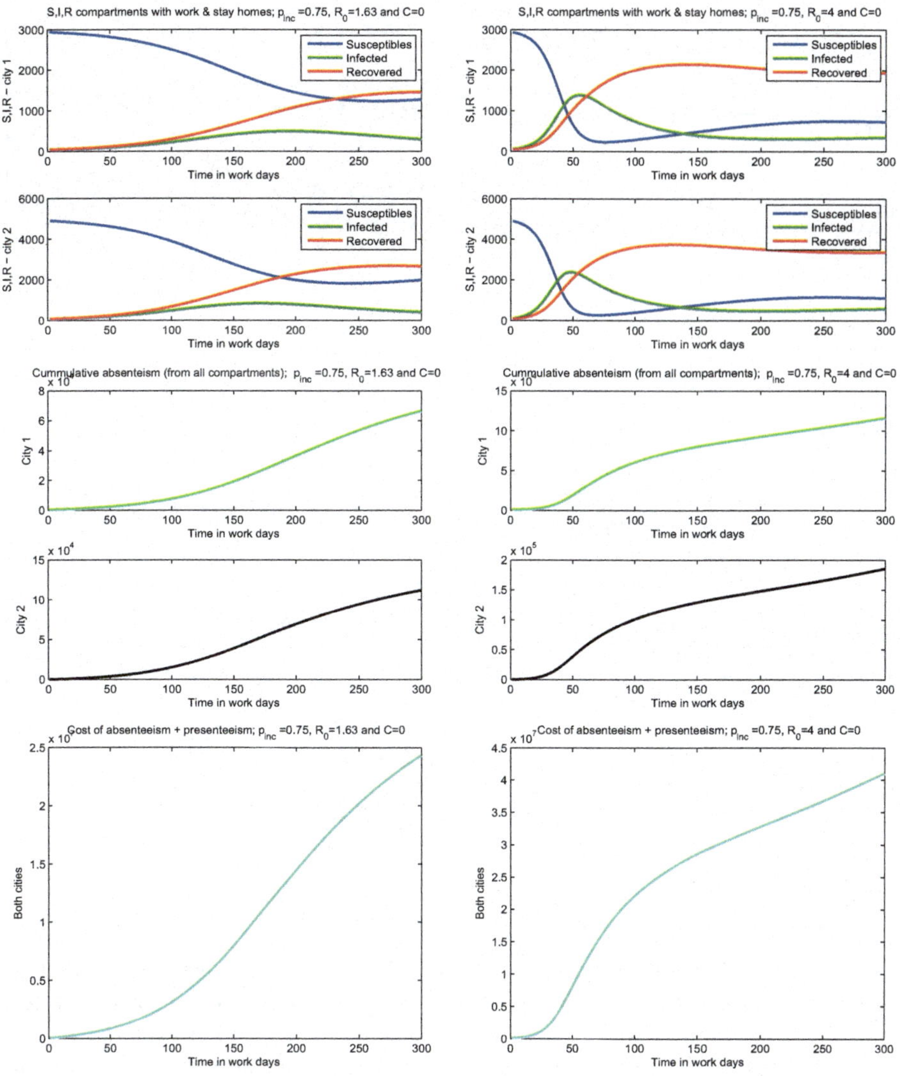

Fig. 4 $p_{inc} = 0.75$

a certain degree of sensitivity to infection levels. We run an experiment where we set $p_{inc} = 0.25$, $R_0 = 1.63$ and we plot the cumulative absenteeism in each city over 300 work days, as well as the cumulative infected individuals who choose to go to work every day, after the same number of 300 work days. We see that as C increases beyond 0 (see Figs. 5 and 6) the proportion of infected who go to work decreases slightly, but the proportion of people staying home increases much more rapidly with the increase of C values. This is consistent with our interpretation of the parameter C as a fear of infection factor. Consequently, the economic cost is monotonically increasing with C, for both types of pandemic.

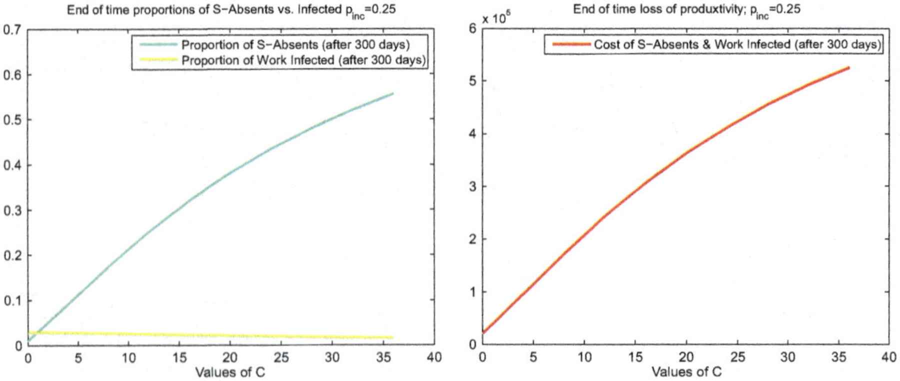

Fig. 5 $p_{inc} = 0.25,\ R_0 = 1.63$

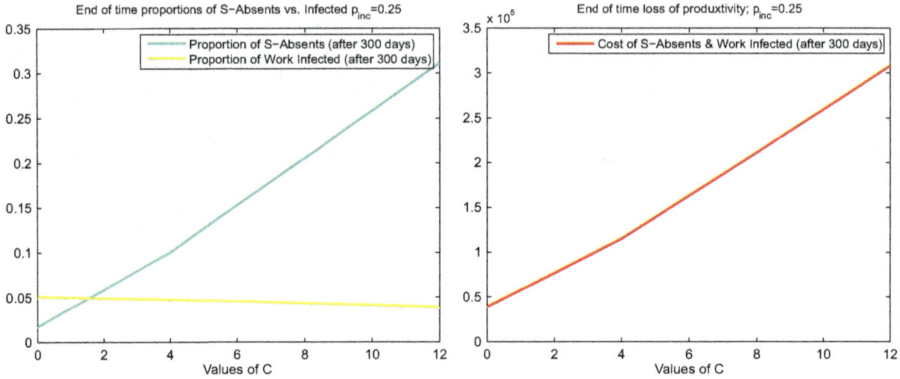

Fig. 6 $p_{inc} = 0.25,\ R_0 = 4$

Thus in these cases, compared with Experiment 2, for instance, we see that the minimal cost to economy is when $C = 0$.[1] Last but not least, we note that the values of C^{max} from Table 2 were found numerically to be $C^{max} \approx 35$ for $R_0 = 1.63$ (Fig. 5), respectively $C^{max} \approx 13$ for $R_0 = 4$ (Fig. 6)—recall C^{max} is such that no susceptible goes to work anymore (state of emergency).

Experiment 4. We are presenting a two parameter sensitivity analysis for our local economy example. We allow $C_1 \neq C_2 \in [0, \delta]$, such that $C_2 = \delta - C_1$ where $\delta \in \{5, 9\}$, respectively, and where $p_{inc} = 0.25$. Note that when $\delta = 4.5$, respectively, $\delta = 2.5$ we regain the cases where $C_1 = C_2$.

[1]Note that the end values of the cost in Fig. 3 lowest panels are the same as the initial values on the graphs in Figs. 5 and 6 right panels, adjusted for the cumulative scale.

Table 2 Parameters used in the example of two cities

Parameter	Notation	Value/range
Population of workers city 1	N_1	30,000
Population of workers city 2	N_2	50,000
Prob. of staying away due infection	p_{inc}	[0,1]
Sensitivity to infected fraction I_i/N_i	C_i	$[0, C_i^{max}], i \in \{1, 2\}$
Initial % of susceptible	$S_i(0)$	$0.98N_i, i \in \{1, 2\}$
Initial % of Infective	$I_i(0)$	$0.01N_i, i \in \{1, 2\}$
Initial % of Recovered	$R_i(0)$	$0.01N_i, i \in \{1, 2\}$
Number of days for disease	N_{days}	[120,300]
Working hours (including travel time)	T^{max}	10 h/day

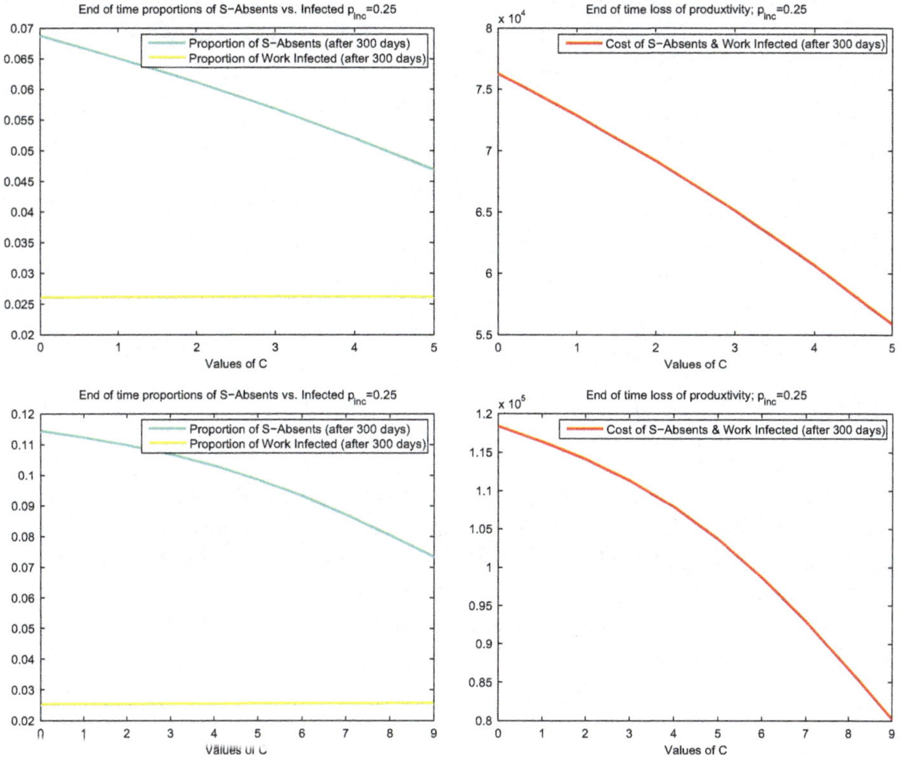

Fig. 7 $p_{inc} = 0.25, C_1 \in [0, \delta], C_2 = \delta - C_1, \delta \in \{5, 9\}, R_0 = 1.63$

We plot the fractions of cumulative absents throughout the 300 work days, and the fraction of infected individuals who decided to go to work throughout the time window simulated (Figs. 7 and 8) below. We run the same analysis for two types of infection, one for $R_0 = 1.63$ (Fig. 7) and the second one for $R_0 = 4$ (Fig. 8). We see that in these cases, there are positive values of C_1 and/or of C_2 which give a

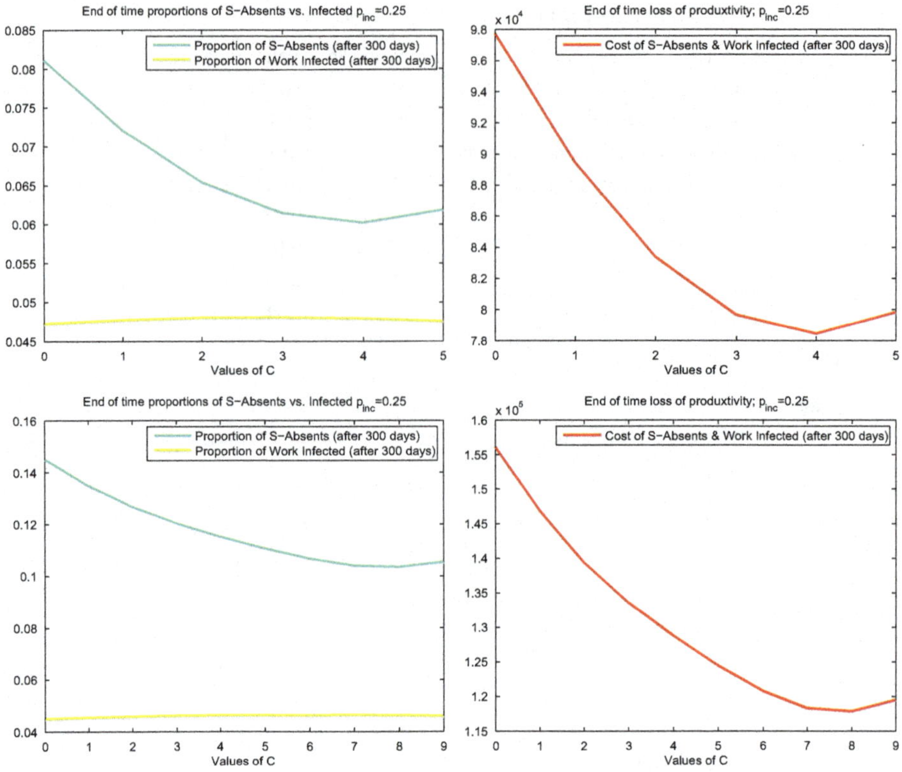

Fig. 8 $p_{inc} = 0.25, C_1 \in [0, \delta], C_2 = \delta - C_1, \delta \in \{5, 9\}, R_0 = 4$

minimal cost on economy for loss of productivity. Specifically (Fig. 7) we see that for $C_1 = 5$, respectively, 9 and $C_2 = 0$ the cost to the economy is minimal, for a pandemic such as H1N1. However, higher values of C_1 will lead to higher cost.

Alternatively (Fig. 8) we see that for values of $C_1 = 4, C_2 = 1$, respectively, $C_1 = 8, C_2 = 1$ the cost to the economy is minimal. Similarly, higher values of C_1 will lead to higher cost.

5 Conclusions and Future Work

The scenario analysis conducted in the paper leads us to a couple of general and interesting conclusions. The first is that, of course, absenteeism costs the local economy, and even if its absence (which is only hypothetically possible), there is loss of productivity in economy due to presenteeism.

The second conclusion is that while allowing for people to miss work, with various degrees of concern for a potential pandemic infection, leads to loss of productivity, this cost needs to be compared, in further analyses, with savings incurred by having a lower percentage of infected in the work force (and in the population at large). The savings consist of more productivity (more individuals are healthy at work) and a reduction in health care costs.

It is interesting to see that heterogeneity in the local population, from the perspective of how sensitive different cities are to their local levels of infection, leads to the emergence of clear positive C-values that lead to a minimal economic cost in the presence of different types of pandemics. Taking the results of Experiment 4 further, it seems that there is a benefit in having the two working populations in the two cities perceive the fear of being infected differently. This suggests that a local policy of "staying at home" from work during a pandemic could be applied, heterogeneously, in different urban areas, with an overall positive result of lowering infected proportions and achieving a minimal economic impact.

As future directions, we plan to extend this model to capture the evolution of the pandemic during non-working hours, once the workers come home. Once this extension is achieved, the results above, especially those of Experiment 4, will be investigated further to understand the possible shapes of the cost functions arising. Last but not least, a scenario with more than two cities would be interesting to develop and explore, so as to expand on the significance of heterogeneity in fear levels of infection among various locations.

Appendix: Computational Method

In this section we give a brief outline of our implementation of the numerical simulations of the projected SIR system described in detail in Sect. 3.

We implemented our code in Matlab 2012 and we wrote original code to approximate numerical solutions to equations given by system (5). We approximate 10 h of each day, for a maximum number of $T = 120$ or $T = 300$ days, as seen in the graphs of Sect. 4. For ease of presentation, we assume here that the transition matrices from (4) are fixed, i.e., C_1, C_2, p_{inc} are fixed parameter values.

1. **Step 0** We initialize the total population numbers in each city N_1, N_2, the data for the disease $\mu, \gamma, \beta, S(0), I(0), R(0), k_1, k_2$, and the transition matrices (4) containing C, p_{inc} as parameters. We also initialize the total number of work days simulated, T, as well as the commuter percentages α_1, α_2.
2. **Step** $k \in [0, T]$: This part has three distinctive substeps:
   (a) Beginning of work day data:
   - Let the initial distribution vectors $(XS_i^k(0), YS_i^k(0))$, $(XI_i^k(0), YI_i^k(0))$, and $(XR_i^k(0), YR_i^k(0))$ denote people going to work vs. staying home in each group S, I and R, in each city $i \in \{1, 2\}$. Using the transition matrices (4) as in relations (7) we deduce the numbers of people going to work from each city, and each group S, I and R;

- We then apply the commuter percentages to determine the numbers of commuters from each city i at beginning of day k in each group, i.e., $(1-\alpha_i)XS_i^k(0)$, $(1-\alpha_i)XI_i^k(0)$ and $(1-\alpha_i)XR_i^k(0)$; these then become the initial conditions of system (5) in beginning of day k;

(b) While at work:

- We simulate the approximate trajectories of the projected system (5) for 100 time steps to represent approximately 10 h work day (including commute time);
- Denoting our 12-dim variables generically as: $x := (S_{11}, I_{11}, R_{11}, S_{12}, I_{12}, R_{12}, S_{22}, I_{22}, R_{22}, S_{21}, I_{21}, R_{21})$; from an initial point $x(0)$ as above, we compute:

$$x_{i+1} := P_K(x_i + \delta F(x_i)), \text{where} i \in \{0, ..., 100\}, \delta := 0.1,$$

  F is the righthand side of (5) and $K := \mathbb{R}^{12} \cap \{\text{All conditions in (6)}\}$.
- The value of the time step $\delta = 0.1$, as well as the convergence of the approximate solution $x_i(t)$ from the step above, follow from the existence proof for solutions of projected systems with a Lipschitz righthand side detailed in Cojocaru and Jonker (2004); since F is L-continuous, if we denote by b its L-continuous constant, then we can assert that there exist approximate solutions of (5) on any interval of length $l \geq \frac{1}{1+b}$ (see Cojocaru and Jonker 2004), thus taking a desired number of steps to be denoted by *steps*, we obtain that the points $x_{i+1} := P_K(x_i + \delta F(x_i))$ with $\delta = \frac{l}{steps}$ form, by interpolation, an approximate trajectory of (5).
- The computation of the projection operator at each intermediate x_i is done modelling the $P_K(x_i + F(x_i))$ as a minimization of the square of the Euclidean distance problem from the point $x_i + F(x_i) \in \mathbb{R}^{12}$ to a point in K; we used a Matlab built-in function for this minimization problem (in the Optimization Toolbox).

(c) End of work day data:

- At end of day k, we recombine each city's population that went to work in day k, in their respective subgroups (S, I and R) as follows:

$$XS_i^k(10\,\text{h}) = (S_{11}(10\,\text{h}) + S_{12}(10\,\text{h}); XI_i^k = I_{11}(10\,\text{h}) + I_{12}(10\,\text{h});$$

$$XR_i^k = R_{11}(10\,\text{h}) + R_{12}(10\,\text{h});$$

$$(YS_i^k, YI_i^k, YR_i^k) := \{\text{those who stayed home in morning of day}$$

$$k = \text{stayed home after } 10\,\text{h}\}.$$

- The vectors $(XS_i^k(10\,\text{h}), YS_i^k(10\,\text{h}))$, $(XI_i^k(10\,\text{h}), YI_i^k(10\,\text{h}))$ and $(XR_i^k(10\,\text{h}), YR_i^k(10\,\text{h}))$ are used in morning of next day in **Step 2**, part (a).

3. Repeat **Step 2** until $k = T$ days initialized in **Step 0**. Then STOP.

When we studied the 1-dimensional variations of parameters p_{inc} and C, we ran a block of code as described above for each value of the parameter under consideration.

References

Arino, J., Van den Driessche, P.: A multi-city epidemic model. Math. Popul. Stud. **10**(3), 175–193 (2003)

Aubin, J.P., Cellina, A.: Differential Inclusions Set-Valued Maps and Viability Theory. Grundlehren der mathematischen Wissenschaften, vol. 264 (1984)

Balcan, D., Hu, H., Goncalves, B., Bajardi, P., et al.: Seasonal transmission potential and activity peaks of the new influenza A(H1N1): a Monte Carlo likelihood analysis based on human mobility. BMC Med. **7**(45), 29 (2009)

Capasso, V., Serio, G.: A generalization of the Kermack-McKendrick deterministic epidemic model. Math. Biosci. **42**(1), 43–61 (1978)

Chowella, G., Ammonb, C.E., Hengartnera, N.W., Hymana, J.M.: Transmission dynamics of the great influenza pandemic of 1918 in Geneva, Switzerland: assessing the effects of hypothetical interventions. J. Theor. Biol. **241**, 193–204 (2006)

Cojocaru, M.-G.: Double-layer dynamics theory and human migration after catastrophic events. In: Nonlinear Analysis with Applications in Economics, Energy and Transportation, pp. 65–86. Bergamo University Press, Bergamo (2007)

Cojocaru, M.-G., Greenhalgh, G.: Dynamic vaccination games and hybrid dynamical systems. Optim. Eng. **13**(3), 505–517 (2012)

Cojocaru, M.-G., Jonker, L.B.: Existence of solutions to projected differential equations on Hilbert spaces. Proc. Am. Math. Soc. **132**(1), 183–193 (2004)

Cojocaru, M.-G., Daniele, P., Nagurney, A.: Projected dynamical systems and evolutionary variational inequalities via Hilbert spaces with applications. J. Optim. Theory Appl. **127**(3), 549–563 (2005)

Cojocaru, M.-G., Bauch, C.T., Johnston, M.D.: Dynamics of vaccination strategies via projected dynamical systems. Bull. Math. Biol. **69**(5), 1453–1476 (2007)

Cojocaru, M.-G., Hawkins, S., Thille, H., Thommes, E.: Double layer and hybrid dynamics of equilibrium problems - applications to markets of environmental products. In: Pardalos, P., Rassias, T.M., Khan, A. (eds.) Nonlinear Analysis and Optimization. Springer, New York (2009)

Davis, K., Collins, S.R., Doty, M.M., Ho, A., Holmgren, A.: Health and productivity among U.S. workers. The Commonwealth Fund, Issue Brief August 2005

Diekmann, O., Heesterbeek, J.A.P.: Mathematical Epidemiology of Infectious Diseases. Wiley, Chichester (2000)

Dupuis, P., Nagurney, A.: Dynamical systems and variational inequalities. Ann. Oper. Res. **44**(1), 7–42 (1993)

Greenwood, M.J.: Human migration: theory, models, and empirical studies. J. Reg. Sci. **25**(4), 521–544 (1985)

Greenwood, M.J., Gary, L.H.: Jobs versus amenities in the analysis of metropolitan migration. J. Urban Econ. **25**(1), 1–16 (1989)

Heesterbeek, J.A.P.: A brief history of R_0 and a recipe for its calculation. Acta Biotheor. **50**(3), 189–204 (2002)

Hethcote, H.W.: The mathematics of infectious diseases. SIAM Rev. **42**(4), 599–653 (2000)

Hyman, J.M., LaForce, T.: Modeling the spread of influenza among cities. In: Banks, H.T., Castillo-Chavez, C. (eds.) Bioterrorism: Mathematical Modeling Applications in Homeland Security. SIAM Frontiers in Applied Mathematics, vol. 28, pp. 211–236 (2003)
Longini, I.M. Jr.: A mathematical model for predicting the geographic spread of new infectious agents. Math. Biosci. **90**(1), 367–383 (1988)
Mitchell, R.J., Bates, P.: Measuring health-related productivity loss. Popul. Health Manag. **14**(2), 93–98 (2011)
Nagurney, A.: Migration equilibrium and variational inequalities. Econ. Lett. **31**(1), 109–112 (1989)
Nagurney, A.: A network model of migration equilibrium with movement costs. Math. Comput. Model. **13**(5), 79–88 (1990)
Nagurney, A.: Network Economics: A Variational Inequality Approach, vol. 10. Springer, Heidelberg (1998)
Nagurney, A., Zhang, D.: Projected Dynamical Systems and Variational Inequalities with Applications. International Series in Operations Research & Management Science, vol. 2. Springer Science & Business Media, New York (2012)
Nagurney, A., Jie, P., Lan, Z.: Human migration networks. Eur. J. Oper. Res. **59**(2), 262–274 (1992)
Nichol, K.L., Tummers, K., Hoyer-Leitzel, A., Marsh, J., Moynihan, M., McKelvey, S.: Modeling seasonal influenza outbreak in a closed college campus: impact of pre-season vaccination, in-season vaccination and holidays/breaks. PLoS ONE **5**(3), e9548 (2010)
Sattenspiel, L., Dietz, K.: A structured epidemic model incorporating geographic mobility among regions. Math. Biosci. **128**(1), 71–91 (1995)
Sattenspiel, L., Herring, D.A.: Structured epidemic models and the spread of influenza in the central Canadian subarctic. Hum. Biol. **70**(1), 91–115 (1998)
Sattenspiel, L., Herring, D.A.: Simulating the effect of quarantine on the spread of the 1918 C19 flu in central Canada. Bull. Math. Biol. **65**(1), 1–26 (2003)
Thommes, E.W., et al.: Examining Ontario's universal influenza immunization program with a multi-strain dynamic model. Vaccine **32**(39), 5098–5117 (2014)
Van Der Schaft, A.: Bisimulation of dynamical systems. In: Hybrid Systems: Computation and Control, pp. 555–569. Springer, Berlin, Heidelberg (2004)
Wency, L.: Going to work sick is bad for business. The Globe and Mail, vol. 13, issue 9 (2011)

Tornado Detection with Kernel-Based Classifiers from WSR-88D Radar Data

Theodore B. Trafalis, Budi Santosa, and Michael B. Richman

Abstract Detection of tornadoes that provides warning times sufficient for evasive action prior to a tornado strike has been a well-established objective of weather forecasters. With modern technology, progress has been made on increasing the average lead time of such warnings, which translates into a number of lives saved. Recently, machine learning (e.g., kernel methods) has been added to the collection of techniques brought to bear on severe weather prediction. In this chapter, we seek to extend this innovation by introducing and applying two types of kernel-based methods, support vector machines and minimax probability machines to detect tornadoes, using attributes from radar derived velocity data. These two approaches utilize kernel methods to address nonlinearity of the data in the input space. The approaches are based on maximizing the margin between two different classes: tornado and no tornado. The use of the Weather Surveillance Radar 1988 Doppler, with continuous data streaming every 6 min, presents a source for a dynamic data driven application system. The results are compared to those produced by neural networks (NN). Findings indicate that these kernel approaches are significantly more accurate than NN for the tornado detection problem.

Keywords Tornado detection • Dynamic data driven application • Generalization error • Feedforward neural networks • Kernel methods

---

T.B. Trafalis (✉)
School of Industrial and Systems Engineering, University of Oklahoma, Stillwater, OK, USA
e-mail: ttrafalis@ou.edu

B. Santosa
Industrial Engineering Department, Institut Teknologi Sepuluh Nopember, Sukolilo, Indonesia
e-mail: budi_s@ ie.its.ac.id

M.B. Richman
School of Meteorology, University of Oklahoma, Norman, OK, USA
e-mail: mrichman@ou.edu

I.S. Kotsireas et al. (eds.), *Dynamics of Disasters—Key Concepts, Models, Algorithms, and Insights*, Springer Proceedings in Mathematics & Statistics 185,
DOI 10.1007/978-3-319-43709-5_16

1 Introduction

Accurate detection of tornadoes with ample warning times has been a longstanding goal of severe weather forecasters. With state-of-the-science weather radar, high speed computing, and advanced signal processing algorithms, steady progress has been made on increasing the average lead time of tornado warnings. Every extra minute of lead time translates into a number of lives saved. One of the severe weather detection algorithms, created by the National Severe Storms Laboratory (NSSL), is the Mesocyclone Detection Algorithm (MDA). This algorithm uses the data stream outputs of the Weather Surveillance Radar 1988 Doppler (WSR-88D) and is designed to detect storm-scale circulations associated with regions of rotation in thunderstorms. The MDA is used by meteorologists as one input in their decision to issue tornado warnings. Marzban and Stumpf (1996) show that the performance of the MDA is improved by neural network (NN) post-processing of the radar data. The present work simplifies the redundancies in the MDA to reduce the amount of information supplied to forecasters, who are barraged by dozens of inputs in the short time interval they have to issue a warning. Streamlining the MDA will help to speed up the detection process so that the forecaster can assimilate information from the data set prior to new data arriving. By identifying patterns associated with tornadoes in a timely fashion, the forecaster can assess the evolution of such patterns.

Kernel-based methods, such as support vector machines (SVM) and minimax probability machines (MPM), are applied to detect tornado circulations sensed by the WSR-88D radar. These methods do not make any assumptions about the data distributions (Vapnik, 1995). Application of the kernel methods is useful to address the problem of nonlinearity of the data in the input space since the data are mapped into a higher dimensional space where there is a high likelihood that the problem becomes linear separable (Vapnik, 1995). By using the state-of-the-art kernel methods, nonlinear classifiers are found that separate the data into tornado and no tornado cases more effectively than traditional linear methods or nonlinear NN (Haupt et al., 2009; Trafalis et al., 2007, 2006).

The current problem is large scale, where a sizeable number of data are required for training and the method of decomposition is an important issue to consider for solution efficiency. Hsu and Lin (2002) proposed a decomposition technique to solve the large-scale SVM training problems. The basic algorithm is a simplification of both sequential minimal optimization by Platt (1999) and SVM^{light} by Joachims (1999). More efficient approaches are provided in recent work by Gilbert and Trafalis (2009). Several other variations of SVM approaches have been developed recently, such as the Analytic Center Machine (Trafalis and Malyscheff, 2002), *p*-Center Machines (Adrianto and Trafalis, 2010), and MPM (Lanckriet et al., 2002).

The present work seeks to quantify the impact of the changing importance of the predictors as a function of time of the year. Previously, a dynamic data driven application systems (DDDAS) approach was used with radar-scan lags to assess the change in attributes from previous scans to improve tornado classification (Trafalis et al., 2004).

This chapter is organized as follows. In Sect. 2.1, the definition of the problem is provided, whereas Sect. 2.2 describes the data. In Sect. 3, the basics of the learning machines used and our methodology are discussed and Sect. 4 describes the experimental setting. Section 5 provides analysis of the results and, finally, Sect. 6 concludes the chapter.

2 Problem Statement and Data Analysis

2.1 Scientific Problem

There are two types of problems addressed in this chapter. One is meteorological, relating to tornado warnings, and the other is methodological. The two are intimately entwined for the prediction of tornadoes.

There are two challenges involved in tornado warnings from the meteorological viewpoint. The first one is tornado detection. Of those tornadoes that do occur, the number of tornadoes detected is smaller. The penalty of missing a tornado that occurs in a populated area is catastrophic. The second challenge is false alarms. This means that the algorithms detect tornado circulations more often than such circulations can be confirmed. False alarms are insidious because warnings have the potential to go unheeded by the public. Accordingly, it is desirable to develop a statistical learning algorithm that will maximize detection and minimize false alarms as a multi-objective optimization problem.

Prediction of tornadoes is a difficult task owing to the small scale of their circulation and their rapid production in the atmosphere. The majority of tornadoes have a circulation smaller than the radar beam width sampling of the atmosphere. Hence, only their parent circulation can be sensed. To complicate matters, tornadoes can form within minutes and disappear just as quickly. Some circulations remain separated from the ground (i.e., funnels) whereas others couple with the ground (i.e., tornadoes). The radar senses both as mesocyclones and combinations of MDA outputs and external information are helpful in distinguishing the two. Without such information, nearly every circulation results in a tornado warning and false alarms. Over time, the public can become apathetic to repeated false alarms. The dynamic nature of this problem requires addressing the time-dependence of this application in an attempt to detect if the circulation is progressing toward the surface. Since tornadoes form so quickly, it is of the utmost importance to sense changes as rapidly as practical within the framework of the radar's limitations to scan a volume of the atmosphere approximately once every 6 min. Forecasts require real time response to observations from radar data, and thus an appropriate problem for the DDDAS model. Once data are collected, they are processed by algorithms that look for signatures of tornadoes in near-real time. The incoming radar data stream can be used for dynamic decision making to increase the lead time in tornado forecasts. However, present day operational radar takes approximately 6 min to complete one volume scan. Furthermore, the spatial resolution averages close to 1/4 km for Doppler radar velocity, and many tornadoes are small in size. Despite these challenges, lead times for tornadoes have increased from a few minutes (a decade ago) to approximately 11 min (with current radar), largely due to improvements in algorithms that use the radar data as inputs (Simmons and Sutter, 2008).

The second research problem is to develop an intelligent system that can analyze data that have a significant noise component. NN are considered robust classifiers in terms of input noise. However, the resulting learning optimization problem is nonconvex. An alternative to NN, with the learning optimization problem being convex, is SVM. Vapnik (1995) shows, based on statistical learning theory, that SVM has better generalization properties than NN. SVM is also robust with bounded noise in the input data (Trafalis and Alwazzi, 2003; Trafalis and Gilbert, 2006, 2007). Recent work by Lanckriet et al. (2002) has shown that MPM provides a classifier that is competitive with SVM.

2.2 Data and Analysis

The MDA data set used for this research is based on the outputs from WSR-88D radar. Any circulation detected on a particular volume scan of the radar can be associated with a report of a tornado. In the severe weather database, supplied by NSSL, there is a label for tornado ground truth that is based on temporal and spatial proximity. If there is a tornado reported between the beginning and ending of the volume scan, and the report is within a reasonable distance of a circulation detection (input manually), then the ground truth value is flagged. The ground truth value is also flagged if a circulation detection falls within the prediction "time window" of -20 to $+6$ min of the ground truth report duration. The key idea behind these timings is to determine whether a circulation will produce a tornado within the next 20 min, a suitable lead time for advanced severe weather warnings by the National Weather Service. Any data, with the aforementioned flagged values, are categorized as tornado cases, with label equal to 1. All other circulations are labeled as 0, corresponding to a no tornado case.

The predictor pool employed in this study consists of 24 attributes of which 23 come from the MDA and are based on Doppler velocity data (Table 1). The 23 MDA attributes have been used successfully by Marzban and Stumpf (1996). Additionally, a month attribute, referring to January through December, was added to account for the strong seasonality exhibited by velocity-based attributes. The database was stratified into seasons and data were sampled from every season to insure generalizable training of the learning machines.

Table 1 List of attributes

Number	Attribute	Unit	Range
1	Base	m	0–12,000
2	Depth	m	0–13,000
3	Strength rank	n/a	0–25
4	Low-level diameter	m	0–15,000
5	Maximum diameter	m	0–15,000
6	Height of maximum diameter	m	0–12,000
7	Low-level rotational velocity	$m\,s^{-1}$	0–65
8	Maximum rotational velocity	$m\,s^{-1}$	0–65
9	Height of maximum rotational velocity	m	0–12,000
10	Low-level shear	$m\,s^{-1}\,km^{-1}$	0–175
11	Maximum shear	$m\,s^{-1}\,km^{-1}$	0–175
12	Height of maximum shear	m	0–12,000
13	Low-level gate-to-gate velocity difference	$m\,s^{-1}$	0–130
14	Maximum gate-to-gate velocity difference	$m\,s^{-1}$	0–130
15	Height of maximum gate-to-gate velocity difference	m	0–12,000
16	Core base	m	0–12,000
17	Core depth	m	0–9000
18	Age	min	0–200
19	Strength index[a] (MSI)	n/a	0–13,000
20	Strength index (MSIr) "rank"	n/a	0–25
21	Relative depth	%	0–100
22	Low-level convergence	$m\,s^{-1}$	0–70
23	Mid-level convergence	$m\,s^{-1}$	0–70

[a] Weighted by average density of integrated layer

3 Methodology

3.1 *Support Vector Machines*

Given a set of data points $\{(\mathbf{x}_i, y_i)\}_{i=1}^{\ell}$ with $\mathbf{x}_i \in \mathbb{R}^n$ and $y_i = \pm 1$, SVM consider a problem where a classifier is sought to separate the two classes of points with maximum margin separation (Fig. 1). The SVM formulation can be written as follows (Haykin, 1999):

$$\min_{\mathbf{w},b,\eta} \left\{ \frac{\|\mathbf{w}\|^2}{2} + C\sum_{i=1}^{\ell} \eta_i : y_i(\langle \mathbf{w}, \mathbf{x}_i \rangle + b) + \eta_i \geqslant 1,\ \eta_i \geqslant 0,\ i = 1, \ldots, \ell \right\}, \quad (1)$$

where C is a parameter to be chosen by the user, $\mathbf{w}$ is referring to the vector perpendicular to the separating hyperplane, η_i refers to the misclassification error variables, and b is the bias of the separating hyperplane. A larger C corresponds to

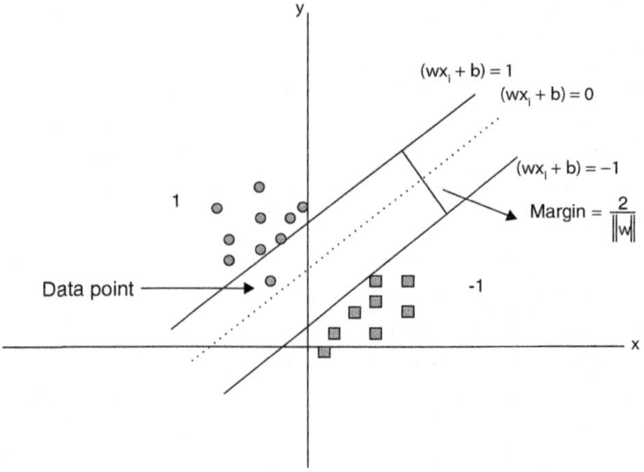

Fig. 1 The geometric illustration of SVM

assigning a larger penalty to errors. Introducing positive Lagrange multipliers α_i, to the inequality constraints in model (1) we obtain the following dual formulation:

$$\min_{\alpha}\left\{\sum_{i,j=1}^{\ell}\alpha_i\alpha_j y_i y_j\frac{\langle\mathbf{x}_i,\mathbf{x}_j\rangle}{2}-\sum_{i=1}^{\ell}\alpha_i:\sum_{i=1}^{\ell}\alpha_i y_i=0,\ 0\leqslant\alpha_i\leqslant C,\ i=1,\ldots,\ell\right\}. \tag{2}$$

The solution of the primal problem is then given by $\mathbf{w}=\sum_{i=1}^{\ell}\alpha_i y_i\mathbf{x}_i$ where $\mathbf{w}$ is the vector that is perpendicular to the separating hyperplane. The free coefficient b can be found from $\alpha_i\big(y_i(\langle\mathbf{w},\mathbf{x}_i\rangle+b)-1\big)=0$, for any i such that α_i is not zero.

SVM map a given set of binary labeled training data into a high-dimensional feature space and separate the two classes of data linearly with a maximum margin hyperplane in the feature space. In the case of nonlinear separability, each data point $\mathbf{x}$ in the input space is mapped into a higher dimensional feature space using a feature map ϕ. In the new space, the dot product $\langle\mathbf{x},\mathbf{y}\rangle$ becomes $\langle\phi(\mathbf{x}),\phi(\mathbf{y})\rangle$. A nonlinear kernel function, $k:(\mathbf{x},\mathbf{y})\mapsto k(\mathbf{x},\mathbf{y})$, can be used to substitute for the dot product $\langle\phi(\mathbf{x}),\phi(\mathbf{y})\rangle$. The use of a kernel function allows the SVM to operate efficiently in nonlinear high-dimensional feature spaces without being adversely affected by the dimensionality of that space. Indeed, it is possible to work with feature spaces of infinite dimension. Moreover, it is possible to learn in the feature space without knowing the mapping ϕ and the feature space $\mathscr{F}$. The matrix $\mathbf{K}_{ij}=\langle\phi(\mathbf{x}_i),\phi(\mathbf{x}_j)\rangle$ is called the *kernel matrix*. In general, the separating hyperplane corresponds to a nonlinear decision boundary in the input space. It can be shown that for each continuous positive semi-definite function $k:(\mathbf{x},\mathbf{y})\mapsto k(\mathbf{x},\mathbf{y})$, there exists a mapping, ϕ, such that $k(\mathbf{x},\mathbf{y})=\langle\phi(\mathbf{x}),\phi(\mathbf{y})\rangle$ for all $(\mathbf{x},\mathbf{y})\in\mathbb{R}^n\times\mathbb{R}^n$, where $\mathbb{R}^n$ is the input space (Mercer's Theorem).

There are three specific kernel functions usually used in the SVM literature: polynomial, radial basis function, and tangent hyperbolic (Haykin, 1999). In this chapter, only polynomial kernel functions ($k:(\mathbf{x},\mathbf{y})\mapsto(\langle\mathbf{x},\mathbf{y}\rangle+1)^p$, where p is the degree of the kernel) are investigated.

3.2 Minimax Probability Machines

The MPM approach was introduced by Lanckriet et al. (2002). The problem of binary classification is solved by minimizing the maximum probability of misclassification of the data points. The problem can be defined as follows. Let $\mathbf{x}$ and $\mathbf{y}$ denote random vectors in a binary class (with $\mathbf{x}$ representing points in one class and $\mathbf{y}$ representing points in the other class), with means and covariance matrices given by $(\bar{\mathbf{x}}, \Sigma_\mathbf{x})$ and $(\bar{\mathbf{y}}, \Sigma_\mathbf{y})$, respectively, where $(\bar{\mathbf{x}}, \bar{\mathbf{y}}) \in \mathbb{R}^n \times \mathbb{R}^n$ and $(\Sigma_\mathbf{x}, \Sigma_\mathbf{y}) \in \mathbb{R}^{n \times n} \times \mathbb{R}^{n \times n}$. Both $\Sigma_\mathbf{x}$ and $\Sigma_\mathbf{y}$ are symmetric and positive semi-definite. Then, a hyperplane $H(\mathbf{a}, b) = \{\mathbf{z} \in \mathbb{R}^n : \langle \mathbf{a}, \mathbf{z} \rangle = b\}$ where $\mathbf{a} \in \mathbb{R}^n \setminus \{\mathbf{0}\}$ and $b \in \mathbb{R}$ that separates the two classes of points with maximal probability with respect to all distributions having these mean and covariance matrices is determined. Mathematically, the resulting optimization problem can be formulated as:

$$\max_{\alpha, \mathbf{a} \neq \mathbf{0}, b} \left\{ \alpha : \inf_{\mathbf{x} \sim (\bar{\mathbf{x}}, \Sigma_\mathbf{x})} P(\{\langle \mathbf{a}, \mathbf{x} \rangle \geqslant b\}) \geqslant \alpha, \inf_{\mathbf{y} \sim (\bar{\mathbf{y}}, \Sigma_\mathbf{y})} P(\{\langle \mathbf{a}, \mathbf{y} \rangle \leqslant b\}) \geqslant \alpha \right\}. \tag{3}$$

The above problem can be simplified as (Lanckriet et al., 2002):

$$\min_{\mathbf{a}} \left\{ \left\| \Sigma_\mathbf{x}^{1/2} \mathbf{a} \right\|_2 + \left\| \Sigma_\mathbf{y}^{1/2} \mathbf{a} \right\|_2 : \langle \mathbf{a}, \bar{\mathbf{x}} - \bar{\mathbf{y}} \rangle = 1 \right\}. \tag{4}$$

In the case of nonlinear separability the kernel method is used. In the high-dimensional feature space we want to find a hyperplane $H(a, b) = \{\phi(\mathbf{z}) : \langle a, \phi(\mathbf{z}) \rangle = b\}$ that corresponds to the nonlinear decision boundary $D(a, b) = \{\mathbf{z} \in \mathbb{R}^n : \langle a, \phi(\mathbf{z}) \rangle = b\}$ in the input space $\mathbb{R}^n$.

3.3 Neural Networks

An NN model is comprised of a large number of processing elements called neurons. Each neuron is connected to other neurons by links, each with an associated weight. Neurons without links toward them are called input neurons and those with no link away from them are called output neurons. The neurons are represented by state variables. State variables are functions of the weighted sum of input variables and other state variables. Neurons perform a simple transformation simultaneously in a parallel-distributed manner. The input–output relation of the transformation in a neuron is characterized by an activation function. The combination of input, output, and links between neurons with associated weights constitute the architecture of the NN (Haykin, 1999). The NN procedure involves learning with training, given a sample of representative data. Testing is performed on data not included in the training set, with the aim of predicting the new outputs to provide the generalization skill. The training process involves different numbers of layers (inputs, output, and hidden), neurons, and links between neurons, with associated weights. The last layer represents the output. The number of hidden layers is user-defined. The number of neurons in each layer can be modified. After the network has been trained and tested, it is ready for use. New sets of input data can be input to the network, and they will produce a forecast, based on what has been learned. A trained NN can be treated as an expert in the category of information it has been given to analyze. This expert can be used to provide predictions given new situations.

3.4 Forecast Evaluation Indices for Tornado Detection

In the detection paradigm, the forecast results are assessed by using a suite of forecast evaluation indices based on a contingency table (also known as a *confusion matrix*). The confusion matrix (defined in Table 2) is a convenient method to assess hypotheses as the forecasts are the row entries and the observations are the column entries.

The cell counts (a, b, c, d) from the confusion matrix can be used to form forecast evaluation indices (Wilks, 2011). In this definition of the confusion matrix, one such index is the probability of detection, POD, which is defined as $a/(a+c)$. POD measures the fraction of observed events that were forecast correctly. Its range is 0 to 1 and a perfect score is 1 (or 100 %). Note that POD is sensitive to hits, and therefore, good for rare events. However, POD ignores false alarms and it can be improved artificially by issuing more "yes" forecasts to increase the number of hits.

False alarm ratio, FAR, is defined as $b/(a+b)$. FAR measures the fraction of "yes" forecasts in which the event did not occur. Its range is 0 to 1, and 0 is a perfect rate. FAR is sensitive to false alarms and it ignores misses. It can be improved artificially by issuing more "no" forecasts to reduce the number of false alarms.

Accuracy is defined as $(a+d)/(a+b+c+d)$. Accuracy measures the fraction of all forecasts that were correct, which makes it a seemingly intuitive measure. The range is 0 to 1, with 1 being best. However, it can be misleading since it is heavily influenced by the most common category, usually "no" event in the case of severe weather.

Bias is defined as $(a+b)/(a+c)$. Bias measures the ratio of the frequency of forecast events to the frequency of observed events. The range is from 0 to infinity. A perfect score is 1. Bias indicates whether the forecast system has a tendency to under-forecast (bias < 1) or over-forecast (bias > 1) events. It does not measure how well the forecast corresponds to the observations. It measures only relative frequencies.

Table 2 Confusion matrix

Pred. vs Obs.[a]	Positive	Negative	Total
Positive	Hit (a)	False alarm (b)	Predicted positives
Negative	Misses (c)	Correct negative (d)	Predicted negatives
Total	Observed positives	Observed negatives	

[a] Rows are for the predicted instances and columns are for the observed instances

Probability of false detection, POFD, is defined as $b/(b+d)$. POFD measures the ratio of false alarms to the total number of "no" observations. The probability of false detection is a measure of inaccuracy with respect to the observations and provides a measure of the extent to which the forecasts provide a false warning for the occurrence of an event. POFD varies from 0 to 1. A perfect score is zero.

The concept of skill is one where a forecast is superior to some known reference forecast (e.g., random chance). Skill ranges from -1 (anti-skill) to 0 (no skill over the reference) to $+1$ (perfect skill). Heidke's skill is commonly utilized in meteorology since it uses all elements in the confusion matrix and works well for rare event forecasting (e.g., tornadoes) (Doswell et al., 1990). Heidke's skill is defined as $2(ad-bc)/\big((a+b)(b+d)+(a+c)(c+d)\big)$.

4 Experiments

The data were split into two sets: *training* and *testing*. All the machine learning algorithms are trained with a fixed training set. Tuning parameters are determined in the training phase and these are used to make predictions for different *testing* sets based on the ratio of tornado to no tornado observations. Increasing the number of no tornado observations while keeping the tornado observations fixed accounts for the various *testing* sets. By investigating this ratio, the performance for each forecast validation statistic can be observed.

For the *training* set, the percentage of tornadic and non-tornadic observations was 50 %–50 %. In the testing sets, the percentage of tornadic observations was varied from 2 % to 10 % in 2 % increments. These small values were selected as the climatological probability of a tornado is only a few percent on any given day. The cases used for training are different from those used in the testing set. The same training and testing sets were applied to all methods.

The SVM, MPM, and NN experiments were performed in the MATLAB environment. OSU SVM Classifier Matlab Toolbox by Ma et al. was used to run experiments using SVM for classification. These codes are MATLAB versions of Chih-Chung Chang and Chih-Jen Lin's LIBSVM algorithm (Chang and Lin, 2001). For the MPM approach, the MATLAB codes from Lanckriet were used. The kernel function used in the experiments is the polynomial kernel. NN experiments were run using MATLAB Neural Networks Toolbox. A feedforward neural network with a hidden layer was selected. Several NN architectures were evaluated and the minimum mean square generalization error was found for five hidden nodes. The network is trained by using a gradient descent algorithm with momentum (Haykin, 1999).

The data structure consists of an m by n matrix, where m refers to observations and n refers to attributes as listed in Table 1. For the training set, m is equal to 749, and for the testing sets it varies from 3938 to 18,202, depending on the percentage of tornado events. The data are preprocessed before each method is applied. For each column, each data point (observation) was divided by the norm of the column. By preprocessing the data, the performance of SVM and MPM improved significantly (Trafalis et al., 2003).

Results show that the behavior of the 23 velocity based attributes have a unique pattern, which is a function of calendar month, as indicated in Fig. 2. The results of investigations of adding the month number as an additional attribute are shown in Tables 3, 4, and 5.

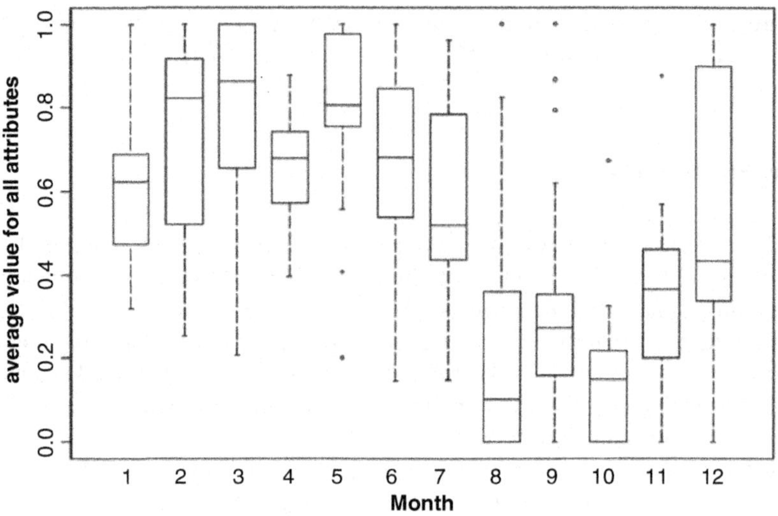

Fig. 2 Box plots of the seasonal variability of the mean of 23 attributes

Table 3 Misclassification rates for SVM

Ratio of tornado/no-tornado	2 %	4 %	6 %	8 %	10 %
Without month	0.017	0.025	0.033	0.037	0.043
With month	0.013	0.021	0.027	0.032	0.037

Table 4 Misclassification rates for MPM

Ratio of tornado/no-tornado	2 %	4 %	6 %	8 %	10 %
Without month	0.013	0.025	0.033	0.038	0.044
With month	0.012	0.020	0.026	0.031	0.036

Table 5 Misclassification rates for NN

Ratio of tornado/no-tornado	2 %	4 %	6 %	8 %	10 %
Without month	0.079	0.078	0.088	0.089	0.092
With month	0.099	0.099	0.110	0.114	0.114

5 Results

Adding month number $(1, 2, \ldots, 12)$ as an attribute has improved the results for SVM and MPM (Table 3). These improvements vary for each testing sample. For SVM, for the testing sample with a 2 % ratio of tornadoes to no tornadoes, the improvement in misclassified error (misses and false alarm) is from 1.7 % to 1.3 % (Table 3). At the other end of the range (10 % ratio), the improvement is from 4.3 % to 3.7 %. Moreover, the results are considerably more accurate than previous research (Trafalis et al., 2003).

Similar improvements were found for MPM (Table 4). For NN, the misclassification error for the testing sets is obtained by taking the average of five runs. The NN results show different patterns of error when a month attribute is added (Table 5).

Figures 3, 4, 5, 6, 7, and 8 depict different performance aspects of the aforementioned forecast evaluation indices for each classification method. The accuracy of MPM and SVM are so similar (Fig. 3) that they are indistinguishable. These methods have highest accuracy with the lowest percentage of tornadoes in the testing set because of the large number of no tornado events. However, it is the prediction of tornado events that is of most interest. The accuracy of NN is about 8 % lower than MPM and SVM for each percentage of tornadoes (Fig. 3).

Much of the behavior can be explained by examination of the FAR (Fig. 4). Again, the SVM values are close to those for the MPM, while FAR values for NN are almost twice as large as for SVM. It is unclear if this is an intrinsic feature of

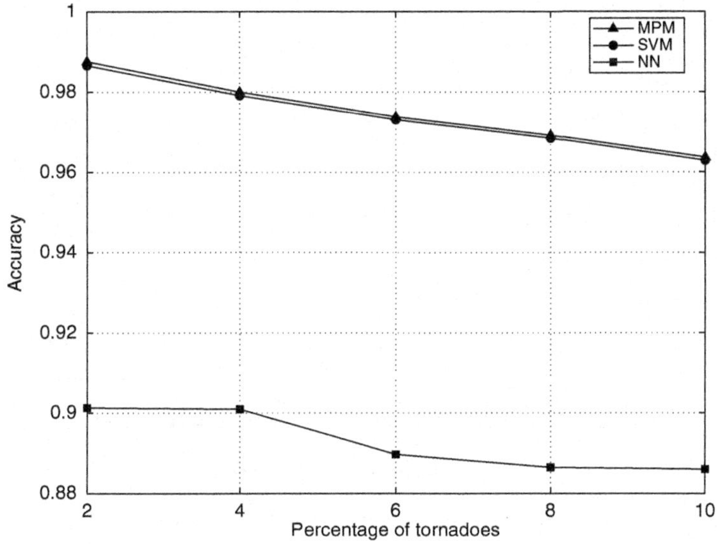

Fig. 3 Accuracy for different methods for different percentages of tornadoes in the test set

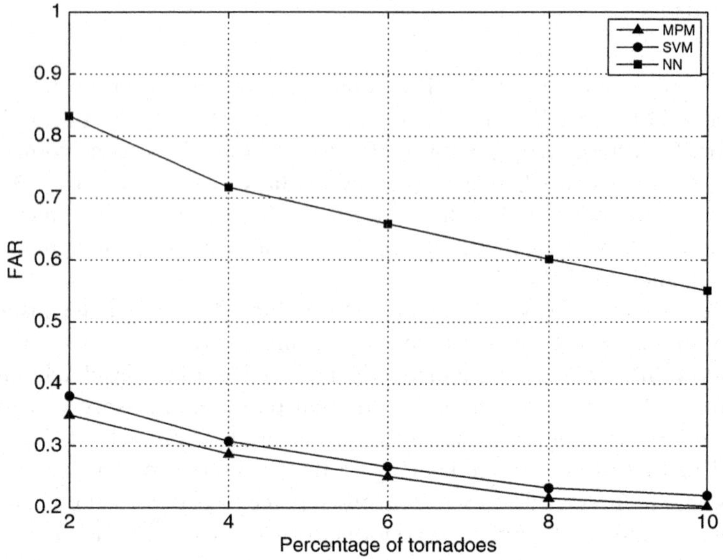

Fig. 4 False alarm ratio (FAR) for different methods for different percentages of tornadoes in the test set

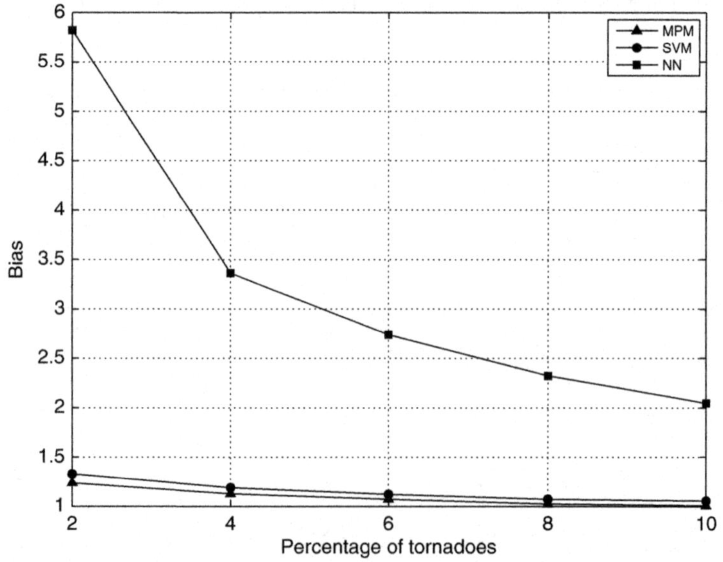

Fig. 5 Bias for different methods for different percentages of tornadoes in the test set

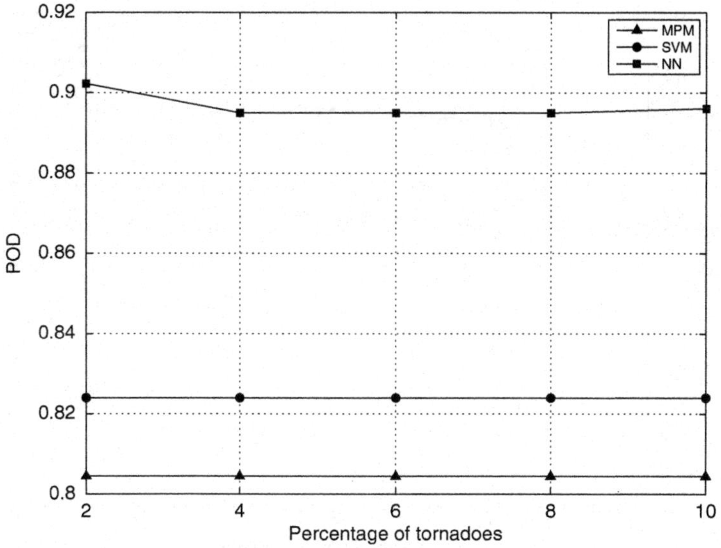

Fig. 6 Probability of detection (POD) for different methods for different percentages of tornadoes in the test set

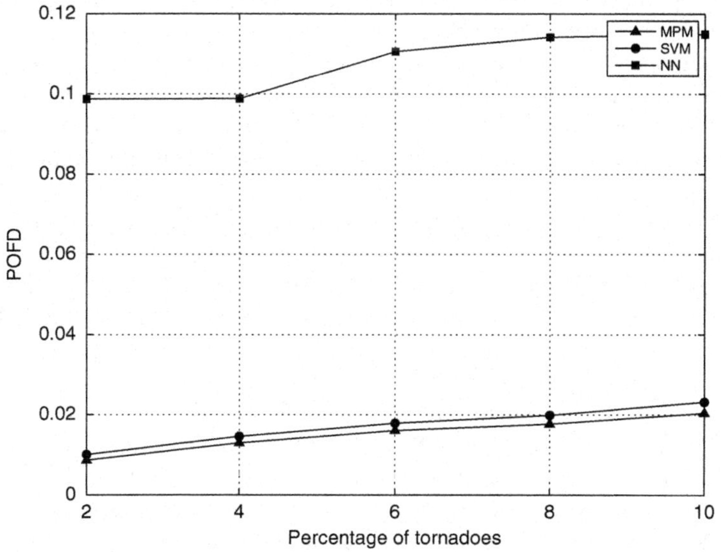

Fig. 7 Probability of false detection (POFD) for different methods for different percentages of tornadoes in the test set

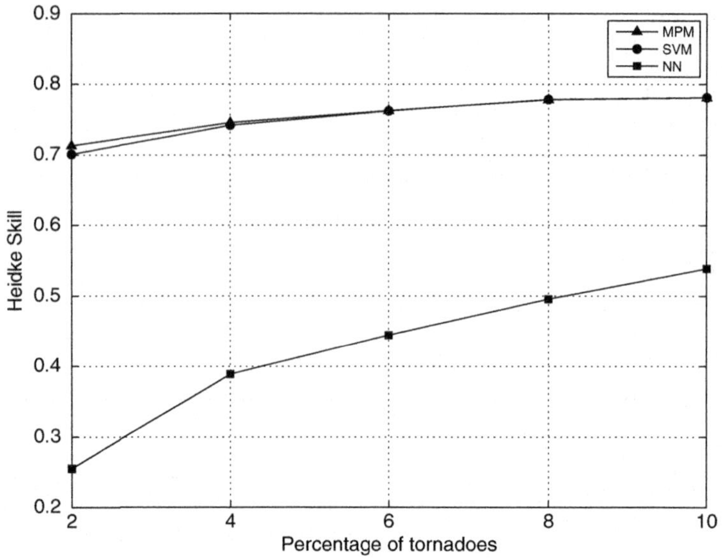

Fig. 8 Heidke's skill for different methods for different percentages of tornadoes in the test set

NN when applied to rare event data, such as tornadoes, or if NN is capturing some local minima and over-fitting noise.

The predisposition to forecast tornadoes when none are observed can be seen in the BIAS values for NN (Fig. 5). Here, the NN has a BIAS of 5.5. The ideal value of 1 is nearly achieved by SVM/MPM. The BIAS values indicate that the SVM and MPM techniques are close to the optimum near the climatological probability of tornadoes.

Another way to assess the sensitivity of the various techniques to detect tornadoes is through the POD (Fig. 6). In this case, the NN had an edge over the other techniques as it could detect over 90 % of the tornadoes in the database, irrespective of the percentage of tornadoes in the testing set. At the 2 % level, the SVM/MPM techniques had approximately 0.8 POD. As the number of tornadoes relative to the number of no tornadoes increased, this value remained the same.

False alarms can be studied further through POFD (Fig. 7). In this case, the "no tornado" observations are isolated, thereby giving a fuller picture, if compared with FAR. The POFD should be as close to zero as possible. SVM and MPM results are quite close to zero, whereas NN has over five times as large as an error.

The Heidke's skill score yields information on how much of an improvement a forecast contains compared to a random guess (or some other reference forecast). Results (Fig. 8) show the NN has lower skill than either SVM or MPM. Since the climatological probability of tornadoes is very low, examination of the 2–4 % range is most revealing. Results in that range suggest relatively low skill compared to a random forecast for NN (0.25) and much larger for MPM and SVM (both approximately 0.70).

Taken collectively, these results suggest that NN behaves in a different way, compared to the kernel methods. This difference can be attributed to NN detecting more tornadoes when there are tornadoes (hits), over-predicting tornadoes in all circumstances (high false alarm rate), and being inferior in detecting no tornado when there is no tornado (correct negative), compared to the kernel methods.

6 Conclusions

In this work shows forecast evaluation indices comparable to SVM. Kernel MPM shows promising performance. Overall, the SVM technique has a slight edge over the MPM method for most forecast evaluations. However, the key finding is that MPM and SVM are more accurate, have a lower false alarm rate, less bias, and more skill than the NN technique. This is particularly noteworthy since the current tornado detection algorithms are processed through NN.

From the meteorological viewpoint, the SVM and MPM techniques need to be tested in an operational setting to assess if lead times are improved. Such experimentation will provide valuable validation that can reveal if these techniques remain stable as prediction tools. In the future, the dynamic use of data collected from multiple radars as sources of predictors is worthy of investigation for obtaining more accurate prediction in real time.

Acknowledgements The present work has been partially supported by the NSF grant EIA-020568. Thanks are extended to Robin C. Gilbert and Weili Zhang for their LaTeX expertise and Jean Shingledecker for her editorial assistance.

References

Adrianto, I., Trafalis, T.B.: The p-center machine for regression analysis. Optim. Methods Softw. **25**(2), 171–183 (2010)

Chang, C.C., Lin, C.J.: LIBSVM: a library for support vector machines. Software available at http://www.csie.ntu.edu.tw/$\sim$cjlin/libsvm (2001)

Doswell, C.A., Davies-Jones, R., Keller, D.: On summary measures of skill in rare event forecasting based on contingency tables. Weather Forecast. **5**, 576–585 (1990)

Gilbert, R.C., Trafalis, T.B.: Quadratic programming formulations for classification and regression. Optim. Methods Softw. **24**(2), 175–185 (2009)

Haupt, S.E., Pasini, A., Marzban, C. (eds.): Artificial Intelligence Methods in the Environmental Sciences. Springer, New York (2009)

Haykin, S.: Neural Networks: A Comprehensive Foundation, 2nd edn. Prentice-Hall, Upper Saddle River, NJ (1999)

Hsu, C.W., Lin, C.J.: A simple decomposition method for support vector machines. Mach. Learn. **46**, 291–314 (2002)

Joachims, T.: Making large-scale SVM learning practical. In: Schölkopf, B., Burges, C., Smola, A. (eds.) Advances in Kernel Methods – Support Vector Learning, pp. 169–184. MIT Press, Cambridge (1999)

Lanckriet, G.R.G., Ghaoui, L.E., Bhattacharyya, C., Jordan, M.I.: A robust minimax approach to classification. J. Mach. Learn. Res. **3**, 555–582 (2002)
Marzban, C., Stumpf, G.J.: A neural network for tornado prediction based on Doppler radar-derived attributes. J. Appl. Meteorol. **35**, 617–626 (1996)
Platt, J.C.: Fast training of support vector machines using sequential minimal optimization. In: Schölkopf, B., Burges, C., Smola, A. (eds.) Advances in Kernel Methods – Support Vector Learning, pp. 185–208. MIT Press, Cambridge (1999)
Simmons, K.M., Sutter, D.: Tornado warnings, lead times, and tornado casualties: an empirical investigation. Weather Forecast. **23**(2), 246–258 (2008)
Trafalis, T.B., Alwazzi, S.A.: Robust optimization in support vector machine training with bounded errors. In: Proceedings of the 2003 IEEE International Joint Conference on Neural Networks, vol. 3, pp. 2039–2042. IEEE, Piscataway, NJ (2003)
Trafalis, T.B., Gilbert, R.C.: Robust classification and regression using support vector machines. Eur. J. Oper. Res. **173**(3), 893–909 (2006)
Trafalis, T.B., Gilbert, R.C.: Robust support vector machines for classification and computational issues. Optim. Methods Softw. **22**(1), 187–198 (2007)
Trafalis, T.B., Malyscheff, A.M.: An analytic center machine. Mach. Learn. **46**, 203–223 (2002)
Trafalis, T.B., Santosa, B., Richman, M.B.: Tornado detection with kernel-based methods. In: Dagli, C.H., Buczak, A.L., Ghösh, J., Embrechts, M., Ersoy, O. (eds.) Intelligent Engineering Systems Through Artificial Neural Networks, vol. 13, pp. 677–682. ASME Press, New York (2003)
Trafalis, T.B., Richman, M.B., Santosa, B.: Dynamic data driven support vector machines applied to tornado detection. In: Dagli, C.H., Buczak, A.L., Ghösh, J., Embrechts, M.J., Ersoy, O., Kercel, S.W. (eds.) Intelligent Engineering Systems Through Artificial Neural Networks, vol. 14, pp. 559–564. ASME Press, New York, NY (2004)
Trafalis, T.B., Richman, M.B., Santosa, B.: Learning networks for tornado detection. Int. J. Gen. Syst. **35**(1), 93–107 (2006)
Trafalis, T.B., Adrianto, I., Richman, M.B.: Active learning with support vector machines for tornado prediction. In: Shi, Y., van Albada, G.D., Dongarra, J., Sloot, P.M.A. (eds.) Computational Science – ICCS 2007. Lecture Notes in Computer Science, vol. 4487, pp. 1130–1137. Springer, Berlin (2007)
Vapnik, V.N.: The Nature of Statistical Learning Theory. Springer, New York, NY (1995)
Wilks, D.S.: Statistical Methods in the Atmospheric Sciences, 3rd edn. Academic Press, London (2011)

Evacuation Modeling and Betweenness Centrality

Chrysafis Vogiatzis and Panos M. Pardalos

Abstract In this chapter, we consider the problem of efficiently evacuating all people in an urban area from danger zones to safe zones. This problem, which has attracted major scientific interest and has been well-studied in literature, is indeed large-scale, and as such difficult to solve. In this work, we propose a solution method based on an islanding scheme. This decomposition approach takes into consideration the betweenness of a set of nodes in the transportation network, and aims to obtain clusters from those nodes that can be easily solved: the idea is to divide the flow more evenly towards multiple paths to safety, leading to a more robust evacuation process. We portray our results on several synthetic and real-life transportation networks. More importantly, we use a very large-scale network representation of the city of Jacksonville, Florida, in the USA to show that our approaches solve the problem, a feat that proved impossible for commercial solvers. We conclude this study with our observations and plans for future work.

Keywords Evacuation • Disaster management • Computational method • Betweenness centrality • Clustering

1 Introduction

Disaster management and evacuation planning are both of utmost importance for the societal welfare of modern countries and states. On top of its humanitarian benefits, proactive planning of a response to a natural or man-made disaster also holds significant cost savings.

C. Vogiatzis (✉)
Department of Industrial and Manufacturing Engineering, North Dakota State University, Fargo, ND, USA
e-mail: chrysafis.vogiatzis@ndsu.edu

P.M. Pardalos
Department of Industrial and Systems Engineering, University of Florida, Gainesville, FL, USA
e-mail: pardalos@ufl.edu

I.S. Kotsireas et al. (eds.), *Dynamics of Disasters—Key Concepts, Models, Algorithms, and Insights*, Springer Proceedings in Mathematics & Statistics 185,
DOI 10.1007/978-3-319-43709-5_17

Most approaches tackling the evacuation planning problem can be categorized based on their underlying methods. Typically, problems of the sort are formulated and tackled as large-scale linear programming problems, with approaches that generalize network flow algorithms. The problem with such methods stems from the large-scale nature of the problem, which does not enable us to solve real-life scenarios. A second approach utilizes simulation and agent-based methods to stochastically imitate the behaviors of drivers and unexpected delays/failures in the infrastructure. Once more, due to the large-scale nature of the problem, these methods are very computationally expensive, especially with the addition of more realistic constraints. Last but not least, there exist many approaches in literature that fall within the general spectrum of heuristics, which try to take advantage of specific characteristics of the evacuation process.

Our work falls within the third category, as we present a heuristic to solve large-scale evacuation problems by decomposing them into a series of smaller, scalable integer linear programming problems. We further incorporate information on the network using notions from graph centrality, in order to ensure a more robust decomposition. This manuscript is organized as follows. In Sect. 2, we present the existing literature on the field with some representative approaches. We also introduce literature from graph centrality that we will be using in the remainder of the paper. We then proceed to introduce our notation, definitions, other preliminaries needed in our work, and the basic mathematical formulation to be used. Section 3 first presents the initial islanding scheme. It further introduces a generalized framework to consider area centralities when decomposing the original problem into subproblems in order to ensure a more evenly distributed flow across the different clusters obtained. Then, in Sect. 4, we present our findings using different network representations of the city of Jacksonville, and other real-life and synthetic networks. Finally, in Sect. 5, we conclude this study and offer insight into future work in the field.

2 Preliminaries

2.1 *Evacuation Planning*

Herein, we will present existing methods to solving the evacuation planning problem, namely routing vehicles to safety in the case of a disaster. Of course this is only part of the overall disaster management problem. The interested reader is referred to Caunhye et al. (2012) for a review of proactive preparedness measures, such as safety center location, transportation of injured people, safety stock location, among others. Furthermore, the ones interested in humanitarian supply chain management in times of distress are referred to the very detailed survey in Richey et al. (2009).

As far as minimum cost flow models (network optimization) for solving evacuation planning problems are concerned, most of the approaches stem from the seminal contribution of Ford and Fulkerson (1956) and the introduction of the maximum dynamic flow problem. This problem, along with extensions such as the earliest arrival flow problem (Gale, 1958), has been the basis of work to come in evacuation planning from both a static and a time dynamic perspective.

As an example, Cova and Johnson (2003) formulate the problem over a time-static network, with the goal of minimizing the conflicts that arise in intersections. This is indeed a serious implication of any evacuation process: who has priority in intersections? Tragically, during the Fukushima prefecture evacuation (Vogiatzis et al., 2013) after the Fukushima Daiichi nuclear disaster, many intersections were overburdened, leading to the entrapment of numerous evacuees. An attempt to consider priorities while solving the overall evacuation process problem can be found in Hamacher and Tufekci (1987) and, more recently, in Chiu and Zheng (2007). Another interesting approach using a static network representation can be found in Lu et al. (2005), with their contribution of using an iterative heuristic method, called Capacity Constrained Route Planner.

Time dynamic approaches (ones that consider a time-expanded network) have many advantages: they offer a schedule of operations and are clearly more realistic. They do, however, come with one important caveat, as they are typically intractable and cannot be easily used to solve large-scale problems, as they arise in urban evacuation routing. In Hamacher and Tjandra (2001), the authors present many of the—then—state-of-the-art in dynamic evacuation modeling. Before them, Hoppe and Tardos (1994) presented a series of evacuation problems that can be indeed solved in polynomial time. More recent developments include the use of robust optimization to tackle the dynamic traffic assignment problem with minimal risks (Ben-Tal et al., 2011), evacuation under a limited budget of emergency personnel (e.g., policemen) (He et al., 2014), and contraflow with lane and/or street reversals.

The contraflow idea is also not new, as it is an intuitive approach towards better utilizing the existing network. However, adding the reversal capabilities to the evacuation model renders the underlying problem NP-hard (Rebennack et al., 2010). Due to the desirability of such solutions though, it has attracted significant interest over the last decade. The interested reader is referred to the following body of work in Kim et al. (2008), Bretschneider and Kimms (2012), Vogiatzis et al. (2013), and the references therein. A later development in the field of considering lane and arc reversals in a time-expanded network can be found in Arulselvan (2014).

2.2 Centrality

In this work we also consider centrality indices within a transportation network. While centrality is not a novel idea (as it has been around since the early 1950s), to the best of our knowledge it has not been applied on the evacuation modeling field. Node centrality represents the importance, or "criticality," of a given node in the grand scheme of the network, and is considered fundamental in network analysis. Numerous studies have been performed over the years starting from the contributions of Bavelas (1948, 1950), Leavitt (1951), and Sabidussi (1966) to the most recent work of Everett and Borgatti (2005) and Borgatti and Everett (2006). More specifically, in their work (Everett and Borgatti, 1999), they propose group centrality, an idea that was adapted to accommodate to clique centrality in Vogiatzis et al. (2015).

Attempts to decompose graphs based on betweenness centrality indices have attracted significant interest since the publication of Newman and Girvan (2004) and later work by Newman (2006). The interested reader is referred to Blondel et al. (2008), Fortunato (2010), Guimera et al. (2005), among many other very interesting works in this vast field.

2.3 *Preliminaries*

Consider a simple, directed graph $G(V, E)$, with $|V| = n$ vertices, and $|E| = m$ edges. Two vertices i and j are said to be connected if there exists a path of vertices beginning at i and ending in j, i.e., $\{v_0, \ldots, v_k\}$ where $v_0 = i$ and $v_k = l$, and for every two consecutive vertices v_l, v_{l+1} in the path, we have that $(v_l, v_{l+1}) \in E$. A graph G is connected if any pair of nodes $i, j \in V$ are connected: it is a common assumption that transportation networks are always connected. Hence, we also make this assumption for simplicity.

As we are considering a time-expanded network, we deal with an instance of the graph G at any given time $t \in T$, where $|T|$ is assumed to be a big enough horizon to cover our evacuation operations. Hence, we have a graph $G^{(t)}$, with its nodeset, $V^{(t)}$, and edgeset, $E^{(t)}$. Every edge is also associated with a parameter m_{ij} that represents the amount of time it would take a vehicle to traverse it. In transportation networks, this can be represented more accurately with a random variable and/or a dependence on the flow currently using the street; however, for our purposes of network decomposition for evacuation problems, it was deemed unnecessary and as such it is treated as a parameter. A subset of nodes $S \subset V$ is described as safe, and is the destination of any vehicle in the transportation network during the evacuation process. We further assume that vehicles do not have a preference on the node $i \in S$ they would like to arrive in.

Moreover, we consider that certain information is readily available on the sets of vertices and edges, before modeling and solving the evacuation problem begins. Namely, each edge $(i, j) \in E^{(t)}$ has a capacity on the number of vehicles it can accommodate at any time $t \in T$, symbolized by $u_{ij}^{(t)}$. For simplicity, it can be generally assumed that $u_{ij}^{(t)} = u_{ij}, \forall t \in T$. The capacity is extended to nodes as well, as we assume that a node $i \in V$ is associated with an upper bound on the number of vehicles that can be there at a given time t, $c_i^{(t)}$. Nodes also have an initial, known demand, $d_i^{(0)}$. Last, we assume that information on the level of danger the node is under at any given time $t \in T$ is readily available. This can be provided to us by simulation of the underlying disaster: as an example, in the case of tropical storms, a tool like CSEVA (Davis et al., 2011) can be used. In order to help us with our mathematical formulation, we assume $l_i^{(t)} = 0$ for all $i \in S$; the remaining nodes $i \in V \setminus S$ take a value $l_i^{(t)} > 0$, with larger values signaling the necessity for immediate evacuation.

For betweenness centrality, we use the definition typically encountered in literature, which is shown in (1) for a node $k \in V$, and in (2) for a set of nodes $N \subseteq V$.

$$C(k) = \sum_{i \in V \setminus \{k\}} \sum_{j \in V \setminus \{k\}: i \neq j} \frac{g_{ij}(k)}{g_{ij}} \tag{1}$$

$$C(N) = \sum_{i \in V \setminus N} \sum_{j \in V \setminus N: i \neq j} \frac{g_{ij}(N)}{g_{ij}} \tag{2}$$

In the above expressions, $g_{ij}(k)$ is the number of geodesic (shortest) paths connecting nodes i and j that pass through k, whereas $g_{ij}(N)$ the same for group of nodes N. The denominator in both cases is the total number of geodesic paths connecting nodes i and j, which can be bigger than 1 in the case of multiple alternate shortest paths. Observe that the group betweenness of a set N cannot be calculated by adding the individual node betweenness of each node members. For a counterexample, we refer the reader to Fig. 1.

In this example, consider the origin-destination pair (s, t). It is easy to see that there are two geodesic paths connecting the pair ($g_{st} = 2$), of which both use node i, and one uses nodes j and k. However, when considering the set of nodes $S = \{i, j\}$, the fraction of shortest paths using *at least one* of the nodes in N is indeed 1.

2.4 Mathematical Formulation

Let $x_{ij}^{(t)}$ denote the number of vehicles traversing edge $(i, j) \in E^{(t)}$ at time t. For every node $i \in V$, let $d_i^{(t)}$ represent the number of vehicles waiting there at time t. Notice that the demands for the nodes at each time $t > 0$ are treated as variables, which enables us to instruct vehicles to wait at their position instead of moving, whenever

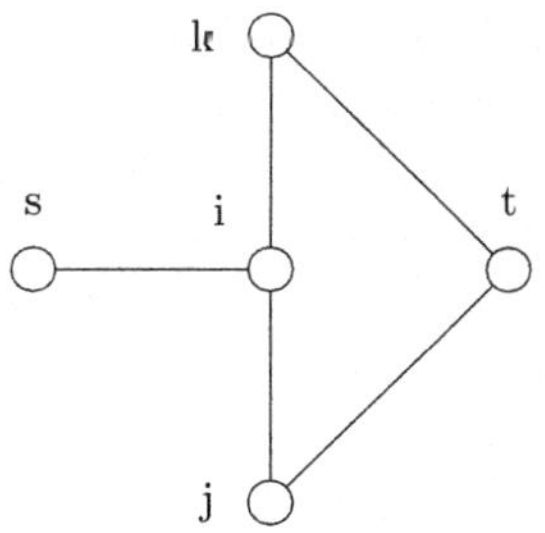

Fig. 1 A counterexample to show that individual node betweenness cannot be used in the calculation of group betweenness

the decision to move them would hinder the success of the evacuation plan. For $t = 0$, $d_i^{(0)}$ is treated as an input parameter. Further, we introduce binary variable y_{ij} as follows:

$$y_{ij} = \begin{cases} 1, & \text{if edge } (i,j) \text{ is reversed during the evacuation process,} \\ 0, & \text{otherwise.} \end{cases}$$

From the definition of the set of y variables, it is clear that we consider only the case that a street can be declared reversed (along with the necessary policing and precautions) from the beginning of the evacuation process and cannot change back to its original sense until the end of operations. Last, we assume we are given a horizon of periods $T = \{1, 2, \ldots, |T|\}$, during which time a maximum number of vehicles are to reach a safe node. Equivalently, the number of vehicles still evacuating can be minimized. The mathematical formulation can now be presented in (3)–(12).

$$\min \sum_{t \in T} \sum_{i \in V} l_i^{(t)} d_i^{(t)} + \sum_{t \in T} \sum_{(i,j) \in E^{(t)}} \frac{l_i^{(t)} + l_j^{(t)}}{2} x_{ij}^{(t)} \tag{3}$$

$$\text{s.t. } x_{ij}^{(t)} \leq (1 - y_{ij}) u_{ij} + y_{ji} u_{ji}, \qquad \forall (i,j) \in E^{(t)}, \forall t \in T \tag{4}$$

$$y_{ij} + y_{ji} \leq 1, \qquad \forall (i,j) \in E \tag{5}$$

$$d_i^{(t)} \leq c_i, \qquad \forall i \in V^{(t)}, \forall t \in T \setminus \{0\} \tag{6}$$

$$d_i^{(t+1)} = d_i^{(t)} - \sum_{j:(i,j) \in E^{(t)}} x_{ij}^{(t)} + \sum_{\substack{j:(j,i) \in E^{(t-m_{ij})} \\ t-m_{ij} \geq 0}} x_{ji}^{(t-m_{ij})}, \qquad \forall i \in V^{(t)}, \forall t \in T \tag{7}$$

$$\sum_{(i,j) \in E} y_{ij} \leq k, \qquad \forall t \in T \tag{8}$$

$$x_{ij}^{(t)} = 0, \qquad \forall i,j \in V^{(t)} : l_i^{(t)} < l_j^{(t)} \tag{9}$$

$$x_{ij}^{(t)} \geq 0, \qquad \forall (i,j) \in E^{(t)}, \forall t \in T \tag{10}$$

$$d_i^{(t)} \geq 0, \qquad \forall i \in V^{(t)}, \forall t \in T \tag{11}$$

$$y_{ij} \in \{0, 1\}, \qquad \forall (i,j) \in E. \tag{12}$$

The objective function in (3) aims to minimize the number of people waiting to be evacuated in the transportation network, either waiting in one of the nodes $i \in V$ or using an edge $(i,j) \in E$. Observe that we employ the simulation-obtained danger factors $l_i^{(t)}$ to derive a weighted sum of the number of demands that remain to be evacuated; for the edges we use an average estimate of the two endpoints at that specific time $t \in T$. That way, evacuees that are further away would take priority in our operations, as they increase the value of the objective function significantly.

Constraints (4) ensure that whenever we have contraflow present in a segment of the evacuation plan, the number of evacuees using the street can increase over its original capacity, as the capacity of the reverse direction can be employed. Keep note that this is not necessarily double the capacity of the original street, as there exist streets where the capacities are asymmetric per direction. Moreover, when a street is reversed, then the constraint ensures that no flow can be sent employing that same street. Constraint (5) guarantees that in an evacuation setting, at most one of the senses of any street can be reversed at a time. Continuing with the constraints of the problem, in (6) we enforce a capacity at every intersection of the transportation network. That way, we can avoid having evacuees accumulate at any intersection, leading to street capacity losses. It can be assumed that the capacity of the safety nodes is much higher than the capacity of the rest of the nodes in the network.

Constraints (7) are the classical flow preservation constraints, as shown in Kim et al. (2008), adapted for the time dynamic network we have at hand. They enforce that the number of evacuees waiting at an intersection i at a time t is as many as there were in the previous time step $t - 1$ when considering also the number of vehicles that are incoming and subtracting the ones outgoing. As a reminder, we have assume that each actual street of the network has been assigned a parameter m_{ij} that stays constant throughout the process which represents the (integer) number of periods it takes to traverse it. Of course, we also need to add a budget constraint, as it is impossible to enable the reversal of every street in our evacuation plan. The budget constraint is shown in (8).

Furthermore, we have constraints (9) that do not allow a vehicle to move from a less dangerous to a more dangerous area. This constraint apart from enabling us with the islanding scheme to be discussed in the following subsection, also guarantees that the objective function is non-increasing from time step to time step. Finally, we guarantee nonnegativity of all variables and the binary nature of the y variables in (10)–(12).

3 Evacuation Modeling

3.1 Simple Islanding Scheme

The above problem is indeed very hard to solve in real-life, large-scale instances. As an example, it would take a commercial solver multiple hours to obtain an optimal solution, when in fact in many situations, it is imperative that an evacuation plan is devised and implemented much faster than that.

To tackle this issue, we propose a decomposition approach that produces a series of connected clusters, called "islands." For the decomposition, we assume that the number of clusters $|N|$ is given as an input. An initial decomposition approach is presented in Algorithm 1. For this algorithm, it is necessary that we know the distance a vehicle needs to traverse from every node $i \in V \setminus S$ to arrive at a node $j \in S$ belonging to the safe zone, without crossing an edge $(i,j) \in E^{(t)}$ such that $l_i^{(t)} < l_j^{(t+m_{ij})}$. Let this function be referred to as *ModifiedDijkstra*(i, S), for a given node $i \in V$ and the set of safe nodes in the network, S.

The modification to the original approach to finding all shortest paths from a node to a set of nodes is pretty straightforward. We perform a preprocessing of the networks $G^{(t)}$ where we remove all edges satisfying $l_i^{(t)} < l_j^{(t+m_{ij})}$; we then proceed to calculate the shortest path from each node to a virtual aggregate node connected to all safe nodes with 0 traversal time. Last, we assume we are given $t_1, t_2, \ldots t_{|N|}$ as threshold distances for a node to belong to cluster $1, 2, \ldots, |N|$, respectively.

3.2 Betweenness Islanding Scheme

An extension of Algorithm 1 is presented in Algorithm 2. This one takes into consideration the betweenness centrality of a cluster C and aims to decompose the central hubs in several, distinct clusters. A simple description of Algorithm 2 can be the following. We first calculate the betweenness of every node in the graph,

Algorithm 1 Decomposition algorithm based on distance from safety and danger factor

```
1: n ← |N|
2: for each node i ∈ V ⊆ S do
3:     dist(i) ← modifiedDijkstra(i, S)
4:     for k ∈ {1, 2, ..., n} do
5:         if dist(i) ≤ t_k then
6:             N_k ← N_k ∪ {i}
7:         end if
8:     end for
9: end for
```

Algorithm 2 Decomposition algorithm based on betweenness

```
max ← 0
N_0 ← ∅
for each set of nodes N ⊆ V do
    if isConnected(N) && |N| = n then
        btn = SafeBetweenness(N)
        if max < btn then
            max ← btn
            N_0 ← N
        end if
    end if
end for
for each node i ∈ V \ N_0 do
    k = argmin{d_ik : k ∈ N_0}
    N_k ← N_k ∪ {i}
end for
```

considering though only the shortest paths from every node $i \in V \setminus S$ to every safe node $s \in S$ (let this procedure be referred to as *SafeBetweenness*(i) for a node $i \in V$ and *SafeBetweenness*(C) for a set of connected nodes $C \subseteq V$). We then calculate the group betweenness of all connected clusters N of size n, using only the shortest paths to safety. Note that in regular betweenness calculations, the shortest paths between every pair of nodes (i, j) in the network are considered; instead, here we only consider the shortest paths between any node $i \in V \setminus S$ and any safe zone $s \in S$. The set of size n with the biggest betweenness centrality is selected and its nodes are now to be divided among the different clusters of the islanding scheme. One way for this to be achieved, is to find the shortest path of every other node in the network to the ones found: then each node can be greedily assigned to the one that it can reach the fastest. A toy example is offered for the reader's convenience in Fig. 2.

The first algorithm is much faster due to the fact that it relies on simple shortest path finding. However, the second algorithm too is tractable for small values of n, as only $\binom{|V|}{n}$ possible set nodes are to be considered. Of course, the detection of the set of nodes N_0 with the highest betweenness can also be formulated as an optimization problem as in Vogiatzis et al. (2015), after removing the clique constraints and replacing them with simple connectivity constraints.

4 Computational Results

All computational experiments were performed on a cluster with two AMD Opteron 6128 Eight-Core CPUs and 12 GB of RAM, with an operating system of Linux x86_64, CentOS 5.9. The clustering schemes were coded in C++, while the integer programs were solved using Gurobi 5.6.3 and CPLEX 12.3. All graph visualizations were performed using networkx 1.9.1 (Hagberg et al., 2008).

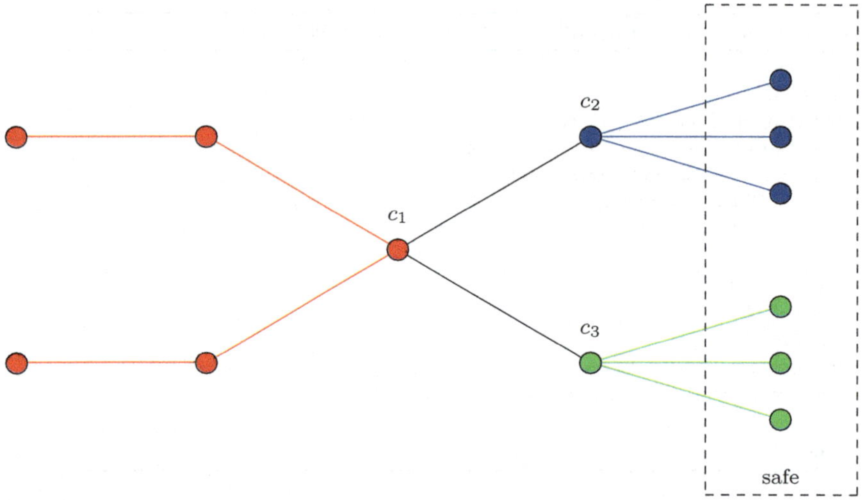

Fig. 2 An example of how Algorithm 2 works. Here, $n = 3$, and the connected set with the highest betweenness centrality is $N = \{c_1, c_2, c_3\}$. The three connected clusters obtained are now shown in *red*, *blue*, and *green*. The set of edges connecting the nodes within N are now to serve as edges connecting supernodes-clusters in the final problem

Our experiments were performed on synthetic networks of varying size (100–10,000 nodes) and real-life transportation network instances, where 10 % of the nodes are selected to be safe zones. Specifically we focus our experiments towards the city of Jacksonville, Florida, which presents a real challenge as it is represented by a huge transportation network. In the Jacksonville network, the safe zones are pre-selected and known. All simulated storms in our experiments are obtained through CSEVA (Davis et al., 2011, 2012).

All instances were solved using a commercial solver (the fastest and/or best solution was only reported), and then by decomposing them using the first and the second clustering scheme and then solving each individual problem by itself. An example of how the two islanding schemes work for the city of Jacksonville and for $n = 5$ clusters is presented in Figs. 3 and 4.

All computational results on synthetic networks are presented in Tables 1 and 2, while the ones for real-life transportation networks are given in Tables 3 and 4. Last, detailed results for the city of Jacksonville are presented in Tables 5 and 6. Our heuristics are represented by H1 for the simple islanding scheme, and H2 for the betweenness-based islanding scheme. Values in bold represent the best computational time or optimality gap, respectively.

From the computational results, we can make the following observations. The first decomposition approach is almost always the fastest: the exception appears in the largest representation of the city of Jacksonville, where the $n = 5$ clusters obtained from the betweenness scheme are obtained (and solved) faster. Another interesting observation is that, as expected, as the number of clusters to be detected

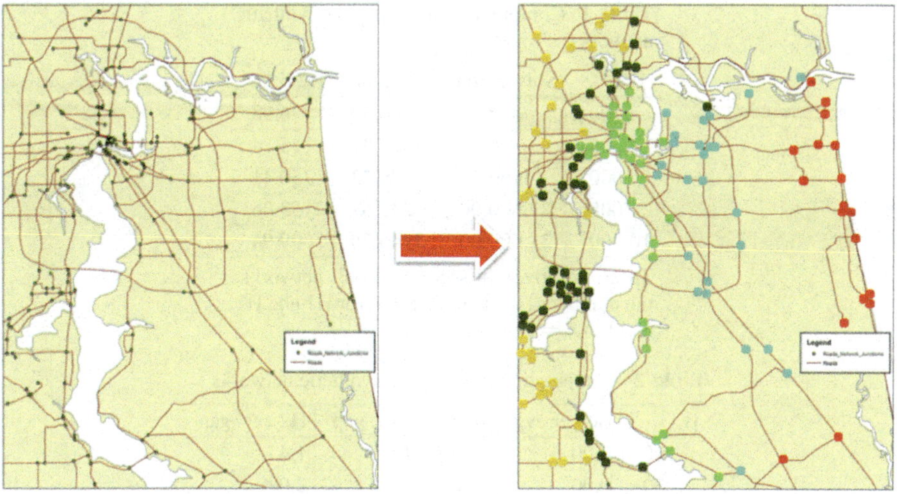

Fig. 3 An example of how the first islanding scheme (Algorithm 1) works for the city of Jacksonville, FL

Fig. 4 An example of how the first islanding scheme (Algorithm 2) works for the city of Jacksonville, FL

increases, the second decomposition scheme does not scale well, as it is always increasing significantly in its runtime. On the other hand, it does lead to higher quality solutions.

Last, we see that solving a large-scale problem, such as the one appearing using the transportation network of Jacksonville, FL, is not a viable option. The time it takes a commercial optimizer to solve the problem is prohibitively big, and also

Table 1 Computational results on synthetic networks

$\|V\|$	Solver time	H1 time	H2 time
100	8.37	**6.52**	6.98
500	78.91	**24.37**	45.74
1000	3600.18	**489.23**	750.41
5000	21,600.00	**1543.67**	2153.06
10,000	21,600.00	**2946.70**	3609.94

Times are given in seconds. In all networks, the number of clusters detected for H1 and H2 is set to $n = 5$

Table 2 Computational results on synthetic networks

$\|V\|$	Solver avg. gap	H1 avg. gap	H2 avg. gap
100	0.00	**0.87**	1.47
500	0.00	2.16	**2.09**
1000	0.00	3.33	**2.85**
5000	N/A	N/A	N/A
10,000	N/A	N/A	N/A

Average optimality gaps are given in %. In all networks, the number of clusters detected for H1 and H2 is set to $n = 5$

Table 3 Computational results on real-life networks

Name ($\|V\|$)	Solver time	H1 time	H2 time
Anaheim (416)	11.52	**3.96**	6.04
Austin (7388)	12,627.90	**1238.81**	2067.01
Philadelphia (13,389)	21,600.00	**3271.31**	3947.25

Times are given in seconds. In all networks, the number of clusters detected for H1 and H2 is set to $n = 5$

Table 4 Computational results on real-life networks

Name ($\|V\|$)	Solver avg. gap	H1 avg. gap	H2 avg. gap
Anaheim (416)	0.00	1.67	**1.08**
Austin (7388)	0.00	2.99	**2.02**
Philadelphia (13,389)	N/A	N/A	N/A

Average optimality gaps are given in %. In all networks, the number of clusters detected for H1 and H2 is set to $n = 5$

contains the risk of running out of memory before it provides us with a valid upper bound as a feasible solution. On the other hand, the first decomposition approach (H1) always terminates with a feasible solution, albeit its optimality gap is significant as the size of the network increases. The second decomposition approach (H2) also provides us with a feasible solution in most instances. On top of that, it provides us with higher quality solutions with a smaller optimality gap than the

Table 5 Computational results on the city of Jacksonville, Florida

Name ($\|V\|$)	Solver time	H1 time	H2 time	Number of clusters (n)
Small (11,724)	60,815.50	**2870.10**	4121.33	5
Small (11,724)	60,815.50	**745.41**	6991.12	10
Medium (∼35,000)	97,254.00	**18,470.10**	28,107.50	5
Medium (∼35,000)	97,254.00	**2917.10**	41,003.20	10
Large (∼90,000)	N/A	**126,491.00**	127,255.00	5
Large (∼90,000)	N/A	**88,267.00**	N/A	10

Times are given in seconds, while optimality gaps in %

Table 6 Computational results on the city of Jacksonville, Florida

Name $\|V\|$	Solver avg. gap	H1 avg. gap	H2 avg. gap	Number of clusters (n)
Small (11,724)	0.00	4.40	**2.77**	5
Small (11,724)	0.00	6.51	**2.41**	10
Medium (∼35,000)	0.00	10.92	**6.80**	5
Medium (∼35,000)	0.00	16.75	**4.02**	10
Large (∼90,000)	N/A	N/A	N/A	5
Large (∼90,000)	N/A	N/A	N/A	10

Average optimality gaps are given in %

first approach. The caveat here though is that the number of clusters that we aim to detect is a significant factor: the more we want to decompose the problem, the bigger the connected set of nodes that we are searching for becomes, which leads to a significant increase in the computational time required to terminate with a solution.

5 Concluding Remarks

In this work, we consider the problem of evacuating large-scale urban areas in the presence of contraflows. The goal is to minimize the number of people who are awaiting rescue, with priority being given to evacuees in more endangered area. Because of the intrinsic difficulty of the problem, which is known to be *NP*-hard, we consider a heuristic approach to decompose the problem into subproblems, based on the danger level, the distance from safety, and a combination of the two. We also propose a new decomposition scheme based on group betweenness. This decomposition can be seen as a way to break down the original problem into subproblems, each containing one of the most important areas during the evacuation process, that is one that multiple evacuation routes use to reach safety.

Our results show that indeed the islanding technique can be used to solve real-life problems, even though there is no theoretical bound to its performance. In practice though, in all computational experiments, the islanding scheme was close to the optimal solution but much faster than using commercial solvers, such as CPLEX and Gurobi. The addition of the clusters obtained by betweenness centrality, on the other hand, leads to a different cluster setup. The problem of detecting such clusters is indeed more computationally expensive, but leads to higher quality solutions. Future work should primarily focus on smartly selecting the number of clusters that should be obtained from a large-scale transportation problem. That way, we could completely control the computational effort required and the quality of the solutions obtained. We are also planning on investigating the convergence of the heuristics presented herein, along with theoretical bounds on their performance. We believe that such a derivation will complement its performance in real-life instances and will prove to be a viable option for practitioners in evacuation and disaster management.

Acknowledgements This research was funded in part by DTRA and the Air Force Research Laboratory Mathematical Modeling and Optimization Institute. Chrysafis Vogiatzis would also like to acknowledge support from ND EPSCoR NSF #1355466.

References

Arulselvan, A., Groß, M., Skutella, M.: Graph orientation and flows over time. In: Algorithms and Computation, pp. 741–752. Springer, Cham (2014)

Bavelas, A.: A mathematical model for group structures. Hum. Organ. **7**(3), 16–30 (1948)

Bavelas, A.: Communication patterns in task-oriented groups. J. Acoust. Soc. Am. **22**(6), 725–730 (1950)

Ben-Tal, A., Do Chung, B., Mandala, S.R., Yao, T.: Robust optimization for emergency logistics planning: risk mitigation in humanitarian relief supply chains. Transp. Res. B: Methodol. **45**(8), 1177–1189 (2011)

Blondel, V.D., Guillaume, J.L., Lambiotte, R., Lefebvre, E.: Fast unfolding of communities in large networks. J. Stat. Mech.: Theory Exp. **2008**(10), P10008 (2008)

Borgatti, S.P., Everett, M.G.: A graph-theoretic perspective on centrality. Soc. Netw. **28**(4), 466–484 (2006)

Bretschneider, S., Kimms, A.: Pattern-based evacuation planning for urban areas. Eur. J. Oper. Res. **216**(1), 57–69 (2012)

Caunhye, A.M., Nie, X., Pokharel, S.: Optimization models in emergency logistics: a literature review. Socio-Econ. Plan. Sci. **46**(1), 4–13 (2012)

Chiu, Y.C., Zheng, H.: Real-time mobilization decisions for multi-priority emergency response resources and evacuation groups: model formulation and solution. Transp. Res. E: Logist. Transp. Rev. **43**(6), 710–736 (2007)

Cova, T.J., Johnson, J.P.: A network flow model for lane-based evacuation routing. Transp. Res. A: Policy Pract. **37**(7), 579–604 (2003)

Davis, J.R., Paramygin, V.A., Figueiredo, R.J., Sheng, Y.P., Vogiatzis, C., Pardalos, P.M.: The coastal science educational virtual appliance (CSEVA). In: Proceedings of the 12th International Conference on Estuarine and Coastal Modeling, St. Augustine, FL, pp. 7–9 (2011)

Davis, J.R., Zheng, Q.P., Paramygin, V.A., Tutak, B., Vogiatzis, C., Sheng, Y.P., Pardalos, P.M., Figueiredo, R.J.: Development of a multimodal transportation educational virtual appliance (MTEVA) to study congestion during extreme tropical events. In: Transportation Research Board 91st Annual Meeting, 12-1119 (2012)

Everett, M.G., Borgatti, S.P.: The centrality of groups and classes. J. Math. Sociol. **23**(3), 181–201 (1999)

Everett, M.G., Borgatti, S.P.: Extending centrality. In: Models and Methods in Social Network Analysis, vol. 35(1), pp. 57–76, Cambridge University Press, New York (2005)
Ford, L.R., Fulkerson, D.R.: Maximal flow through a network. Can. J. Math. **8**(3), 399–404 (1956)
Fortunato, S.: Community detection in graphs. Phys. Rep. **486**(3), 75–174 (2010)
Gale, D.: Transient flows in networks. Tech. Rep., DTIC Document (1958)
Glenn Richey, R. Jr., Natarajarathinam, M., Capar, I., Narayanan, A.: Managing supply chains in times of crisis: a review of literature and insights. Int. J. Phys. Distrib. Logist. Manag. **39**(7), 535–573 (2009)
Guimera, R., Mossa, S., Turtschi, A., Amaral, L.N.: The worldwide air transportation network: anomalous centrality, community structure, and cities' global roles. Proc. Natl. Acad. Sci. **102**(22), 7794–7799 (2005)
Hagberg, A.A., Schult, D.A., Swart, P.J.: Exploring network structure, dynamics, and function using NetworkX. In: Proceedings of the 7th Python in Science Conference (SciPy2008), Pasadena, CA, pp. 11–15 (2008)
Hamacher, H., Tufekci, S.: On the use of lexicographic min cost flows in evacuation modeling. Nav. Res. Logist. **34**(4), 487–503 (1987)
Hamacher, H.W., Tjandra, S.A.: Mathematical modelling of evacuation problems: a state of art. Fraunhofer-Institut für Techno-und Wirtschaftsmathematik, Fraunhofer (ITWM) (2001)
He, Y., Wang, Y., Shi, J., Liu, Z.: An evacuation network flow optimization model for city transportation systems with policemen resource allocation. In: 2014 International Conference on Informative and Cybernetics for Computational Social Systems (ICCSS), pp. 45–50. IEEE, New York (2014)
Hoppe, B., Tardos, É.: Polynomial time algorithms for some evacuation problems. In: SODA, vol. 94, pp. 433–441 (1994)
Kim, S., Shekhar, S., Min, M.: Contraflow transportation network reconfiguration for evacuation route planning. IEEE Trans. Knowl. Data Eng. **20**(8), 1115–1129 (2008)
Leavitt, H.J.: Some effects of certain communication patterns on group performance. J. Abnorm. Soc. Psychol. **46**(1), 38 (1951)
Lu, Q., George, B., Shekhar, S.: Capacity constrained routing algorithms for evacuation planning: a summary of results. In: Advances in Spatial and Temporal Databases, pp. 291–307. Springer, Berlin (2005)
Newman, M.E.: Modularity and community structure in networks. Proc. Natl. Acad. Sci. **103**(23), 8577–8582 (2006)
Newman, M.E., Girvan, M.: Finding and evaluating community structure in networks. Phys. Rev. E **69**(2), 026113 (2004)
Rebennack, S., Arulselvan, A., Elefteriadou, L., Pardalos, P.M.: Complexity analysis for maximum flow problems with arc reversals. J. Comb. Optim. **19**(2), 200–216 (2010)
Sabidussi, G.: The centrality index of a graph. Psychometrika **31**(4), 581–603 (1966)
Vogiatzis, C., Walteros, J.L., Pardalos, P.M.: Evacuation through clustering techniques. In: Models, Algorithms, and Technologies for Network Analysis, pp. 185–198. Springer, New York (2013)
Vogiatzis, C., Yoshida, R., Aviles-Spadoni, I., Imamoto, S., Pardalos, P.M.: Livestock evacuation planning for natural and man-made emergencies. Int. J. Mass Emerg. Dis. **31**(1), 25–37 (2013)
Vogiatzis, C., Veremyev, A., Pasiliao, E., Pardalos, P.: An integer programming approach for finding the most and the least central cliques. Optim. Lett. **9**(4), 615–633 (2015). doi:10.1007/s11590-014-0782-2. http://dx.doi.org/10.1007/s11590-014-0782-2

Ode to the Humanitarian Logistician: Humanistic Logistics Through a Nurse's Eye

Deborah Wilson

Abstract The following chapter is a description of the role of the humanitarian logistician as viewed through the eyes of a medical professional while working on the same team in Liberia during the Ebola crisis in 2014. Rather than being a technical explanation of logistical tasks, this chapter describes the heroic work the logisticians performed and how their work was critical in enabling the medical team to function, save lives, and survive.

Keywords Ebola • Humanitarian logistics • Supply chain • Humanitarian relief

1 Introduction

I have it on good authority (namely, logisticians themselves) that in any humanitarian mission the nurses are the most demanding "customer." Our trump card, which we play with abandon, is to tell the logistician that we need certain materials so that we can save patients' lives and, by the way, we need those supplies *yesterday*. I plead guilty to this, yet, in all sincerity, I know that without the logisticians I and all nurses and medical team members could never be able to do our job. In recognition and honor of logisticians, or logs, as they are called in the field, I wanted to paint a picture of the extraordinary work I have seen them perform under extreme conditions during the 2014 Ebola outbreak in Liberia. This piece is by no means a technical description of the role of the logistician; rather, it is the personal observations of a nurse who has witnessed the tireless work the logs have done, which has allowed me to safely and effectively care for the victims of a terrible virus.

In September 2014, I was deployed to Foya, Liberia, with the international humanitarian organization Medecins Sans Frontieres (MSF; known in English as Doctors Without Borders). I was slated to be there for 6 weeks managing a 120-bed Ebola Treatment Unit (ETU) (Fig. 1). To give some sense of the scale of the disaster, at the height of the Ebola outbreak MSF had deployed 325 international staff and

D. Wilson (✉)
RN, BSN, CRNI
e-mail: debbiew.49@gmail.com

I.S. Kotsireas et al. (eds.), *Dynamics of Disasters—Key Concepts, Models, Algorithms, and Insights*, Springer Proceedings in Mathematics & Statistics 185,
DOI 10.1007/978-3-319-43709-5_18

Fig. 1 The Ebola Treatment Unit (ETU) (credit Martin Zinggl)

hired 4000 locally trained staff to deal with the outbreak in Liberia, Guinea, and Sierra Leone (MSF 2016). MSF set up seven ETUs and two transit centers treating a total of 10,376 patients (MSF 2016). MSF also reports that it shipped more than 1305 tonnes (800 tons) of supplies to the affected countries from March to November 2014 (MSF 2016). By the end of the outbreak, MSF had spent more than 96 million euros combating the outbreak.

All these supplies needed to be effectively managed, recorded, transported, and set up by the logisticians.

On the MSF website, the job description of a logistician is described as:

> Whether coordinating the purchase and transport of supplies, locally or internationally; organizing the transport of vaccines; restoring damaged hospitals; setting up feeding centers; managing national staff; or maintaining vehicles and communication systems, logisticians keep MSF's programs running smoothly.
>
> Logisticians normally have frequent interaction with local authorities and organizations, and often oversee the implementation of security protocols. In some cases, the positions of logistician and administrator are combined. Individuals ranging in professional background from construction managers to film location scouts, from supply chain managers to engineers have joined the organization.

This description seems rather dry given what I have witnessed the logs manage and coordinate in the field. What follows is a description of some of the heroic things I have witnessed the logs execute and achieve during one of the humanitarian emergencies I have worked in.

In any MSF mission, the logisticians manage different areas. There is the log base, who oversees where central communications and administration are housed, as well as staff accommodations (either at base or in the local town). The log base also organizes the security and safe transportation of all staff. Then there is the logistician who oversees the hospital, or in this case, the ETU. This person looks after construction involved, supplies, the managing of local staff, and the coordination and communication of the different teams that make up the ETU. The log supply is the one who manages the warehouse where all the supplies from Geneva and Monrovia are housed, tracked, and distributed. Although pharmacy supplies were managed by nursing, it was the log supply who made sure that all orders were shipped from Geneva and delivered to the hospital. Water sanitation is a specialized logistics role, and at the ETU in Foya the water sanitation crew needed to provide 15,000 gallons of clean water a day so that it could function.

2 The Roles of the Logisticians from a Nurse's Eye

In subsequent subsections, I present further details as to the observed logisticians' roles in the field in the context of an ETU in Liberia.

2.1 Log Base

The log base was the person who made our lives livable. During a crisis, accommodations are stark and rudimentary due to the sudden emergency arrival of a lot of staff. The log base made sure that a Jeep would pick us up at the tiny dirt airstrip in Kissidougou, Guinea, where we landed aboard a UN plane. Left standing in the hot sun as the plane took off, it was entirely possible that we might have never been seen or heard from again. However, sure enough, a Jeep materialized, taking us and our supplies on the 3-hour journey through the jungle to the MSF facility in Gueckedou, Guinea, for a quick meal and dropping off some staff and supplies. We continued by Jeep and then canoe across the Mankano River, arriving safely in Liberia (Fig. 2). Even before my mission had begun, I had been taken care of by the logs. To fill out the picture of what a logistician has to do to make this one job of picking us up happen, Balcik et al. (2009) explain that simply dealing with transportation requires the logistician to have to find maps of the area, cope with poor infrastructure and poor roads, hire local drivers and vehicles (often scarce and in bad repair), assess security, and manage the sheer bulk of materials that have to be transported. A lot had gone into simply getting a Jeep to pick us up.

In envisioning arrival in the field, especially in an emergency such as the Ebola outbreak, one expects things to be rudimentary, and this was certainly the case when we got there. Our Internet access was spotty, and usually had to be reserved for administration tasks. Not being able to be instantly in touch with home leads to

Fig. 2 Crossing the Mankano River (credit Martin Zinggl)

unhappy expatriates and worried family members. I watched as the log base worked with local staff to create a functioning satellite communication system, successfully getting the Internet up and running. Although the connection was patchy, the log made it possible for us to be in touch with our loved ones. Other extraordinary things I noticed: the log base kept an eye on our accommodation and the staff. Did we have supplies? Was there enough food? Over at base one day I saw all these closets being built. The log had noticed that we had nowhere to put our personal items and was having closets constructed for us. Something like that means a lot when you come home after a desperate day watching people die, working your hardest to give everyone a chance to live.

I witnessed our log base pull off the seemingly impossible: When the Foya ETU commenced operations, all blood samples from patients suspected of having Ebola needed to be transported by motorbike across the river (and border) to the MSF facility in Gueckedou, where they could be tested. It would take 5 days to receive the results, leaving people waiting in isolation areas, not knowing if they had Ebola or not. This increased the risk of cross infection, which was a real concern during the waiting period. In mid-September, the European Mobile Laboratory Consortium offered to send equipment and a team of volunteers to set up and run a lab. The equipment was brought over from Europe and was to arrive in Foya by helicopter. The coordination between the logs in Monrovia, the capital, and in Foya was phenomenal. Our log base negotiated with the local village chiefs to establish a helicopter pad so that our equipment could arrive. Our log hired local workers to put together a dust-free room where all the equipment could be set up. Anyone who has been to Africa knows that "dust-free" is an oxymoron. Yet, the logs made it happen. Because of this, wait times for results dropped from 5 days to 6 h. Lives were saved, all because of the logs.

2.2 *The ETU Logistician*

So much was happening at the ETU that we had our own dedicated logistician. The largest ETU built prior to this outbreak held 30 beds. At the height of the crisis, the ETU in Foya was dealing with more than 120 patients a day. The urgent need for rapid expansion of isolation areas, triage, and laundry was met with the hiring of construction workers, hasty meetings, and planning late into the night. Even our minor request for a trolley to push equipment from patient to patient was honored. In a couple of days, we had a trolley with wooden wheels that saved many trips back and forth—no small feat when one is gowned in full protective equipment in 100-degree heat. Establishing a communications person turned out to be a huge time-saver: One person manning a transmitter radio kept track of us and allowed communications to keep flowing, which was critical when so many decisions had to be made quickly. Scores of workers showed up every week looking for employment and job applications, and references deluged the ETU log. Choosing who could get work and who could not must have been tough. Once the ETU log mentioned that he had finally found a job for a man who had lost an arm and a leg. He mentioned that this man had kept applying for every job posted and it had been difficult to keep giving it to someone else. Then the right job came along that would be safe for this gentleman and he became part of the Foya team.

The ETU log also oversaw the destruction of all infectious waste. Huge pits were dug behind the isolation area, where all waste was disposed. Before my arrival, it had been decided to set fire to the existing waste as the pits were filling up. This turned out to be unexpectedly explosive, setting fire to one of the hospital areas. Apparently, the logs were up all night containing the fire. No one was injured and, after that, the pits were kept alight and smoldering 24/7 (Fig. 3).

One evening, I was speaking with the ETU logistician as he getting ready to leave his mission. He told me how the long hours and endless demands had nearly finished him; he had been counting the days for his mission to come to an end. But then he saw three Ebola patients across the fence in the isolation area, walking out into the courtyard amid cheers. At first, he thought they had escaped the isolation area but then he realized that they had actually beaten the virus, that they were cured. At that moment, he remembered why he was doing what he was doing. He signed up to come back after the mandatory 21-day break.

Finally, one other logistical nightmare was that generators ran power for the entire ETU. At one point at night, the generators kept malfunctioning and fuses would blow. The logs were called out of bed night after night to fix them and could not find the reason why. Then one of them worked it out. All the nurses and hygienists would plug in their phones and electrical devices to charge them.

Fig. 3 The fire (credit Martin Zinggl)

The local market where people went to charge their devices was closed because of the outbreak. Instead of forcing everyone to stop charging their devices, the log got another generator delivered from Monrovia.

2.3 Log Supply

Overseeing the nursing staff of the ETU involved managing pharmacy supplies. Being in charge of this meant that I spent some time every few days over at the warehouse ordering supplies and liaising with the logistician in charge of the materials needed to run the ETU. Walton et al. (2011) describe clearly the challenges faced by the logistician in charge of the global supply chain; the amount of close communication with field requestors (i.e., the likes of me) that was essential and the difficulty of this simply because of the heavy workload and manual communication tools. During times of disasters, it is often only one full-time employee who has to manage the requests, following through from initial contact to delivery in the country. Walton et al. (2011) included this quote that describes exactly what I witnessed our log supply having to cope with on a daily basis:

> . . . with multiple shipments every day, it's a challenge to finish each shipment and keep up with the next one. They rotate on top of each other, and you're trying to keep up. You have to stay focused, and sometimes it can be confusing what shipment we're talking about.—*emergency logistics team*

3 Water Sanitation

The water sanitation engineer, or WatSan as they are referred to in the field, is responsible for supplying clean water to all parts of the project. It was fascinating to watch the trucks with the water bladders drive off to the river to collect water (Fig. 4). In the West, we take clean running water for granted, but in Africa finding clean water is a huge project in and of itself. When the Foya ETU first started, everyone was provided with bottled water until the WatSans had safe water available. Once this was established, and water stations were set up, the nurses refused to drink the water because it was yellowish in color. A battle ensued. It was discovered that the nurses were using the bottled water that was for the patients. Administration was planning punitive measures. However, discussion with the WatSan resulted in a different outcome. She met with the nurses and described the whole process of how the water was cleaned and purified. She also listened to the nurses' concerns. I learned that all water was being taken from the river, leaving the town's well water and water table safe. After talking with the nurses, the WatSan rethought the whole plan, met with the village chiefs, and was able to get permission to obtain well water for drinking only. River water was also purified, but used for laundry, spraying, and cleaning. The WatSan went above and beyond her duties to make sure that everyone was happy. In any global humanitarian mission, power

Fig. 4 The water sanitation system (credit Deborah Wilson)

imbalances can be steep, and any efforts as expatriates to acknowledge and seek to address it allows for a relationship with other health professionals to become one of colleagues and partners, working together to meet local health needs (Hunt et al. 2014).

4 Special Mention of the Logisticians I Did Not See

Special mention must go the logistics teams in Geneva and Brussels who monitored everything. For example, I heard that they were developing opaque non-permeable body bags so that family members could see that it was their loved one inside. This innovation was needed because the family members of Ebola victims would not always believe that it was their relative in the white body bags typically being used (Fink et al. 2014). Even worse, the families feared that the organs of their loved ones were being harvested. Because a corpse contains a high level of active Ebola virus and can do so for up to 7 days after death, there was too much risk to open the body bags to show a family who was inside (Prescott et al. 2015). Hence, an opaque body bag would relieve a lot of concern, fear, and stress. Brussels also sent an architect to meet with all the staff in small groups to find out what was working and what was not, so he could design a safer, more efficient ETU.

5 Conclusions

There are many more stories that could be told about the bravery and tenacity of the logisticians working in a humanitarian emergency. I hope this snapshot gives a sense of the extraordinary, life-saving work of the logisticians I encountered in Liberia. They are the foundation and backbone to any project, the unsung heroes of disaster relief, without whom, we medics would not be able to function.

References

Balcik, B., Beamon, B.M., Krejci, C.C., Muramatsu, K.M., Ramriez, M.: Coordination in humanitarian relief chains: practices, challenges and opportunities. Int. J. Prod. Econ. **126**(2010), 22–34 (2009)

Fink, S., Nossiter, A., Kanter, J.: Doctors Without Borders Evolves as It Forms the Vanguard in Ebola Fight. New York Times. Retrieved from http://www.nytimes.com/2014/10/11/world/africa/doctors-without-borders-evolves-as-it-forms-the-vanguard-in-ebola-fight-.html (2014)

Hunt, M.R., Schwartz, L., Sinding, C., Elit, L.: The ethics of engaged presence: a framework for health professionals in humanitarian assistance and development work. Dev. World Bioeth. **14**, 47–55 (2014). doi:10.1111/dewb.12013

Médecins Sans Frontières/Doctors Without Borders: Ebola Crisis Update—14th January 2016. London Médecins Sans Frontières UK; 2016 Jan. http://www.msf.org/article/ebola-crisis-update-14-january-2016
Médecins Sans Frontières/Doctors Without Borders: General Logisticians: 2016 Sept. http://www.doctorswithoutborders.org/work-us/work-field/who-we-need/general-logisticians
Médecins Sans Frontières/Doctors Without Borders Medical issues/Ebola 2016, Jan. http://www.doctorswithoutborders.org/our-work/medical-issues/ebola
Prescott, J., Bushmaker, T., Fischer, R., Miazgowicz, K., Judson, S., Munster, V.J.: Postmortem stability of Ebola virus. Emerg. Infect. Dis. **21**(5), 856–859 (2015). doi:10.3201/eid2105.150041
Walton, R., Hayes, R., Haselkorn, M.: Defining 'fast': experience of speed humanitarian logistics. In: Proceedings of 8th international ISCRAM conference, Lisbon, Portugal, May 2011

CPSIA information can be obtained
at www.ICGtesting.com
Printed in the USA
LVOW05*1717270417
532413LV00001BA/1/P

9 783319 437071